Chemistry of Variable Charge Soils

Topics in Sustainable Agronomy

SERIES EDITORS

Rattan Lal
Pedro Sanchez
Malcolm Sumner
M. E. Swisher
P. B. Tinker
Robert E. White

VOLUMES

Chemistry of Variable Charge Soils

Edited by
T. R. YU

with the assistance of G.L. Ji
and other members of the staff of
the Department of Soil Electrochemistry
Institute of Soil Science
Chinese Academy of Sciences

New York   Oxford
Oxford University Press
1997

Oxford University Press

Oxford   New York

Athens   Auckland   Bangkok   Bogota   Bombay   Buenos Aires
Calcutta   Cape Town   Dar es Salaam   Delhi   Florence   Hong Kong
Istanbul   Karachi   Kuala Lumpur   Madras   Madrid   Melbourne
Mexico City   Nairobi   Paris   Singapore   Taipei   Tokyo   Toronto

and associated companies in
Berlin   Ibadan

Library of Congress Cataloging-in-Publication Data
Yu, T. R.
Chemistry of variable charge soils / T.R. Yu.
p.   cm. — (Topics in sustainable agronomy)
Includes bibliographical references and index.
ISBN 0-19-509745-9
1. Variable charge soils.   I. Title.   II. Series.
S592.53.Y85   1996
631.4'1—dc20   95-41882

1 3 5 7 9 8 6 4 2

Printed in the United States of America
on acid-free paper

PREFACE

In the vast areas of tropical and subtropical regions there is a category of widely distributed, highly weathered soils known as "Red Earths," also referred to as Ferralsols, Acrisols, Oxisols, Ultisols, Latosols, Lateritic Soils, Krasnozems, Ferrisols, and Ferrallitic Soils. Because of the abundant rainfall and high temperature, this type of soil is considered one of the most important agricultural soils in the world. However, because of some unfavorable properties inherent in these soils, the productivity is generally low. Proper utilization and melioration of these soils remain an important problem in agricultural production.

Besides agriculture, soil pollution is becoming an increasingly important problem following the industrialization in these regions. Since "Red Earths" are generally low in their buffering capacity against acid precipitation and other pollutants, the problem presents a series of special features.

From a scientific standpoint, just as most of the characteristics of paddy soils are related to "water," red earths' characteristic features are closely related to "red," a visible indication of the presence of the large amounts of iron oxides. These iron oxides, together with the large amounts of aluminum oxides, cause the surface charge carried by the clay to become markedly variable. This is the reason for the name *variable charge soils*. Because of this feature, the soils possess certain properties different from those of the constant charge soils in temperate regions. It would be helpful to have a comprehensive understanding of the chemical properties of these soils, and the results from research on these soils can enrich our present knowledge that comes primarily from research on constant charge soils.

Beginning in 1962 when establishing its electrochemical research, the Department of Soil Electrochemistry in the Institute of Soil Science of the Chinese Academy of Sciences chose "Red Earths" as the principal subject of study. In particular, after the completion of the monograph *Physical Chemistry of Paddy Soils* (Yu, 1985; Science Press/Springer-Verlag) in 1982, the department has concentrated most of its efforts on systematic research of the chemical properties of variable charge soils.

During the research it was realized that in order to thoroughly study the chemical phenomena of these soils it was necessary to develop appropriate research techniques. Therefore, focusing on the characteristics of variable charge soils, we developed a series of new electrochemical methods that were described in the monograph, *Electrochemical Methods in Soil and*

Water Research (Yu and Ji, 1993; Pergamon Press). In retrospect it is clear that the application of these methods played an important role in helping us to explore new research fields and to study certain chemical phenomena not understood previously.

This book summarizes the experience of our research during the past 30 years. The materials cited in the book, except for those referred to in the literature at the end of each chapter, are previously unpublished data from our research.

In this book, emphasis is placed on the interactions among charged particles (clay, ions, protons and electrons) and their chemical consequences in soils. This is the essence of soil chemistry. Actually, the topics cover most areas of modern soil chemistry.

It is my hope that this book may help the reader to understand the basic characteristics of variable charge soils and the fundamental difference between this kind of soil and constant charge soils. Such an understanding, when applied to research, may be helpful for further development in this field.

I wish to express my sincere thanks to Prof. G. Sposito of University of California at Berkeley and Prof. M. E. Sumner of the University of Georgia for their help in facilitating the publication of this book. I am thankful to my colleagues in the Department of Soil Electrochemistry, particularly Prof. G. L. Ji, currently the head of the department, for their painstaking work in assisting me to complete the manuscript. I am also grateful to my wife, Prof. B. H. Li of Nanjing Medical University, for her help in preparing the typescript. The research was supported by the National Natural Science Foundation of China.

T. R. Yu

ACKNOWLEDGMENTS

Permission from the following publishers to reproduce copyrighted material is gratefully acknowledged:

American Chemical Society for Table 3.1, which appeared in *J. Phys. Chem.*, 63:1831-1837 (1959).

Blackwell Scientific Publications for Figs. 6.3, 6.10, 6.13, 6.15, 6.16, 13.6, 13.7, which appeared in *Journal of Soil Science*, 35:471 (Fig. 3), 474 (Fig. 8) (1984); 38:31 (Figs. 1, 2), 32 (Fig. 3), 33 (Figs. 4, 5) (1987).

Elsevier Science Publishers for Figs. 2.17, 11.21, 13.20, 13.21, 13.22, 13.23, 13.24, 13.25, which appeared in *Geoderma*, 32:290 (Figs. 1, 2), 291 (Fig. 3), 292 (Fig. 4), 293 (Fig. 5), 294 (Fig. 6) (1984) ; 44:282 (Fig. 5), 283 (Fig. 6) (1989).

International Society of Soil Science for Figs. 3.23, 13.13, 13.15, 13.16, 13.17, 13.18, 13.19, which appeared in *Transactions of 14th International Congress of Soil Science*, pp. II-64 (Fig. 1), II-65 (Figs. 2A, 2B), II-66 (Fig. 3), II-72 (Fig. 2) (1990).

Springer-Verlag for Figs. 11.23, 11.29, 13.2, 13.8, 13.10, 13.11, 13.12, 14.1, 14.4, 14.6, 14.7, 14.10, which appeared in *Physical Chemistry of Paddy Soils* (Yu Tian-ren, ed.) (1985), pp. 14 (Fig. 1.9), 15 (Fig. 1.10), 29 (Fig. 2.2), 31 (Fig. 2.3), 35 (Fig. 2.8), 39 (Fig. 2.14), 76 (Fig. 4.1), 78 (Fig. 4.2), 81 (Fig. 4.5), 88 (Fig. 4.8), 149 (Fig. 7.17), 152 (Fig. 7.19).

VCH Verlagsgesellschaft mbH for Figs. 5.1, 5.4, 5.5, 12.1, 12.2, 12.3, 12.4, 12.5, 12.10, 12.15, which appeared in *Z. für Pflanzenernährung und Bodenkunde*, 144:518 (Fig. 1), 519 (Fig. 2) (1981); 149:602 (Fig. 4) , 603 (Fig. 5), 604 (Fig. Fig. 7) (1986); 150:19 (Figs. 1, 2) (1986).

Williams & Wilkins for Figs. 8.17, 8.18, 8.24, 8.25, 8.26, 9.1, 9.2, 9.3, 9.4, 9.5, 9.6, 9.7, 9.8, 9.9, 9.10, 9.11, 9.12, 9.13, 10.1, 10.2, 10.3, 10.4, 10.5, 10.6, 10.7, 10.9, 10.10, 10.11, 10.12, 10.13, 10.14, 10.15, 10.16, 10.17, 10.18, 11.20, which appeared in *Soil Science* 139:169 (Fig. 7) (1985); 144:405 (Fig. 3) (1987); 147:38 (Figs. 4, 5), 92 (Figs. 1, 2, 3), 93 (Fig. 4), 175 (Fig. 1), 176 (Fig. 2) (1989); 150:832 (Fig. 2), 833 (Fig. 3) (1990); 151:438 (Figs. 1, 2), 439 (Figs. 3, 4), 440 (Figs. 5, 6, 7) (1991); 152:27 (Fig. 1), 28 (Fig. 2), 29 (Figs. 3, 4, 5), 30 (Figs. 6, 7, 8, 9), 31 (Fig. 10) (1991).

CONTENTS

Chemistry of Variable Charge Soils

1

INTRODUCTION

T. R. Yu

1.1 SCOPE OF SOIL CHEMISTRY

The constitution and properties of soils have their macroscopic and microscopic aspects. Macroscopically, the profile of a soil consists of several horizons, each containing numerous aggregates and blocks of soil particles of different sizes. These structures are visible to the naked eye. Microscopically, a soil is composed of many kinds of minerals and organic matter interlinked in a complex manner. In addition, a soil is always inhabited by numerous microorganisms which can be observed by modern scientific instruments. To study these various aspects, several branches of soil science, such as soil geography, soil mineralogy, and soil microbiology, have been developed. If examined on a more minute scale, it can be found that most of the chemical reactions in a soil occur at the interface between soil colloidal surface and solution or in the solution adjacent to this interface. This is because these colloidal surfaces carry negative as well as positive charges, thus reacting with ions, protons, and electrons of the solution. The presence of surface charge is the basic cause of the fertility of a soil and is also the principal criterion that distinguishes soil from pure sand. The chief objective of soil chemical research is to deal with the interactions among charged particles (colloids, ions, protons, electrons) and their chemical consequences in soils. As depicted in Fig. 1.1, these charged particles are closely interrelated.

The surface charge of soil colloids is the basic reason that a soil possesses a series of chemical properties. At present, considerable knowledge has been accumulated about the permanent charge of soils. On the other hand, our understanding is still at an early stage about the mechanisms and the affecting factors of variable charge. The quantity of surface charge determines the amount of ions that a soil can adsorb, whereas the surface charge density is the determining factor of adsorbing strength for these ions. Because of the complexities in the composition of soils, the distribution of

3

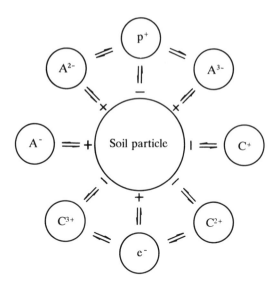

Fig. 1.1. Interactions among charged particles in soils. A, anion; C, cation; p, proton; e, electron.

positive and negative charges is uneven on the surface of soil colloidal particles. Insight into the origin and the distribution of these charges should contribute to a sound foundation of the surface chemistry of soils.

Interactions between charged colloidal surface and ions are the core of soil chemical research. These interactions manifest themselves chiefly in the structure and the changing pattern of the electric double layer. However, this electric double layer is submicroscopic and is practically unaccessible to direct measurement. In order to study these interactions, certain comparatively macroscopical methods must be employed. Some of these methods are static in nature, and some are dynamic. The static condition of these reactions is in reality a dynamic equilibrium determined mainly by the balance between electric energy and thermal energy. The distribution of ions between the solid phase and the solution phase after adsorption and/or negative adsorption is one reflection of this dynamic equilibrium. Soil colloidal surfaces adsorb cations as well as anions; during adsorption both electrostatic force and specific force are involved. In particular, the electric double layers around both positive and negative charge sites at the surface of soil colloids are not independent of one another, but can overlap to a certain extent. Besides, in soils there are generally many kinds of ions present together. These ions may be competitive or cooperative when adsorbed by soil colloidal surfaces. Therefore, the interactions between the surface of soil colloids and various kinds of ions constitute a broad research field in soil chemistry.

In addition to static methods under equilibrium conditions, some dynamic methods, in which one component of the soil is forced to migrate in an electric field or force field, can also disclose the mechanisms of the interactions between soil colloids and ions. One can observe the migration of colloidal particles in an electric field (electrophoresis) or the movement of ions together with colloidal particles in an electric field (electric conductance). Alternatively, one can measure the diffusion rate of ions caused by the microregional difference in chemical potential due to concentration gradient. Clay particles and ions have mutual influences during the relative displacement, with the extent of influence determined by the reacting force between the two charged species. For instance, in an alternating current (AC) electric field of different frequencies the electric conductivity of ions adsorbed by the soil with different adsorbing strengths may differ remarkably.

Protons and electrons in the soil cannot exist in a free state for a long time. Instead, they undergo transfer reactions very actively. During the transfer both the solution phase and the solid phase may be involved. The acceptance and release of protons by and from the solid phase of the soil is the basis for the production of variable charges. Aluminum ions released from the clay due to reactions of protons with the solid phase is the principal cause for the occurrence of acid reaction in a soil. Electrons liberated from the decomposition products of soil organic matter can react with variable-valency elements, inducing oxidation-reduction reactions. It is now known that these oxidation-reduction reactions not only take place actively in submerged soils, but can also occur widely in soils under aerated conditions.

In the history of soil science, the scope of soil chemistry was viewed quite differently during various developing stages and by different soil chemists. Broadly speaking, at the early stage of development all chemical composition and properties of soils were included in soil chemistry. Later, following the wide application of chemical methods in soil research and the establishment of various branches of soil science, soil chemical research was confined to basic principles of chemical phenomena. In this book, the interactions among charged particles and their chemical consequences are treated. These interactions are electrical in nature. Their consequences may be reflected in various chemical phenomena, most of which occur at or near the surface of soil particles. Therefore, soil chemistry may be viewed as soil electrochemistry or, more exactly, soil surface electrochemistry.

Up to the present, soil chemical research focuses its scope mainly on the soil colloid-solution interface. Very little attention has been paid to the soil colloid-solution-plant roots interfaces. Research on the soil colloid-solution-microorganisms interfaces remains a nearly unexplored field.

1.2 CHARACTERISTICS OF VARIABLE CHARGE SOILS

Soils on earth are numerous in type. However, they may be broadly classified into two basic categories: constant charge soils and variable charge soils.

Soil components that can carry surface charges include three major kinds, layer silicate minerals, free oxides, and humus. Silicate minerals carry permanent negative charge under ordinary environmental conditions. Oxides of iron, aluminum, and manganese carry variable negative charge as well as variable positive charge, whereas humus carries variable negative charge and a small amount of variable positive charge.

Now, let us examine the characteristics of the two categories of soils.

In soils of temperate regions, the content of iron oxides is generally less than 50 g kg^{-1}. The principal clay minerals are hydrous mica, montmorillonite, and vermiculite. These three minerals carry permanent negative surface charge in quantities of 10–40, 80–120 and 100–150 cmol kg^{-1}, respectively. For these minerals the variable negative surface charge component cannot manifest itself markedly under ordinary pH conditions. Humus and oxides can contribute to variable charge of the soil to a certain extent. However, this contribution to variable charge generally is small when compared to permanent charge carried by clay minerals. Therefore, for soils of these regions it is mainly the permanent negative charge that determines the surface chemical properties. Thus, it would be appropriate to refer to this category of soil as *permanent charge soil*.

The dominant clay mineral in soils of a large part of tropical and subtropical regions is kaolinite. This mineral carries only 3–10 cmol kg^{-1} of permanent negative surface charge. On the other hand, it carries substantial quantities of variable positive charge and variable negative charge. In particular, the iron oxide content in soil is 50–250 g kg^{-1} or even more and is accompanied by a large amount of aluminum oxides. In addition, because the quantity of surface charge carried by clay minerals is low, the relative contribution of humus to surface charge of the whole soil is much more important compared to soils of temperate regions. Thus, for this category of soil the quantities of surface charges are markedly variable, and the positive surface charge may play a very important role in affecting the surface properties. Therefore, it would be appropriate to refer to this category of soil with kaolinite as the predominant clay mineral and containing large amounts of iron and aluminum oxides as *variable charge soil*. Of course, constant charge soils may carry a certain quantity of variable charge, and variable charge soils also carry permanent negative surface charge and even permanent positive surface charge (Tessens and Zauyah, 1981). Hence, strictly speaking, it is difficult to have a clear–cut definition between the two categories of soils.

The development of soil science came mainly from research on soils of temperate regions of northern Europe and North America. It was observed that for each soil in these regions there was a nearly constant cation-exchange capacity. The concept of cation-exchange capacity of a soil became well established during the 1930s and the 1940s, because it was found that the clay fraction of a soil is composed chiefly of various kinds of layer silicate minerals and that the negative surface charges of these minerals originated from isomorphous substitutions in the crystal lattice. Although it was also known that the cation-exchange capacity of a soil is in reality not a constant, but varies to a considerable extent with experimental conditions, particularly with pH, most of the researchers tried to find the cause of this phenomenon from the method of determination. Since then, the concept of cation-exchange capacity was accepted widely in soil science and has remained until the present time. It should be noted that as early as the beginning of the 1930s Mattson already regarded soil colloids as amphoteric colloids and used variable charge soils of America as his subject of study, and Schofield, taking a variable charge soil of Africa as an example, illustrated the variability of negative and positive surface charges of the soil in the early 1950s.

On the other hand, beginning in the 1960s, because of the increase in research on soils of tropical and subtropical regions, it was gradually recognized that soils of these regions possess a series of properties quite different from those of temperate regions. A fundamental difference was that the quantities of surface charge of these soils were remarkably variable and that this characteristic feature had a profound influence on many other properties of the soil, such as the ability to adsorb anions. To exchange knowledge in this field, an International Conference on Soils with Variable Charge was held in New Zealand in 1981, sponsored by the International Society of Soil Science. During the conference many soil chemists reevaluated the contribution of Mattson to soil science. This conference, especially the monograph entitled *Soils with Variable Charge*, edited by B. K. G. Theng, greatly stimulated the worldwide development in research on variable charge soils. At present, the importance of the variability of surface charge in soils has been widely recognized by soil scientists, including those who have paid special attention to clay minerals.

One question closely related to the variability of surface charge is the significance of iron and aluminum oxides in soils. Based on the concept that clay minerals are the key materials in determining the basic properties of the soil, these oxides were generally regarded as inert materials. Because these oxides may mask the surface of clay minerals in the form of colloidal coatings and thus make the surface charge of clay minerals unable to manifest themselves entirely, it was a frequent practice to remove these "free" materials from the soil by chemical means when studying the chemical properties of soils. It is now known that it is these oxides that are the major cause for the soil to carry variable charges. This point is of particular significance for variable charge soils, because, owing the low quantity of surface charge carried by kaolinite, it is frequently these oxides that determine the surface properties of the soil.

1.3 TYPES OF VARIABLE CHARGE SOILS

There are many types of variable charge soils in the world. The most important ones are Rhodic and Orthic Ferralsols, Nitosols, Ferric, Ferrali-, and Ali- Acrisols and Xanthic soils. In the Soil Taxonomy system, they include Oxisols and Ultisols. In Tables 1.1 and 1.2, the names and properties of some representative soils dealt with in this book are given.

The Rhodic Ferralsol derived from basalt and distributed in the Leizhou Peninsula and Hainan Island may be taken as the representative of Rhodic Ferralsols of China. The contents of clay and iron oxides are 600–750 g kg^{-1} and 150–200 g kg^{-1}, respectively. The predominant clay minerals in the clay fraction are kaolinite, gibbsite, and hematite (Fig. 1.2). The soil does not carry noticeable amount of permanent negative charge. One peculiar feature of this type of soil is that the organic matter content of the subsoil may frequently be 7–10 g kg^{-1}, higher than that of Ferrali–Haplic Acrisols and Ali–Haplic Acrisols (about 5 g kg^{-1}).

Table 1.1 Names of Some Representative Variable Charge Soils

Name	Location	Soil Taxonomy	Local Name
Hyper-Rhodic Ferralsol	Kunming, China	Rhodic Hapludox	Latosol
Rhodic Ferralsol	Xuwen, China	Rhodic Hapludox	Latosol
Rhodic Ferralsol	Queensland, Australia	Rhodic Hapludox	Krasnozem
Rhodic Ferralsol	Hawaii, U.S.A.	Rhodic Hapludox	Oxisol
Rhodic Ferralsol	Brasilia, Brazil	Acrustox	Dark Red Latosol
Xanthic Ferralsol	Manaus, Brazil	Acrorthox	Pale Yellow Latosol
Ferric Acrisol	Tully, Australia	Rhodudult	Red Earth
Ferrali-Haplic Acrisol	Guangzhou, China	Kandiudult	Lateritic Red Soil
Ali-Haplic Acrisol	Jiangxi, China	Hapludult	Red Earth

Table 1.2 Properties of Representative Variable Charge Soils

Soil	Location	Fe$_2$O$_3$ (g kg^{-1})	Clay (g kg^{-1})	Clay Minerals[a]	Parent Material
Hyper–Rhodic Ferralsol	Kunming	250	600-800	K,G,H (V,A,Go)	Basalt
Rhodic Ferralsol	Xuwen	186	500-750	K,G,H(Go)	Basalt
Rhodic Ferralsol	Queensland	–	600	K,Go(G)	Basalt
Rhodic Ferralsol	Hawaii	145	–	K,F(I)	?
Rhodic Ferralsol	Brasilia	100	660	K(H,Go)	Diluvial
Xanthic Ferralsol	Manaus	53	470	K(Go)	Diluvial
Ferric Acrisol	Tully	–	510	K(G,H/Go)	Sedimentary
Ferrali–Haplic Acrisol	Guangzhou	50–80	200-500	K(I,V,F)	Granite
Ali–Haplic Acrisol	Jiangxi	50–80	300-500	K,I(V,F)	Quaternary diluvial

[a] A, anatase; F, ferric oxides; G, gibbsite; Go, goethite; H, haemitite; I, hydrous mica; K, kaolinite; V, vermiculite

In the Kunming region of Yunnan Province, there is a Ferralsol derived from basalt or paleosol and can be called Hyper–Rhodic Ferralsol. The iron oxide content in soil is 250 g kg^{-1} or more, and the titanium content is 30-60 g kg^{-1}. The clay fraction is dominated by kaolinite, gibbsite and hematite, with certain amounts of vermiculite and anatase (Fig. 1.3). This soil is not only regarded as the most highly weathered soil in China, although it contains vermiculite, but is also one of the rarely encountered

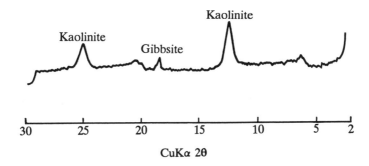

Fig. 1.2. X–ray diffraction pattern of the clay fraction of Rhodic Ferralsol of Xuwen (free Fe$_2$O$_3$ removed).

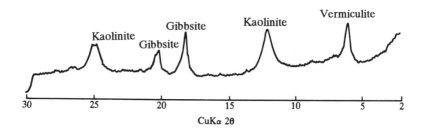

Fig. 1.3. X-ray diffraction pattern of the clay fraction of Hyper-Rhodic Ferralsol (Kunming) (Fe$_2$O$_3$ removed).

cases in the world insofar as the distinctively characteristic feature of variable charge soils is concerned. Its chemical properties (Yu, 1950) and surface electrochemical properties (Zhang et al., 1989) have been investigated in detail. At present, we do not know how such a high titanium oxide content in the soil affects the chemical properties of the soil, although research with pure minerals has shown that these minerals may adsorb cations as well as anions (Milnes and Fitzpatrick, 1989).

The Rhodic Ferralsol (Dark Red Latosol) collected from Brasilia is developed from Quaternary diluvial deposit distributed on an undulating low plateau. This type of soil occupies large areas in central Brazil, and is an important agricultural soil there (Macedo and Bryant, 1987). The soil contains 500–700 g kg^{-1} of clay, and the dominant clay mineral is kaolinite (Fig. 1.4). Although the content of iron oxides is not very high (about 100 g kg^{-1}), the soil possesses a series of surface properties similar to those of

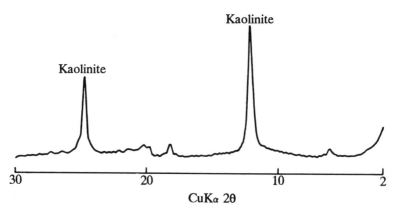

Fig. 1.4. X-ray diffraction pattern of the clay fraction of Rhodic Ferralsol of Brasilia (free Fe$_2$O$_3$ removed).

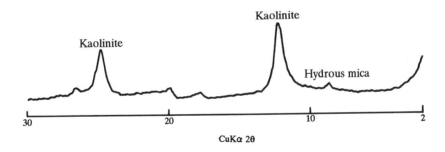

Fig. 1.5. X-ray diffraction pattern of the clay fraction of Rhodic Ferralsol of Hawaii (free Fe_2O_3 removed).

the Hyper-Rhodic Ferralsol of Kunming, as can be seen in later chapters of this book.

The Rhodic Ferralsol (Oxisol) of Hawaii was supplied by the University of Hawaii. The soil contains 145 g kg^{-1} of iron oxides. The clay minerals are dominated by kaolinite and accompanied by a small amount of hydrous mica (Fig. 1.5).

The Rhodic Ferralsol (Krasnozem) of Queensland collected by Dr. G. P. Gillman was derived from basalt on a slightly undulating plain, containing 600-650 g kg^{-1} of clay, with kaolinite, geothite, and gibbsite as the principal clay minerals.

The Ferric Acrisol (Red earth) also collected by Dr. G. P. Gillman contains 300-500 g kg^{-1} of clay. The predominant clay minerals of this soil are kaolinite, gibbsite, and hematite.

The chemical properties of the two Australian soils have been studied in detail by Gillman and Bell (1976).

Among Ferrali-Haplic Acrisols widely distributed south of the Nanling Mountain in China the one derived from granite is the most important as far as the distributed area is concerned. Because the parent material has been intensively weathered, the soil does not contain primary minerals such as mica and feldspar, except for quartz. Therefore, it is composed chiefly of quartz sand and clay, with a clay content ranging from 200 to 500 g kg^{-1}. The soil contains 50-80 g kg^{-1} of iron oxides. The clay fraction is dominated by kaolinite, with only small amounts of hydrous mica and vermiculite (Fig. 1.6).

For soils distributed in the Xishuangbanna region of China and derived from sandstone, shale, phyllite, granite, or Quaternary deposit, although geographically the region should be dominated by Ferralsols, they belong to the type Ferrali-Haplic Acrisols because, in addition to kaolinite, the clay fraction contains considerable amounts of hydrous mica and vermiculite (Fig. 1.7). These soils contain 60-80 g kg^{-1} of iron oxides.

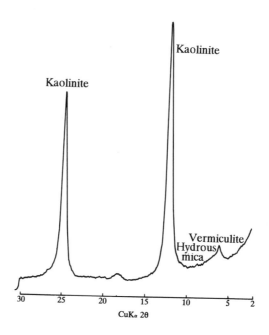

Fig. 1.6. X-ray diffraction pattern of the clay fraction of Ferrali–Haplic Acrisol of Guangzhou (free Fe_2O_3 removed).

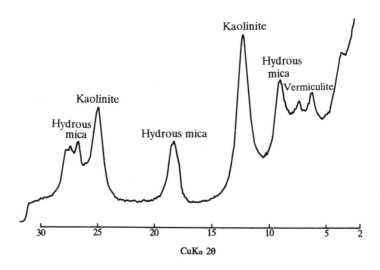

Fig. 1.7. X-ray diffraction pattern of the clay fraction of Ferrali–Haplic Acrisol of Xishuangbanna (free Fe_2O_3 removed).

Ali–Haplic Acrisols may derive from various kinds of parent materials. In China, those derived from Quaternary red clay occupy the largest area and are the most important in agricultural production. Soils widely distributed in the central part of the Jiangxi Province may be taken as the typical example of this type of soil. The soil contains 300–500 g kg^{-1} of clay and 50–80 g kg^{-1} of iron oxides. The clay minerals are dominated by kaolinite and hydrous mica and also contain a certain amount of vermiculite (Fig. 1.8). Because of this feature in mineralogical composition, the whole soil can carry 7–11 cmol kg^{-1} of negative surface charge at pH 7, due to the presence of a large proportion of permanent negative charge.

Xanthic soils can be subdivided into two subtypes. One is the soil developed under humid climatic conditions or in local wet areas in regions of Ferralsols or Acrisols. For this subtype of soil, except the form of iron–containing minerals, the chemical composition does not differ greatly from the zonal variable charge soils of the same region. For example, for a Xanthic–Haplic Acrisol derived from Quaternary red clay in the Pingba region of the Guizhou Province, the clay minerals are dominated by kaolinite and hydrous mica and are accompanied by certain amounts of vermiculite and gibbsite. The Xanthic Ferralsol (Pale Yellow Latosol) collected from Manaus of Brazil, a region not far from the Equator, is developed on a diluvial plain under the rain forest. The clay content and iron oxide content are 470 g kg^{-1} and 53 g kg^{-1}, respectively. The clay mineralogy is almost exclusively composed of kaolinite. For another subtype of xanthic soils, distributed at a higher elevation above sea level, the weathering is comparatively weak due to low temperature. As a consequence, the clay fraction contains, in addition to kaolinite, large amounts of

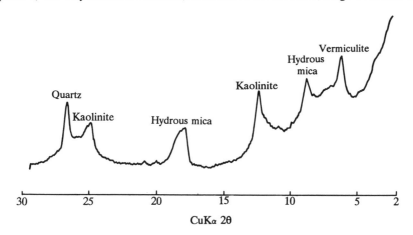

Fig. 1.8. X–ray diffraction pattern of the clay fraction of Ali–Haplic Acrisol (Jiangxi) (free Fe$_2$O$_3$ removed).

2:1–type clay minerals such as hydrous mica and vermiculite. Thus, xanthic soils as a whole belong to a transitional type between constant charge soils and variable charge soils, and whether such a soil can be regarded as a variable charge soil is dependent on the actual conditions of the given soil.

The experimental data cited in this book are mainly concerned with three representative variable charge soils of China: Rhodic Ferralsol of Xuwen, Ferrali–Haplic Acrisol of Guangzhou, and Ali–Haplic Acrisol of central Jiangxi Province. The comprehensive overview of these materials presented here may reflect the basic status of the chemical properties of variable charge soils of the world, with the exception of some azonal soils (e.g., Andosols derived from volcanic ash).

1.4 SIGNIFICANCE OF CHEMISTRY OF VARIABLE CHARGE SOILS

For a long time our knowledge about the chemical properties of soils came chiefly from studies on constant charge soils of temperate regions. The results of research on variable charge soils can supplement or modify our present understanding about the chemical properties of the soil, and thus they are helpful in compiling a more comprehensive theory in soil science.

Variable charge soils occupy vast areas in the world, and at present they are one of the main kinds of soils that deserve reclamation. Therefore, in addition to theoretical importance, research on these soils is of practical significance. For example, although these soils have a disadvantage in that the quantities of nutrient cations such as potassium, ammonium, calcium, and magnesium that a soil can hold are comparatively low, the holding capacity can be easily regulated by artificial means. It has been estimated that by raising the pH from its original value of 5 to 7 simply through the application of alkaline materials, such as lime, the holding capacity of some Ferralsols for cations can be increased by 70% with an accompanying increase in holding intensity (Yu and Zhang, 1990). An estimated increase in CEC for some soils from their natural pH to pH 7 may be as large as five–fold (Theng, 1980). Variable charge soils possess the ability to hold nutrient anions such as nitrate and sulfate to protect them from leaching loss, whereas constant charge soils do not. Therefore, knowledge about the regularities of the changing patterns in surface charge and their relations to various kinds of nutrient ions may be useful in adopting appropriate soil management measures. Except for nutrient environment, a suitable acidity environment and a suitable oxidation–reduction environment of the plant roots are also important. For instance, without the functioning of buffering ability of the soil, the pH of soil solution may change drastically and constantly due to changes in environmental factors and thus would be unfavorable to the growth of plants. Too strong an oxidizing condition can lead to the disturbance of the equilibria among nutrients, whereas too strong a reducing condition can frequently cause the production of toxic substances.. In variable charge soils both acidification–alkalinization reactions and oxidation–reduction reactions proceed actively because of high temperature and ample rainfall. Therefore, it is necessary to regulate these reactions based on a thorough understanding about the various factors affecting them. Besides, the significance of soil structure in soil fertility is well known. The phenomenon of flocculation caused by interactions among soil particles is the first necessary step for the formation of soil structure. It may be frequently observed that the structure of many variable charge soils is not so bad under field conditions as might be suspected, although the clay content is as high as 500 g kg^{-1} or more. In this respect, the peculiar structural status caused by the presence of large amounts of iron oxides is an important reason. From the above–mentioned examples it is evident that many chemical properties of variable charge soils are very important factors in affecting the fertility of the soil.

Chemical properties of variable charge soils are also of significance in soil genesis. It may be observed that the surface layer of well–developed variable charge soils under natural vegetation is less red in color than that of the underlying horizons; if the soils are cultivated, the color becomes gray after the decomposition of the large amount of organic matter. In soil science, there is the term "podzolized red soils." Actually, this layer is the result of the leaching loss of ferrous iron, in the form of chelate, after the reduction of iron oxides under periodical reduction conditions. Besides, the texture of this layer is usually lighter than that of the lower horizons. This is closely related to the decomposition of aluminum–containing minerals caused by the periodical change in acidity of the soil as a result of seasonal change in oxidation–reduction conditions as well as by the complexation effect of aluminum with organic ligands.

It is also possible to subdivide soil types among variable charge soils based on the differences in their surface charge properties.

As mentioned in Section 1.2, our knowledge at present regarding the properties of soils comes chiefly from studies on constant charge soils of temperate regions. There are reasons to believe that, with the further development in research on chemical properties of variable charge soils, a comprehensive soil chemistry system integrating characteristics of both constant charge soils and variable charge soils will be established.

BIBLIOGRAPHY

Gillman, G. P. and Bell, L. C. (1976) *Aust. J. Soil Res.*, 14:351–360.

Gillman, G. P. and Yu, T. R. (1986) in *Proceedings of the International Symposium on Red Soils* (Institute of Soil Science, ed.). Science Press/

Elsevier, Beijing/Amsterdam. pp. 251-261.

Macedo, J. and Bryant, R. B. (1987) *Soil Sci. Soc. Am. J.*, 51:690-698.

Milnes, A. R. and Fitzpatrick, R. W. (1989) in *Minerals in Soil Environments* (J. B. Dixon and S. B. Weed, eds.). Soil Science Society of America, Madison, WI. pp. 1131-1205.

Tessens, E. and Zauyah, S. (1981) *Soil Sci. Soc. Am. J.*, 46:1103-1106.

Theng, B. K. G. (1980) *Soils with Variable Charge*. New Zealand Society of Soil Science, Lower Hutt.

Yu, T. R. (1950) *Bul. Soil Sci. Soc. China* (ch.), 1:177-186.

Yu, T. R. (1976) *Electrochemical Properties of Soils and Their Research Methods*. Science Press, Beijing.

Yu, T. R. (1992) *Adv. Agron.*, 48:205-250.

Yu, T. R. and Ji, G. L. (1993) *Electrochemical Methods in Soil and Water Research*. Pergamon, Oxford.

Yu, T. R. and Zhang, X. N. (1990) in *Soils of China* (Institute of Soil Science, ed.). Science Press, Beijing, pp. 494-513.

Zhang, X. N., Zhang, G. Y., Zhao, A. Z. and Yu, T. R. (1989) *Geoderma*, 44:275-286.

2

SURFACE CHARGE

X. N. Zhang and A. Z. Zhao

The surface of soil colloids carries electric charges, and these surface charges are the basic cause for soil to possess a series of surface properties. Soil surface charges affect the chemical properties of the soil through varying the quantity of electric charge and the surface charge density. For example, adsorptions of cations and anions are caused by negative and positive surface charges of the soil, respectively. The amount of ions adsorbed is determined by the quantity of surface charge, whereas the tightness of adsorption is related to charge density. In addition, the migration of ions in soil, the formation of organo-mineral complexes, and the dispersion, flocculation, swelling, and shrinkage are all affected by surface charge properties of the soil. Therefore, surface charge properties have an important bearing on soil structure and plant nutrition.

Variable charge soils are characterized by the high content of iron and aluminum oxides. The clay mineralogical composition is dominated by 1:1-type minerals, such as kaolinite. These two factors make the surface charge properties of variable charge soils distinctly different from those of constant charge soils of temperate regions which chiefly containin 2:1-type clay minerals. However, unlike the case for pure variable charge minerals, in variable charge soils there is generally the presence of a certain amount of 2:1-type clay minerals. Therefore, as a mixture of variable charge minerals and constant charge minerals, the surface charge properties of variable charge soils is more complicated.

In this chapter, the origin and factors affecting surface charges of the soil as well as the relationship between these charges and soil type will be discussed.

2.1 CHARGED SURFACE

Despite the complexity in composition, a soil may be regarded as a mixed system consisting of constant charge surface materials and constant

17

potential surface materials in different ratios (Anderson and Sposito, 1992; Gillman and Uehara, 1980). Examples of the former type such as montmorillonite and vermiculite carry permanent negative charges, while those of the latter type such as iron oxide and aluminum oxide carry variable charges.

2.1.1 Permanent Charge Surface

Commonly found constant charge clay minerals in soils include those layer silicates such as hydrous mica, vermiculite, montmorillonite, and chlorite. The characteristic feature of their structure is that one aluminum–oxygen or magnesium–oxygen octahedral sheet is sandwiched by two silicon–oxygen tetrahedral sheets, forming a structural unit. Both sides of the crystal layer of this 2:1–type mineral are oxygen atoms connected to the silicon atom of the tetrahedron, and the functional group of the exposed basal plane is siloxane ($\equiv$Si–O–Si$\equiv$). Since kaolinite and other 1:1–type clay minerals are constructed of one silicon–oxygen tetrahedral sheet and one a-luminum–oxygen octahedral sheet, only one–half of the basal plane is a siloxane type surface. Silicon atoms firmly attach to oxygen atoms, and thus the siloxane surface is a hydrophobic surface.

The electric charge of the siloxane surface is a permanent charge, produced by isomorphous substitution in layer silicate minerals. If the silicon atom in the silicon–oxygen tetrahedron is substituted by an aluminum atom with a lower valence, or the aluminum atom in the aluminum–oxygen octahedron is substituted by a magnesium atom or a ferrous atom, a surplus negative charge would be produced.

The origins of permanent negative surface charge in various 2:1–type layer silicate minerals differ considerably. The negative surface charge of montmorillonite comes chiefly from the substitution of a part of aluminum in the aluminum–oxygen octahedron by divalent magnesium atoms, and in each substitution one negative charge is produced. There is also the possibility that a small portion of negative charges is produced by the substitution of silicon atoms in the silicon–oxygen tetrahedra with aluminum atoms. However, this kind of substitution generally does not account for more than 15% of the total substitution. In montmorillonite, negative charges are partially compensated for internally, leaving about a 0.666 surplus negative charge for each unit cell. The cation–exchange capacity of montmorillonite is 80–130 cmol kg^{-1}, in which more than 80% is permanent negative surface charge. In hydrous mica, about one–sixth of the silicon atoms in the silicon–oxygen tetrahedra are substituted by aluminum atoms, producing about 1.3–1.5 surplus negative charges for each unit of crystal micelle. There is also a small amount of substitutions of trivalent aluminum in the octahedra by divalent atoms, producing surplus negative charges. In hydrous mica, most of the negative charges are compensated for by interlayer nonexchangeable potassium ions, and a small portion of them are compensated for by nonexchangeable calcium and magnesium ions, leaving only a small portion exposed to the outside. The cation-exchange capacity of hydrous mica is 10–40 cmol kg^{-1}. In vermiculite, negative charges originate chiefly from the partial substitution of silicon atoms in the silicon-oxygen tetrahedra with aluminum atoms. Except for the internal compensation of a portion of negative charges, there are still 1–1.4 surplus negative charges compensated for by exchangeable cations such as calcium and magnesium for each unit of crystal micelle. The cation-exchange capacity of large-sized vermiculite may be as large as 100–150 cmol kg^{-1}. In acid soils, however, because of the presence of nonexchangeable polymerized hydroxyl aluminum ions in interlayers, the cation-exchange capacity is much smaller than that of the large-sized mineral. The specific surface area of vermiculite is also comparatively small.

Permanent negative charges originating from the isomorphous substitution in tetrahedra are close to the plane of the siloxane group, and therefore their field strength is relatively strong. By contrast, negative charges originating from the isomorphous substitution in octahedra are far from the plane of the siloxane group, and therefore they possess a relatively weak field strength. Hence, the tightness in the retention of cations by negative charge sites is different for these two origins.

Espinoza et al. (1975) found that volcanic ash soils also carry a large quantity of permanent negative surface charge. They suggested that in the ionic substitution within amorphous and hydrated oxides and in the crystal lattice vacancies there is also the possibility of the production of permanent negative surface charge. For iron oxide minerals in Oxisols, permanent positive surface charges may be produced through the isomorphous substitution of trivalent Fe^{3+} ions by tetravalent Ti^{4+} ions (Tessens and Zauyah, 1982).

Since permanent negative surface charges originate from isomorphous substitution in crystal lattices of minerals, and since this substitution occurs during the formation of minerals, they are not affected by environmental factors such as the pH of the medium or electrolyte concentration.

According to the electric double layer theory, the surface charge density σ and surface potential ψ_0 are related to the electrolyte concentration C of the liquid phase following the Gouy-Chapman equation:

$$\sigma = \left(\frac{2C\epsilon RT}{\pi} \right)^{\frac{1}{2}} \sinh \frac{zF\psi_0}{2RT} \qquad (2\text{-}1)$$

For constant charge surface, σ is a constant, and therefore we have

$$\left(\frac{2C\varepsilon RT}{\pi} \right)^{\frac{1}{2}} \sinh \frac{zF\psi_0}{2RT} = \text{constant} \qquad (2\text{-}2)$$

It can be seen from the equation that when the electrolyte concentration C, dielectric constant ε, valence z of compensating ions, or temperature T is changed, the surface potential ψ_0 will also change, so that the left side of equation (2-1), σ, can remain constant. An increase in concentration and valence of compensating ions induces a decrease in surface potential and a corresponding contraction of the electric double layer. A decrease in dielectric constant of the medium also induces a decrease in thickness of the electric double layer.

2.1.2 Variable Charge Surface

Humus is an important source of variable charge of most soils, especially for soils with a high content of organic matter. For variable charge soils, the principal sources of variable charge are the oxides and hydrated oxides of iron and aluminum. In addition, edge surfaces of clay minerals of soil are also variable charge surfaces.

2.1.2.1 *Humus*

Humus contains a large quantity of acid groups. Variable negative charges are produced after the dissociation of these acid groups. It has been estimated that the quantity of negative charge carried by humus is within the range of 200–500 cmol kg^{-1}. These negative charges originate chiefly from carboxyl groups with a pK_a value of 4–5. Hydroxyl groups including phenolic hydroxyl, quinonic hydroxyl, and enolic hydroxyl groups can also make important contributions to negative charge. The extent of dissociation of quinonic hydroxyl groups is close to that of carboxyl groups. On the other hand, the pK_a value of phenolic hydroxyl groups is larger than that of carboxyl groups, and the variation range is also wider. Under alkaline conditions the contribution of phenolic hydroxyl groups to negative charge is even more important. In soil humus, carboxyl groups and hydroxyl groups account for about 50% and 30% of the total functional groups, respectively (Yu, 1976). In addition, at high pH, amino groups can also contribute to negative charge.

2.1.2.2 *Hydrated Oxides*

The principal oxides and hydrated oxides in soils include those of iron, aluminum, manganese, titanium, and silicon. The oxides of these elements such as hematite (α-Fe_2O_3), corundum (α-Al_2O_3), and quartz (SiO_2) are easily hydrated under moist conditions, changing their surface to a state identical to that of hydrated oxides such as goethite (α-$FeO\cdot OH$), gibbsite [$Al(OH)_3$], and silica gel ($SiO_2\cdot nH_2O$). Hence, the surface of all these oxides and hydrated oxides contains hydroxyl groups. These surface hydroxyl groups can bond directly with iron, aluminum, or silicon ions, and they can also bond with adsorbed water molecules through hydrogen bonding. This is the distinct difference between these groups and the inert oxygen of siloxane surface. The silanol surface ($\equiv Si$-OH) can change into hydrophobic siloxane surface with a very low charge density. However, this kind of surface differs from the siloxane surface of 2:1-type clay minerals in that it can react with water vapor, restoring it to a silanol surface.

Hydrated oxides of iron and aluminum have different crystal structures. In addition, the number of iron or aluminum atoms attached to the hydroxyl group of different crystal planes is also different. The number may be one, two, or three. Therefore, the number and activity of hydroxyl groups on different crystal planes vary considerably.

Electric charges at the surface of hydrated oxides are produced through the dissociation of H^+ ions of surface hydroxyl groups and the combination of these hydroxyl groups with H^+ ions of the solution. This can be expressed as follows:

$$\begin{array}{ccc} +H^+ & & -H^+ \\ M\text{-}OH_2^+ \rightleftharpoons & M\text{-}OH \rightleftharpoons & M\text{-}O^- \\ -H^+ & & +H^+ \end{array} \qquad (2\text{-}3)$$

where M represents a metal atom or a silicon atom. Al-OH and Fe-OH are called an aluminol group and a ferrol group, respectively. The surface will carry negative charge when the hydroxyl group loses one H^+ ion, and it will carry positive charge when the hydroxyl group adsorbs one H^+ ion.

Allophane is a group of hydrated aluminum silicates with different silicon-to-aluminum ratios. All allophanes are short-order amorphous materials possessing Si-O-Al bonding. Allophane carries a negative surface charge at high pH through the dissociation of H^+ ions from the siloxane group ($\equiv Si$-OH), and it carries a positive surface charge at low pH owing to the acceptance of protons by the aluminol group (Al-OH).

2.1.2.3 *Edge Face of Layer Silicates*

The edge face of kaolinite is the most typical site of variable charge surfaces among layer silicates. The change in electric charge may be expressed as follows:

Si-O	-1	Si-OH		Si-OH	
$\|$		$\|$		$\|$	
O	-½	OH	+½	OH	+½
⋮		⋮		⋮	
Al		Al		Al	
⋮		⋮		⋮	
OH	-½	OH	-½	OH_2	+½

-2	0	+1
(a)	(b)	(c)

Under nearly neutral conditions (b), each of the two oxygens of aluminum–oxygen octahedron connects to one hydrogen, while at the same time it connects to aluminum through one half of the bond. Since one of the oxygens also connects to silicon, this oxygen will carry 1/2 positive charge. Similarly, another oxygen connecting to aluminum carries 1/2 negative charge. In this case, the net charge at the edge surface is zero. When kaolinite is under an acid condition (c), the oxygen originally connected to aluminum and carrying a 1/2 negative charge accepts one proton and thus turns to carry a 1/2 positive charge, making one unit of edge surface carry one positive charge. Under alkaline conditions (a), one unit of edge will carry two negative charges due to the dissociation of H^+ ions from two OH groups connected to silicon.

On the edge face of other layer silicates there can also be the production of variable charge, caused by the dissociation of H from or the adsorption of proton by exposed OH groups of broken bonds. However, because 2:1-type layer silicates carry a large quantity of permanent charge, this kind of variable charge component generally does not manifest itself markedly.

2.1.3 Surface Charge and Surface Potential of a Variable Charge Surface

The surface potential of variable charge systems is controlled by adsorbed potential–determining ions. Hence, it is controlled by the activity of this ion species in the equilibrium solution and is independent of the presence of supporting electrolytes. According to the diffuse double layer theory, the surface charge density σ_v in this system is related to surface potential ψ_0 as

$$\sigma_v = \left(\frac{2C\varepsilon RT}{\pi} \right)^{\frac{1}{2}} \sinh \frac{zF}{2RT} \, (\text{constant } \psi_0) \qquad (2\text{-}4)$$

If potential-determining ions are H^+ or OH^-, the constant surface potential of this electrochemically reversible surface should obey the Nernst equation, that is,

$$\psi_0 = \frac{RT}{zF} \ln \frac{H^+}{H_0^+} \tag{2-5}$$

or

$$\psi_0 = \frac{RT}{F} 2.303 \, (pH_0 - pH) \tag{2-6}$$

where H^+ is the activity of hydrogen ions, and H_0^+ is the activity of hydrogen ions when $\psi_0 = 0$.

When ψ_0 is substituted into equation (2-4), we get:

$$\sigma_V = \left(\frac{2C\epsilon RT}{\pi} \right)^{1/2} \sinh 1.15z \, (pH_0 - pH) \tag{2-7}$$

It is thus seen that the surface charge density of a variable charge system is affected by the valence (z) of compensating ions, dielectric constant (ϵ), temperature (T), electrolyte concentration (C), pH of the solution, and zero point of charge (pH_0) of the variable charge surface.

For 1:1-type supporting electrolytes, when the difference between pH and pH_0 does not exceed one pH unit, $\sinh 1.15z(pH_0 - pH) \approx 1.15(pH_0 - pH)$ at 20°C. Then, the above equation can be simplified to

$$\sigma_V = 0.135 C^{1/2} \, (pH_0 - pH) \tag{2-8}$$

The above mathematic equation derived from the diffuse double layer theory is valid only for electrolytes that are not adsorbed specifically. If there are specifically adsorbed ions in the solution, both pH_0 and σ_V would change. When studying changes of surface charge and zero point of charge of hematite in the presence of various specifically adsorbed ions, Breeuwsma and Lyklema (1973) found that the zero point of charge in 10^{-3} N KNO_3 solution was 8.5 and that it decreased to 6.5 after the addition of 10^{-3} N $Ca(NO_3)_2$. It increased to 9.6 after the addition of K_2SO_4. The quantity of surface charge was also different at different pH.

2.1.4 Systems Consisting of Two Kinds of Surfaces

Except in some occasional cases, variable charge soils consist of both variable charge minerals and constant charge minerals. Therefore, most of them are systems in which variable and permanent charge surfaces coexist. For these systems, it is necessary to describe them with the model of mixed systems (Uehara and Gillman, 1980).

In such mixed systems, the total net surface charge density σ_T should be the sum of permanent surface charge density σ_P and variable surface charge density σ_V, that is,

$$\sigma_T = \sigma_P + \sigma_V \qquad (2\text{-}9)$$

When equation (2-7) is substituted into the above equation, we get:

$$\sigma_T = \sigma_P + \left(\frac{2C\varepsilon RT}{\pi} \right)^{\frac{1}{2}} \sinh 1.15z\,(\,pH_0 - pH\,) \qquad (2\text{-}10)$$

If the simplified equation (2-8) is substituted into equation (2-9), the resulting equation would be

$$\sigma_T = \sigma_P + 0.135\,C^{\frac{1}{2}}\,(\,pH_0 - pH\,) \qquad (2\text{-}11)$$

It can be seen from equations (2-10) and (2-11) that $\sigma_T = \sigma_P$ when $\sigma_V = 0$. This is the pure permanent charge system. When $\sigma_T = 0$, the pH is called the *zero point of net charge* (ZPNC), at which $\sigma_P + \sigma_V = 0$.

When the pH is equal to ZPNC, we get from equation (2-11) that

$$\sigma_P + 0.135\,C^{\frac{1}{2}}\,(\,pH_0 - ZPNC\,) = 0 \qquad (2\text{-}12)$$

Thus, for a mixed system there are two zero points of charge, namely, pH_0 and ZPNC. The relationship between these two parameters is

$$ZPNC = pH_0 + \frac{\sigma_P}{0.135\,C^{\frac{1}{2}}} \qquad (2\text{-}13)$$

where pH_0 is the zero point of charge of the variable charge component. Because this pH_0 does not change with the change in concentration of supporting electrolytes, it is also called the *zero point of salt effect*. On the contrary, ZPNC is dependent on the concentration of electrolytes and will approach pH_0 gradually with the increase in electrolyte concentration. If the quantity of permanent charge is zero, ZPNC would be equal to pH_0, and in this case the system is actually a pure variable charge system.

The relationships of ZPNC, pH_0, and σ_P with the concentration of electrolyte C are shown in Fig. 2.1.

It is known from the above discussion that pH_0 is the pH at which the quantity of net surface charge of the variable charge component in the mixed system equals zero. This kind of charge is produced by the protonation or deprotonation of hydroxyl surfaces of the solid phase. This pH is not affected by the concentration of a supporting electrolyte. The parameter pH_0 is a characteristic feature inherent in a system and is independent of the permanent charge of the system. However, if there are specifically

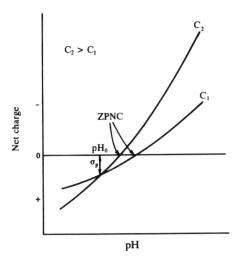

Fig. 2.1. ZPNC, pH_0 and σ_p in relation to electrolyte concentration C.

adsorbed cations or anions in the electrolyte solution, this pH_0 may change. On the other hand, ZPNC is a parameter characterizing the relationship between net surface charge composed of electric charges of all origins and pH in a mixed system. It is affected by both the nature and the concentration of the electrolyte. Colloids with a ZPNC lower than pH_0 carry permanent negative surface charge, whereas those with a ZPNC higher than pH_0 carry permanent positive surface charge. The larger the difference between ZPNC and pH_0, the larger the quantity of permanent surface charge.

Iso–electric point (IEP) is also a parameter for characterizing the surface properties of colloids. IEP of a colloid is the pH at which the net charge on the slip surface (plane of shear) of the electric double layer is zero, and thus the (electrokinetic) potential equals zero. This IEP is affected by quantities of variable and permanent surface charges as well as by specific adsorption of ions. For pure variable charge systems not carrying a permanent surface charge, under conditions with no specific adsorption of ions, surface charge density should equal electrokinetic charge density on the slip surface, and therefore IEP should be identical to pH_0. For mixed systems carrying permanent surface charge, when there is no specific adsorption of ions, IEP should be lower than pH_0 and closer to ZPNC. For variable charge soils, IEP is an important parameter for characterizing their surface properties. This will be discussed in detail in Chapter 7.

The model for describing mixed systems mentioned above is a simplified model derived from the Gouy–Chapman equation and the Nernst equation.

The practical applicability of this model may be limited. Following the increase in electrolyte concentration, the model tends to become inapplicable gradually. It is also inapplicable when there is specific adsorption of ions. The main drawback of this model is that the effect of permanent surface charge has not been considered when describing variable surface charge using the diffuse double layer equation. Actually, electric charges of ions in the double layer are compensated for by both variable and permanent surface charges of the solid, and therefore the assumption of neutrality principle in the derivation cannot hold. Besides, when expressing surface potential with the Nernst equation the effect of permanent surface charge has also been ignored (Madrid et al., 1984). Therefore, the model for mixed systems can only be applied to systems in which the proportion of permanent surface charge is very small.

In mixed systems, the numerical value of pH_0 is not a constant, but is affected by the nature of the electrolyte, although, by definition, pH_0 is the zero point of charge of the variable charge component. It will change when there is specific adsorption of ions (Breeuwsma and Lyklema, 1973; Gallez et al., 1976; Keng and Uehara, 1974; Li, 1985; Pyman et al., 1979a,b). Besides, the quantity of permanent surface charge can also affect the pH_0 value.

The difference between pH_0 and titration zero point is generally regarded as a reflection of the quantity of permanent surface charge. However, some studies showed that the technique of determination and the pretreatment could affect the numerical value of σ_P (Hendershot et al., 1978, 1979; Sakurai et al., 1988).

In view of the above-mentioned reasons, Parker et al. (1979) called the intersection point of proton titration curves in different concentrations of electrolyte solutions the *zero point of salt effect*.

2.1.5 Determination of Surface Charge

Methods for the determination of surface charge may be classified into two categories: ion-adsorption method and potentiometric titration method.

In ion-adsorption method, the soil is allowed to adsorb cations and anions in a salt solution (generally NH_4Cl) of a certain concentration at different pH, and the amounts of adsorbed cations and anions are taken as the quantities of negative and positive surface charge at that pH, respectively. The pH at which the quantities of two adsorbed ion species equal each other is called the *zero point of net charge*. This method was established by Schofield (1949). The basic assumption in this method is that the adsorption of these two ion species is caused solely by electrostatic force, and therefore one ammonium ion neutralizes one negative surface charge while one chloride ion neutralizes one positive surface charge. However, as shall be seen in the subsequent chapters of this book, in the adsorption of chloride ions by variable charge soils, besides electrostatic force, a specific force may also be involved and the quantity of negative surface charge of the soil may increase to a certain extent after adsorption. Thus, the quantities of both positive and negative surface charges as calculated based on this method would be higher than the actual values. Besides, whether the adsorption of ammonium ions by variable charge soils is solely electrostatic in nature cannot be said with certainty. Another complicating factor is that, owing to the extremely strong adsorption energy of aluminum ions, monovalent cations such as ammonium may not be able to replace all of the ions originally adsorbed by the soil (Gallez et al., 1976). Therefore, the results using ion–adsorption method would be conditional. Other methods of this kind, such as the cesium chloride method (Greenland, 1974), seems to be even less feasible.

Potentiometric titration method was introduced to soil research by Van Raij and Peech (1972). In this method, the soil is titrated with acid or alkali in the presence of different concentrations of an electrolyte. Since hydrogen ions and hydroxyl ions are assumed to be potential–determining ions, the quantity of hydrogen ions adsorbed by the soil at a given pH may be taken as the quantity of positive surface charge, and that of hydroxyl ions may be taken as the quantity of negative surface charge at that pH. The intersection point of several titration curves (sometimes they do not overlap entirely) is taken as the zero point of charge of the soil. Thus, the potentiometric titration method is actually one alternative of the ion–adsorption method. From the discussions in Section 2.1.4 it should be noted that this intersection point merely represents the point of zero charge of the variable charge component, that is, pH_0. For the majority of variable charge soils carrying both permanent and variable charge surface, this pH_0 should be different from their ZPNC. Besides, the situation in the relationship between adsorbed hydrogen ions and hydroxyl ions on the one hand and surface charge on the other hand may be rather complex. In this case, the two ion species may not behave merely as potential–determining ions when reacting with the surface of the soil as in the case for ideal reversible surfaces. In particular, when the concentration of hydrogen ions is high, as will be seen in Chapters 10 and 11, these ions can also react with the amorphous aluminum of the solid surface or even the aluminum of clay minerals. Thus, the quantity of hydrogen ions consumed in the reaction may greatly exceed the amount of positive surface charge of the soil (Espinoza et al., 1975; Parker et al., 1979; Sposito, 1992). One possible way to improve the potentiometric titration method is to correct for these secondary reactions (Duquette and Hendershot, 1993a; Schulthess and Sparks, 1986, 1987, 1988).

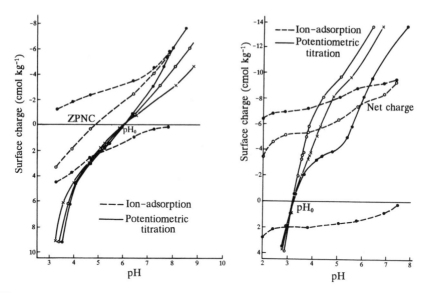

Fig. 2.2. (Left) Comparison of method of determination of surface charge for Hyper-Rhodic Ferralsol.

Fig. 2.3. (Right) Comparison of method of determination of surface charge for Ali-Haplic Acrisol.

Comparisons between the two methods for the determination of surface charge of a Hyper-Rhodic Ferralsol and an Ali-Haplic Acrisol are shown in Figs. 2.2 and 2.3, respectively. It can be seen from the figures that for the Hyper-Rhodic Ferralsol with little permanent surface charge and containing large amounts of iron and aluminum oxides, the zero points of charge as determined by the two methods differ very little. Except at very low pH, the potentiometric titration curve lies close to the ion-adsorption curve. On the contrary, for the Ali-Haplic Acrisol containing a substantial amount of 2:1-type clay minerals, owing to the presence of a large quantity of permanent surface charges, the results determined by the two methods differ markedly.

It can be concluded from the above discussions that both the quantities of positive and negative surface charges and the zero point of charge will be different when determined by the two methods, with the magnitude of difference depending on the composition of the soil (El-Swaify and Sayegh, 1975; Espinoza et al., 1975; Parker et al., 1979; Sposito, 1981; Van Raij and Peech, 1972). At present, no method of quantifying surface charge of soils has universal applicability (Sposito, 1992).

In soil science, the quantity of negative surface charge is commonly called the *cation-exchange capacity* (CEC) of the soil. This concept comes principally from studies on clay minerals and constant charge soils of temperate regions. For variable charge soils, however, because the quantity of negative surface charge is dependent on pH of the medium and is also affected by other factors in determination, the concept CEC in reality does not have strictly scientific meaning. In practical use, it is only possible to refer to the result determined under a specified condition such as at pH 7 as "apparent CEC" for the purpose of comparison among various soils.

2.2 CONTRIBUTION OF SOIL COMPONENTS TO SURFACE CHARGE

2.2.1 General Discussion

Soil components can be broadly classified into three types: clay minerals, oxides, and organic matter. These three components coexist together in soils in a complicated manner and may also have mutual influences. For variable charge soils, various oxides are of special significance, because their amounts are high, the variability of their surface charge is great, and they are the basic cause of the difference in surface charge properties between variable charge soils and constant charge soils. These oxides, including the oxides of iron, aluminum, manganese, titanium and silicon, generally exist in the form of hydrous oxides. All of them are amphoteric, and therefore they have a zero point of charge (ZPC). The approximate ZPC of various oxides in soils are as follows: SiO_2, pH 2; TiO_2, pH 4.5; MnO_2, pH 4; Fe_2O_3, pH 6.5-8; Al_2O_3, pH 7.5-9.5 (Parks, 1965). These differences are caused by the difference in affinity of various coordinating ions with electrons.

Hydrous oxides can exist in different forms in soils. The contributions of various forms of oxides to surface charge of the soil differ considerably. Besides, the difference in interactions between amorphous hydrous oxides and crystalline minerals may also have different effects on surface charge.

2.2.2 Iron Oxides

Iron oxides are the principal materials in producing positive surface charge in variable charge soils. This is because the Fe-OH groups on their surface can adsorb H^+ ions from the solution when the pH is lower than their zero point of charge. The contribution of different forms of iron oxides to positive surface charge differs. In various soils both the content and the form of iron oxides are different. In addition, a considerable part of these oxides is combined with clay minerals in various ways. Therefore, iron oxides may have a complex effect on surface charge of the soil.

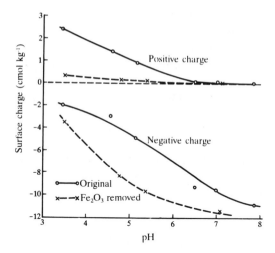

Fig. 2.4. Change in surface charge of a Xanthic–Haplic Acrisol with pH before and after removal of free iron oxides (Zhao and Zhang, 1991).

The variations of positive and negative surface charges with pH before and after the removal of iron oxides for four variable charge soils are shown in Figs. 2.4 to 2.7. It can be seen that for all the soils the quantity of

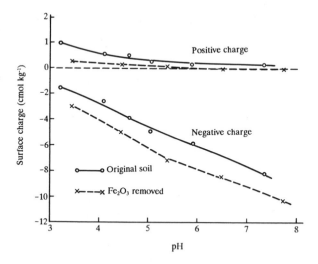

Fig. 2.5. Change in surface charge of Ferrali–Haplic Acrisol (Xishuangbanna) with pH before and after removal of free iron oxides (Zhao and Zhang, 1991).

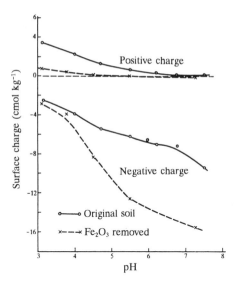

Fig. 2.6. Change in surface charge of Ferric-Haplic Acrisol with pH before and after removal of free iron oxides (Zhao and Zhang, 1991).

positive surface charge increases with the decrease in pH. At pH 3.5 the quantities for the Xanthic-Haplic Acrisol, Ferrali-Haplic Acrisol, Ferric-Haplic Acrisol and Hyper-Rhodic Ferralsol are 2.3, 0.8, 3.0, and 3.6

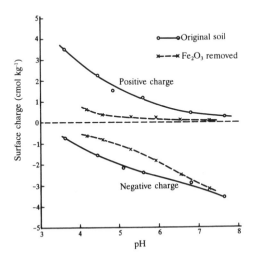

Fig. 2.7. Change in surface charge of Hyper-Rhodic Ferralsol with pH before and after removal of free iron oxides (Zhao and Zhang, 1991).

cmol kg^{-1}, respectively. At pH 7, the soils practically do not carry positive surface charge, except for the Hyper–Rhodic Ferralsol which still carries about 0.5 cmol kg^{-1} of positive charge. After the removal of iron oxides, except at very acid conditions, positive surface charge disappears nearly completely. The free Fe_2O_3 contents of the four soils are 9.58%, 7.29%, 15.4%, and 21.2%, respectively. The decrease in quantity of positive surface charge after the removal of iron oxides coincides with the content of free Fe_2O_3, suggesting that free iron oxides are the principal carriers of positive surface charge in soils.

However, the contribution of free Fe_2O_3 to positive surface charge in various soils differs greatly when calculated on an unit weight basis, especially at low pH. As can be seen in Fig. 2.8, for the Ferrali–Haplic Acrisol containing the lowest amount of iron oxides, the contribution to positive surface charge for one gram of Fe_2O_3 is only 55 μmol at pH 4, and it decreases slightly with the rise in pH. For the Xanthic–Haplic Acrisol with an intermediate amount of iron oxides the contribution is as high as 170 μmol at pH 4, and it decreases to 20 μmol at pH 7. Conversely, for the Hyper–Rhodic Ferralsol with the highest amount of iron oxides the contribution of one gram of Fe_2O_3 is not the largest. This difference in contribution to positive surface charge for various soils is related to the form of iron oxides, because it has been known that the amount of positive charge carried by amorphous iron oxides is larger than that carried by the crystalline form (Shao and Wang, 1991).

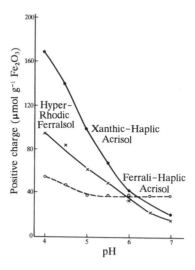

Fig. 2.8. Contribution of Fe_2O_3 to positive surface charge in various soils (Zhao and Zhang, 1991).

One part of hydrous iron oxides in soils is combined with clay minerals. This combination may affect the negative surface charge of the soil. For the Xanthic–Haplic Acrisol, Ferrali–Haplic Acrisol and Ferric–Haplic Acrisol, the quantity of negative surface charge increases, with the increment differing with the type of soil, while for the Hyper–Rhodic Ferralsol the quantity decreases after the removal of iron oxides.

The sign and quantity of net surface charge of a soil is determined by the algebraic sum of positive and negative charges. When the net charge is positive, its quantity increases with the decrease in pH, whereas the quantity increases with the rise in pH if the net charge is negative. The ZPNC may reflect the net charge status of a soil to a certain extent. For the Ferrali–Haplic Acrisol, no ZPNC can be observed. The ZPNC of Xanthic–Haplic Acrisol and Ferric–Haplic Acrisol is about pH 3.5, and it disappears after the removal of iron oxides. The ZPNC of the Hyper–Rhodic Ferralsol decreases to pH 4.2 from an original value of pH 4.8 when iron oxides are removed.

It is generally supposed that free iron oxides affect the quantity of negative surface charge in two ways. The positive charge carried by iron oxides can compensate for a portion of negative surface charge sites. Also, iron oxides can exert a physical masking effect on negative surface charge sites. The increment of net negative surface charge after the removal of iron oxides may be regarded as the liberation of negative surface charge sites as a result of the elimination of the above two factors. Since the decrement of positive surface charge after the removal of free iron oxides is approximately equivalent to the quantity of positive charge carried by iron oxides, the difference between the increment of net negative surface charge and the

Table 2.1   Effect of Free Iron Oxides on Net Negative Surface Charge[a] of Soils (Zhao and Zhang, 1991)

pH	Xanthic Acrisol		Ferrrali–Haplic Acrisol		Ferric Acrisol		Hyper–Rhodic Ferralsol	
	E.N.[b]	P.M.	E.N.	P.M.	E.N.	P.M.	E.N.	P.M.
4	1.65	2.57	0.4	1.57	2.01	1.16	1.8	0.6
5	0.95	4.03	0.26	1.71	0.95	5.07	1.6	0.9
6	0.40	2.98	0.26	1.56	0.45	6.50	1.3	0.7
7	0.20	2.39	0.27	1.55	0.30	6.65	0.8	-0.3

[a] Unit: cmol kg^{-1}
[b] E.N., electrostatic neutralization; P.M., physical masking.

decrement of positive surface charge after the removal of free iron oxides may be regarded as the effect of physical masking on negative surface charge. The results of such an estimation are given in Table 2.1. For the Hyper-Rhodic Ferralsol the effect of free iron oxides on net negative surface charge is caused chiefly through electric neutralization, whereas for the other three soils the effect of physical masking greatly exceeds the effect of electric neutralization. However, it should be noted that during the removal of iron oxides by the DCB method a part of active aluminum may also be removed, and therefore the effect of a part of active aluminum should also be included in the effect of free iron oxides discussed above.

2.2.3 Aluminum Oxides

In addition to alumino-silicates, there is a certain amount of free aluminum oxides present in variable charge soils. These aluminum oxides are present in forms of minerals such as gibbsite, boehmite, diaspore, and amorphous aluminum oxides. Their surfaces belong to the hydroxyl type, and they can carry positive charge or negative charge due to protonation or deprotonation, respectively.

The changes in surface charge of three soils after the coating with aluminum oxides in the form of gibbsite are shown in Figs. 2.9 to 2.11, respectively. For all the soils the quantity of positive surface charge increases

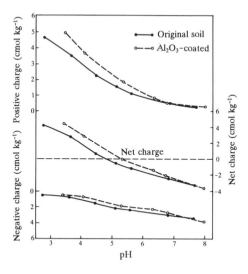

Fig. 2.9. Change in the surface charge of Hyper-Rhodic Ferralsol after coating with aluminum oxides (Zhao and Zhang, 1992).

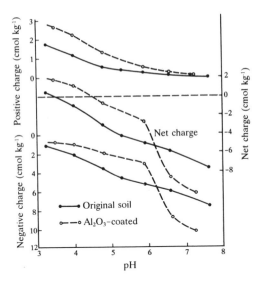

Fig. 2.10. Change in the surface charge of Ali–Haplic Acrisol (Guilin) after coating with aluminum oxides (Zhao and Zhang, 1992).

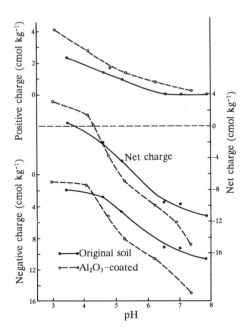

Fig. 2.11. Change in the surface charge of Xanthic–Haplic Acrisol after coating with aluminum oxides (Zhao and Zhang, 1992).

markedly after coating with aluminum oxides. The increase is enhanced by the lowering of pH. For the Hyper–Rhodic Ferralsol, the quantity of negative surface charge increases slightly after the addition of aluminum oxides. For the Ali–Haplic Acrisol, the quantity of negative surface charge decreases markedly when the pH is lower than 6.2 but increases remarkably when the pH is higher than 6.2. For the Xanthic–haplic Acrisol, the quantity of negative surface charge increases markedly when the pH is higher than 4.2. Based on these data, it can be suggested that aluminum oxides precipitated in the soil may carry positive charges under acid conditions. However, their effect on negative surface charge varies with the soil. For the Hyper–Rhodic Ferralsol, aluminum oxides only exert a light physical masking effect. For the Ali–Haplic Acrisol and Xanthic–Haplic Acrisol, they have a marked masking effect on negative surface charge sites at low pH but cause an increase in negative surface charge at higher pH, because in the latter case the pH is perhaps above the zero point of charge of these oxides.

The zero point of charge of aluminum oxides is generally higher than the ZPNC of soils. Therefore, the ZPNC of soils increases after coating with aluminum oxides. Such a situation is shown in Table 2.2 for several soils. As seen in the table, the magnitude of increase in ZPNC varies considerably for different soils, caused presumably by the difference in original surface conditions among soils and thus the difference in properties of aluminum oxides precipitated on them.

2.2.4 Silicon Oxides

Because the zero point of charge of SiO_2 lies at about pH 2, silicon oxides carry negative charge in soils under ordinary pH conditions. However,

Table 2.2 Change in Zero Point of Net Charge (ZPNC) of Soils After Coating with Aluminum Oxides (Zhao and Zhang, 1991)

Soil	Free Fe_2O_3 (g kg^{-1})	ZPNC Original	ZPNC Coated
Xanthic–Haplic Acrisol (Pingba)	95.8	3.6	4.3
Ali–Haplic Acrisol (Guilin)	78.0	3.5	4.5
Ferrali–Haplic Acrisol (Guangzhou)	68.2	3.3	3.7
Ferrali–Haplic Acrisol (Yunnan)	67.2	3.3	4.3
Hyper–Rhodic Ferralsol (Kunming)	212	4.8	5.3

the quantity of negative charge carried by amorphous silicon oxides is not large within the pH range of 2-6 (Perrot, 1977; Pyman et al., 1979a,b). Greenland and Mott (1978) showed that the charge density of silica gel at pH 4 was 1.67 charges nm^{-2}, corresponding to only one-third of the theoretical maximum value, 5 charges. This implies that within the ordinary pH range of variable charge soils the surface of the majority of silicon oxides is present in the form of uncharged SiOH group, leaving only a small portion in the form of negatively charged $Si-O^-$ group. Therefore, the significance of silicon oxides in variable charge soils is small when compared with iron and aluminum oxides.

2.2.5 Layer Silicates

Layer silicates are a group of clay minerals constructed of silicon–oxygen tetrahedra and aluminum–oxygen octahedra arranged with a certain regularity. Owing to the isomorphous substitution of silicon in the tetrahedron and/or aluminum in the octahedron by cations with a lower valence, permanent negative charges are produced, with the quantity of the charge varying with the extent of substitution. On the edge surface of layer silicate crystals there are both Si-OH groups and Al-OH groups. Under ordinary soil pH conditions the former group dissociates H^+ ions and thus becomes negatively charged while the latter group adsorbs H^+ ions and thus becomes positively charged.

Since the quantity of negative charge carried by various clay minerals differs greatly, the quantities of negative surface charge of various soils constituted of different clay minerals are different. It can be seen from

Table 2.3 Negative Surface Charge of Soil Colloids (<1 μm) with Different Clay Mineralogical Compositions at pH 7

| Soil | pH | | Neg. Charge | Dominant |
	H_2O	KCl	(cmol kg^{-1})	Minerals[a]
Rhodic Ferralsol (Hainan)	5.0	4.8	5.2	Kl.,Gb.
Ferrali-Haplic Acrisol (Guangdong)	4.8	4.2	12	Kl.
Ali–Haplic Acrisol (Jiangxi)	4.8	4.1	22	Kl.,Il,Vm.
Cambisol (Nanjing)	7.0	6.1	40	Il.,Vm.,Mt.
Kastanozem (Mongolia)	–	–	91	Mt.

[a] Kl., kaolinite; Gb., gibbsite; Il., hydrous mica;
Vm., vermiculite; Mt., montmorillonite.

Table 2.3 that the quantity of negative surface charge carried by different soil colloids is closely related to the clay mineralogical composition. For variable charge soils dominated by kaolinite, the quantity of negative surface charge is much smaller than that of constant charge soils such as Cambisol and Kastanozem.

The relationship between the net charge–pH curve and clay mineralogical composition of four variable charge soils is shown in Fig. 2.12. For these soil samples from which iron oxides have been removed, the quantity of net negative surface charge carried by Rhodic Ferralsol consisting principally of kaolinite and gibbsite is the smallest. The quantity of net negative surface charge carried by Ferrali–Haplic Acrisol with kaolinite as the predominant clay mineral is comparatively larger. The Ferric–Haplic Acrisol derived from limestone and containing a large amount of vermiculite besides kaolinite carries the largest quantity of net negative surface charge. For the Xanthic–Haplic Acrisol, the quantity of net negative surface charge is also quite large because it contains considerable amounts of vermiculite and hydrous mica, although the mineralogical composition is dominated by kaolinite.

For mixed systems consisting of permanent charge minerals and variable charge minerals, the value σ_p on the potentiometric titration curve can reflect the quantity of permanent surface charge in a broad way. Therefore, the value σ_p is closely related to the mineralogical composition of the soil (Hendershot and Lavkulich, 1978; Laverdiere and Weaver, 1977).

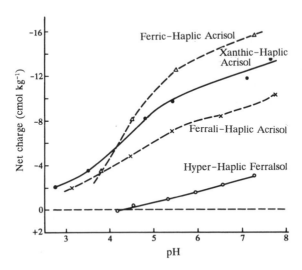

Fig. 2.12. Net surface charge–pH curves of variable charge soils (free Fe_2O_3 removed) with different clay mineralogical composition.

2.2.6 Organic Matter

Humus carries 200–500 cmol kg^{-1} of negative charge. Therefore, it may make an important contribution to negative surface charge of soil colloids. For Acrisols of south China, the increase of humus by 1% in soil colloids induces an increase in negative surface charge by about 1 cmol per kilogram of colloid, whereas for soils at the lower reaches of the Yangtze River the corresponding figure is 2 cmol or larger (Zhang and Jiang, 1964). Despite the fact that the contribution of humus to negative surface charge of variable charge soils is smaller than that of constant charge soils in absolute quantity, its relative contribution is larger than the latter type of soil, because the mineral part of this type of soil carries only a small quantity of negative surface charge. For paddy soils derived from Acrisols, the contribution of organic matter to total negative surface charge of the soil amounts to 5–42%, averaging 21% (Yu, 1976). For the surface layer of some soils of Africa, the quantity of negative surface charge is linearly correlated with the content of organic carbon (Greenland, 1986).

In soils a considerable part of organic component is combined with mineral component through various bonding forces. Because of this combination, the amount of negative surface charge of the organo–mineral complex is smaller than the sum of charges carried by the two components separately. This phenomenon is called nonadditivity of negative surface charge of soil colloids. The mechanism of this nonadditivity is not known. Two cases may be involved. Negatively charged organic colloids may combine with positively charged iron and aluminum oxides or with the positive sites at edges of clay minerals. Another possibility may be that humus is precipitated on the surface of mineral colloids through coagulation by polyvalent cations, thus exerting a masking effect on the negative surface charge sites of mineral colloids. From the change in positive and negative surface charges of the clay fraction of Rhodic Ferralsol after the removal of organic matter shown in Fig. 2.13, it seems that, for such soil colloid containing a large amount of free iron oxides, the decrease in amount of net negative surface charge at pH higher than the ZPNC after the removal of organic matter is caused chiefly by the increase in positive charge rather than by the decrease in negative charge. This indicates the presence of both of the above–mentioned cases, and it also shows the important role of iron oxides in the formation of organo–mineral complexes in soils.

When the ratio of the decrement of negative surface charge after the removal of organic matter to the content of organic matter is taken as the amount of "apparent negative charge" of soil organic matter, it has been estimated that the quantity of this apparent negative charge of humus in the clay fraction of Acrisols in south China is in the range of 80–120 cmol kg^{-1}, whereas the corresponding value for Cambisols of central China is about

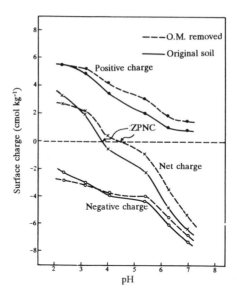

Fig. 2.13. Change in surface charge of the clay fraction of Rhodic Ferralsol after removal of organic matter (Zhang and Jiang, 1964).

250 cmol kg^{-1}. For paddy soils, the apparent negative charge of those soils derived from Acrisols is also smaller in quantity than those derived from neutral alluvial soils. These phenomena imply that iron and aluminum oxides are important factors in inducing the nonadditivity of surface charge in the formation of organo–mineral complexes in soils.

Organic matter can also affect zero point of charge of the soil, because it is an important variable charge component. The zero point of charge of the A horizon is generally lower than that of the B horizon (Gallez et al., 1976; Hendershot et al., 1979; Laverdier and Weaver, 1977; Van Raij and Peech, 1972). The zero point of charge of soils increases to various degrees after the removal of organic matter (Hendershot and Lavkulich, 1979). The zero point of charge of soils may be linearly correlated with the content of organic matter (Basu et al., 1986).

2.3 INFLUENCE OF ENVIRONMENTAL FACTORS ON SURFACE CHARGE

2.3.1 pH

Among various environmental factors, pH is the most important one in affecting the surface charge of soils. The data cited in the previous sections, as shown in Figs. 2.1 to 2.13, indicate that the quantities of surface charge of either variable charge soils or their components are affected by pH. As shown in equation (2–3), in the production of surface charge both the dissociation of H^+ ions from the hydroxylated surface and the adsorption of H^+ ions by the hydroxylated surface are dependent on the pH of the medium. This is the reason why variable charge minerals such as iron and aluminum oxides carry negative charge when the pH is higher than their zero point of charge and carry positive charge when the pH is lower than their zero point of charge. Therefore, when examining the surface charge of a given soil, it is necessary to consider the pH of that soil at the same time.

2.3.2 Concentration of Electrolyte

The quantity of surface charge of variable charge soils varies with the concentration of electrolyte. This is another characteristic feature of variable charge soils as compared to constant charge soils. The mechanisms of the effect of electrolyte on surface charge of soils are rather complex. According to the theory of diffuse double layer, two mechanisms may exist. The surface of variable charge minerals is one kind of reversible constant potential surface. The charge density σ on this type of surface is proportional to the square root of ion concentration C of the solution, that is, $\sigma = K$ $(C)^{1/2}$, where K is a constant (van Olphen, 1977) [see equation (2–7)]. This is the principal cause of the effect of electrolyte concentration on the quantity of surface charge of variable charge colloids. Besides, on the surface of variable charge colloids there are positive and negative charge sites existing independently. These sites have their diffuse double layers of oppositely charged ions. Because the thickness of the diffuse double layer is inversely proportional to the square root of the concentration of ions, the decrease in ion concentration would induce an increase in thickness of the double layer. When the thickness extends to a certain extent, two adjacent diffuse double layers may overlap partly, causing mutual compensation of positive and negative charges. This would result in the decrease in adsorption of anions and cations. This is another possible reason for the effect of electrolyte concentration on the quantity of surface charge of variable charge colloids. In soils, both of these two mechanisms may operate (Barber and Rowell, 1972). Practical measurements showed that for various variable charge soils the slope of the linear correlation line between the quantity of positive, negative, or net surface charge and the concentration of electrolyte differs with the type of the soil (Okamura and Wada, 1983; Uehara and Gillman, 1981; Wada and Okamura, 1983; Wada and Wada, 1985).

Variable charge soils generally carry a certain amount of permanent charge. Therefore, such soils are mixed systems in which both variable charge and permanent charge coexist. According to the model for describing the surface charge of mixed systems, the relationships between ZPNC, pH_0, and permanent surface charge density σ_p on one hand and electrolyte concentration C on another hand are as follows:

$$\text{ZPNC} = pH_0 + \frac{\sigma_P}{\text{constant} \times C^{\frac{1}{2}}} \tag{2-13a}$$

In the equation, pH_0 is a parameter characteristic of a given system and is independent of either the quantity of permanent surface charge or the concentration of electrolyte, while ZPNC varies with the change in C. If the colloid carries permanent negative surface charges, ZPNC would be lower than pH_0. Conversely, ZPNC would be higher than pH_0 when the colloid carries permanent positive surface charges.

In Table 2.4, the ZPNC of Rhodic Ferralsol decreases while the quantity of surface charge at ZPNC increases with the increase in electrolyte concentration. Conversely, the ZPNC of Hyper–Rhodic Ferralsol remains nearly unchanged (at about pH 4.8) within the electrolyte concentration range of $0.005-0.1$ mol L^{-1}, although the quantity of surface charge increases with the increase in electrolyte concentration. It has been shown that the pH_0 of this soil is also 4.8 and that this soil carries very little permanent surface charge. According to equation (2–13a), this means that the term $(\sigma_p)/[\text{constant} \times (C)^{\frac{1}{2}}]$ equals zero. This is an extreme case for mixed systems in which the soil only carries variable surface charges, or more probably the quantities of permanent negative charge and permanent positive charge equals each other, so that the quantity of net permanent surface charge equals zero (Zhang et al., 1989). This extreme case seems quite rare.

Table 2.4 Effect of Electrolyte (KCl) Concentration on ZPNC of Soils

Concentration (M)	Rhodic Ferralsol		Hyper–Rhodic Ferralsol	
	ZPNC	Charge (cmol kg^{-1})	ZPNC	Charge (cmol kg^{-1})
0.005	4.3	0.6	4.7	1.3
0.01	4.1	1.1	4.8	1.6
0.05	3.9	1.6	4.8	2.2
0.1	3.6	2.3	4.9	2.2

2.3.3 Kind of Electrolyte

The valence, ionic radius, and thickness of hydrated layer of cations and anions of various electrolytes are different. These would induce the difference in interactions between these ions and the surface of soil colloids. As a consequence, the surface charge properties of a given soil would be affected differently by different electrolytes. In this respect, the data in the literature are quite inconsistent. For example, the quantities of negative surface charge and net surface charge of seven tropical soils of Brazil showed a certain correlation with the kind of the electrolyte, with the zero point of charge in $MgCl_2$, $MgSO_4$, and K_2SO_4 solutions lower than that in KCl solution (Morais et al., 1976). The zero points of charge of four Nigerian tropical soils in $MgCl_2$ solution were slightly higher than those in NaCl solution, whereas the values in K_2SO_4 solution were much higher than those in NaCl or $MgCl_2$ solution (Gallez et al., 1976). The pH_0 values of two Hawaiian soils in $CaCl_2$ solution and in $MgCl_2$ solution were close to those in NaCl solution, but they were much higher in Na_2SO_4, $MgSO_4$, and $CaSO_4$ solutions (Li, 1985). The zero points of charge of Brazilian Oxisols in $CaCl_2$ solution and in $MgSO_4$ solution were lower and higher than those in NaCl solution, respectively (Van Raij and Peech, 1972). The pH_0 value of Oxisols decreased after the adsorption of phosphate (Stoop, 1980).

The potentiometric titration curves of the colloid fraction of Rhodic Ferralsol in two electrolyte solutions are shown in Figs. 2.14 and 2.15,

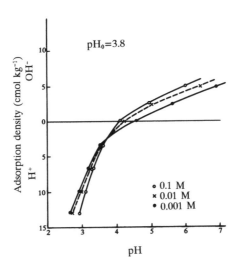

Fig. 2.14. Zero point of charge pH_0 of the colloid fraction of Rhodic Ferralsol in KCl solution (Zhang and Zhao, 1988).

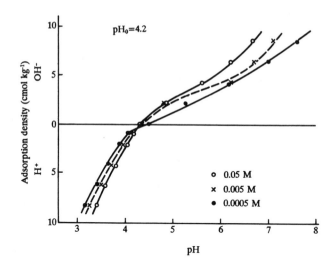

Fig. 2.15. Zero point of charge pH_0 of the colloid fraction of Rhodic Ferralsol in $CaCl_2$ solution (Zhang and Zhao, 1988).

respectively. The zero points of charge of the soil in KCl solution is 3.8, while in $CaCl_2$ solution it is 4.2.

It appears that the quantities of positive and negative surface charges as well as the zero point of charge of a soil in an electrolyte solution are determined chiefly by the relative affinities of the two constituent ion species of the electrolyte with the soil. In particular, the specific adsorption of one ion species can affect the surface charge of the soil markedly. This will be discussed in the next section.

2.3.4 Specific Adsorption of Anions

Variable charge soils can adsorb some cations and anions specifically. These specific adsorption would induce the change in surface charge properties of the soil. The effect of specific adsorption of cations shall be discussed in chapter 4. In this chapter, the effect of specific adsorption of anions on surface charge will be examined.

Some anions can exchange with the ligand OH or OH_2 of the coordination shell of metal atoms on the surface of soil colloids, and thus they are adsorbed on the surface. This reaction occurs within the Helmholtz layer of the electric double layer and is called *coordination adsorption* or *specific adsorption*. The mechanisms of coordination adsorption of anions shall be discussed in detail in Chapter 6. Suffice it to say that this specific adsorption of anions can result in the decrease in quantity of positive surface charge and/or the increase in quantity of negative surface charge. For variable charge soils, specific adsorptions of sulfate, silicate, and particularly phosphate are of greater significance.

According to an experiment, the treatment of the clay fraction of a Rhodic Ferralsol with silicate led to the decrease of pH_0 from 3.8 to 3.5. Treatment with phosphate resulted in the change of pH_0 from 3.8 to 3.2 and the change of ZPNC from 3.7 to 3.4. The change in surface charge would induce the change in zeta potential and thus the change in IEP. Such a change in IEP for this colloid is shown in Fig. 2.16. As seen in the figure, after treatment with phosphate or silicate, the IEP decreased to 3.75 and 4.2, respectively, from an original value of 4.4, indicating that the effect of phosphate ions on surface charge of soil colloid is greater than that of silicate ions.

The extent of the effect of specific adsorption of anions on surface charge of soils is related to the amount of the anions. It can be seen from Fig. 2.17 that, following the increase in sulfate ions, the quantity of positive surface charge of Rhodic Ferralsol decreased sharply at first and then remained practically unchanged when the amount of sulfate exceeded 5 cmol kg⁻¹, whereas the quantity of negative surface charge increased gradually to a maximum value of about 15 cmol kg⁻¹. Generally speaking, the decrement of positive surface charge exceeded the increment of negative surface charge, especially when the amount of sulfate added was low. Thus, it appears that when the concentration of sulfate was low the

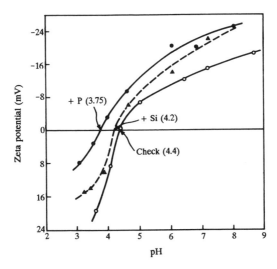

Fig. 2.16. Effect of phosphate and silicate on IEP of the colloid fraction of Rhodic Ferralsol (Zhang and Zhao, 1988).

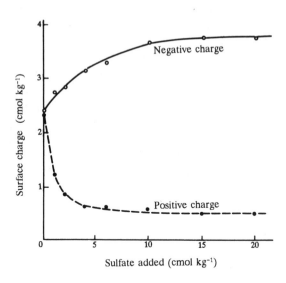

Fig. 2.17. Effect of sulfate on surface charge of Rhodic Ferralsol at pH 5 (Zhang et al., 1989).

effect of sulfate ions was chiefly to neutralize positive charges on the soil colloid surface.

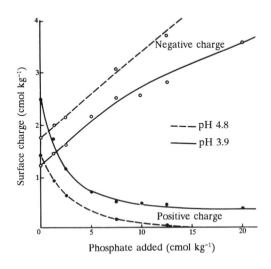

Fig. 2.18. Effect of phosphate on surface charge of Rhodic Ferralsol at different pH.

For phosphate ions, the quantity of positive surface charge of the Rhodic Ferralsol decreased in a parabolic fashion, while the quantity of negative surface charge increased gradually with the increase in phosphate. The extent of change in positive and negative surface charges as well as the net increment in negative surface charge were related to pH (Fig. 2.18). The net increment in negative charge was smaller than that caused by the adsorption of sulfate ions, indicating that in this case the transfer of electric charge was less than that during sulfate adsorption. Hingston et al. (1968) observed a similar phenomenon. This can be explained by closer proximity of phosphate ions to clay surface as compared to sulfate ions (Barrow, 1985).

The extent of the effect of phosphate ions on surface charge of soils differs with the type of the soil. As seen in Fig. 2.19, for the Ferric–Haplic Acrisol derived from limestone and the Ferrali–Haplic Acrisol derived from Quaternary deposit, because the quantity of positive surface charge is small, the magnitude of change is not large, while the quantity of negative charge increases remarkably with the increase in phosphate.

Despite the increased effect of anions on surface charge of the soil with the increase in anion adsorption, the increment of net negative charge per unit of adsorbed anions decreases with the increase in adsorption, as shown in Fig. 2.20 for phosphate.

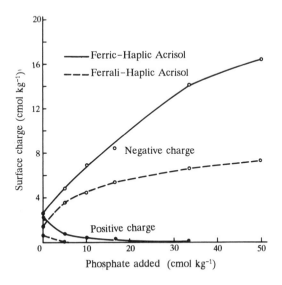

Fig. 2.19. Effect of phosphate on surface charge of Ferric–Haplic Acrisol and Ferrali–Haplic Acrisol (Xishuangbanna) at pH 4.

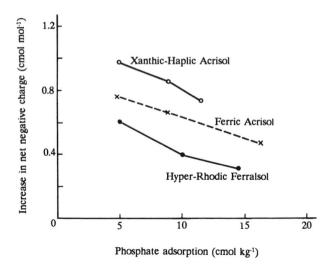

Fig. 2.20. Change in net negative surface charge of soils at pH 4 after adsorption of different amounts of phosphate.

All the experimental data presented in Figs. 2.16 to 2.20 were obtained when the pH was kept nearly constant. Actually, for variable charge soils, the effect of specific adsorption of anions on the surface charge of soils is closely related to pH. As shall be shown in Chapter 6, the specific adsorption of anions is profoundly affected by pH. On the other hand, after anion adsorption the surface charge property varies at different pH.

The increases in net negative surface charge of three variable charge soils after the adsorption of phosphate at different pH are shown in Fig. 2.21. The overall tendency is that the higher the pH the larger the increment of net negative surface charge induced by unit phosphate. This is the result of the combined effect of several factors. Generally speaking, the lower the pH, the more the adsorption of anions. On the other hand, the more the adsorption, the smaller the increment of net negative surface charge per unit of anions adsorbed (Fig. 2.20). For phosphate ions in particular, because in the chemical equilibrium $H_2PO_4^- \rightleftharpoons PO_4^{3-}$ a high pH favors the rightward reaction and thus the existence of more PO_4^{3-} ions, the adsorption would be enhanced at high pH. Besides, for a given soil the higher the pH, the larger the quantity of net negative surface charge. In Fig. 2.21, the increases in net negative surface charge for the three soils are more remarkable when the pH is higher than about 5. This highlights the important role of the last two factors. The role of the last factor can be evidenced in Fig. 2.22. Despite the reduced increment of positive surface charge after phosphate adsorption when the pH is higher, the increment of

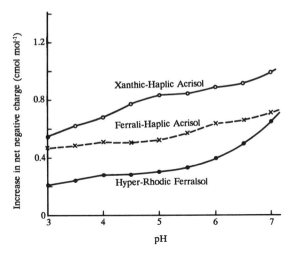

Fig. 2.21. Increase in net negative surface charge of variable charge soils after phosphate adsorption at different pH values.

negative surface charge exceeds the decrement of positive surface charge. Hence, the overall result is that the quantity of negative surface charge increases with the increase in pH.

Owing to the decrease in positive surface charge and the increase in negative surface charge after specific adsorption of anions, both zero point of charge and ZPNC would decrease. However, because the estimation of

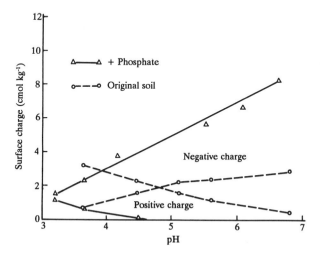

Fig. 2.22. Change in surface charge of Hyper–Rhodic Ferralsol at different pH after phosphate adsorption.

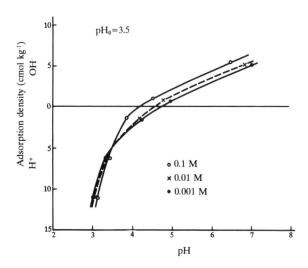

Fig. 2.23. Effect of sulfate ions (0.005 M K_2SO_4) in KCl solution on pH_0 of the colloid fraction of Rhodic Ferralsol (Zhang and Zhao, 1988).

pH_0 is based on the adsorption of H^+ ions and OH^- ions, the situation in the effect of specific anions may be complicated. For example, if a comparison between Fig. 2.14 and Fig. 2.23 is made it can be found that when a small amount of K_2SO_4 is present in the KCl solution the specific adsorption of sulfate ions leads to a decrease in pH_0 of the colloid fraction of Rhodic Ferralsol from the original value of 3.8 to 3.5, and the pH_0 increases to 4.2 when the electrolyte is solely K_2SO_4 (Fig. 2.24). In the literature, reports are also quite inconsistent. In the majority of cases the pH_0 decreased after specific adsorption of anions (Hingston et al., 1972; Parks, 1965, 1967; Wann and Uehara, 1978a,b). There are also reports in which the pH_0 increased (Arnold, 1977; Breeuwsma and Lyklema, 1973; Gallez et al., 1976; Keng and Uehara, 1974). In order to explain the latter phenomenon, some authors (Breeumsma and Lyklema, 1973; Uehara and Gillman, 1981) supposed that, in the presence of a large amount of sulfate, SO_4^{2-} ions are adsorbed in the inner layer of the electric double layer. This would induce the adsorption of H^+ ions or the desorption of OH^- ions by or from the solid surface, resulting in an increase in positive surface charge. In this case, for the compensation of positive surface charge to achieve the zero point of charge, the required amount of OH^- ions would be increased or that of H^+ ions decreased. This is the reason why the pH_0 of the colloid increases when K_2SO_4 is present as the electrolyte. Pyman et al. (1979a,b) thought that when the electrolyte solution contains a small portion of specific ions, pH_0 corresponds to the actual zero point of charge. On the

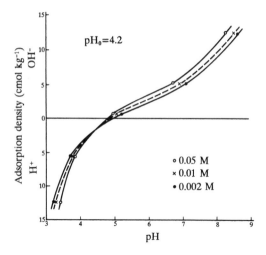

Fig. 2.24. Zero point of charge of the colloid fraction of Rhodic Ferralsol (Zhang and Zhao, 1988).

other hand, when specific ions are the only anion species the intersection point of several titration curves is not the real zero point of charge of the colloid. In the latter case, the surface of the colloid still carries negative charges.

From the above discussions it can be concluded that, in addition to the basic factor hydrogen ions, other ions, particularly the specifically adsorbed anions such as phosphate and sulfate, can also affect the surface charge properties of soils. Thus, variable charge soils and ions are mutually affected. This is different from constant charge soils in which it is the surface charge that determines the relationship between the soil and various ions. Under field conditions, soil pH as well as ionic composition of soil solution are constantly changing. Therefore, the surface charge status of variable charge soils is in a state of constant change.

2.4 SURFACE CHARGE AND SOIL TYPE

2.4.1 Quantity and Variability of Surface Charge

The types of variable charge soils are numerous. Owing to the differences in clay mineralogical composition and particularly in the kind, form, and amount of iron and aluminum oxides, the quantities of surface charges as well as the changing pattern of these charges with the change in pH are remarkably different.

The changes in surface positive and negative charges of the clay fraction

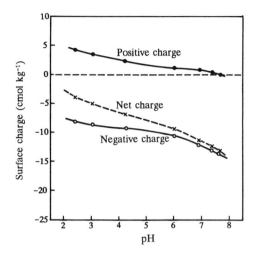

Fig. 2.25. Change in surface charge with pH for the clay fraction of Ali–Haplic Acrisol (Zhang and Jiang, 1964).

of the subsoil of three representative variable charge soils of China are shown in Figs. 2.25 to 2.27, respectively. For the Ali–Haplic Acrisol derived from Quaternary red clay containing kaolinite and hydrous mica as the predominant clay minerals and 9.1% of free Fe_2O_3, the clay fraction carries 16 cmol kg^{-1} of permanent negative surface charge and 7 cmol kg^{-1} of variable negative surface charge at pH 7.7. The quantity of positive surface charge at pH 3 is 3 cmol kg^{-1}. The Ferrali–Haplic Acrisol derived from granite contains well–crystallized kaolinite as the predominant clay mineral. It contains only a small amount of 2:1–type clay minerals. The free Fe_2O_3 content of the clay fraction is 10.8%. The clay fraction of this soil carries 10 cmol kg^{-1} of permanent negative surface charge, 3 cmol kg^{-1} of variable negative surface charge at pH 7.7, and 3.5 cmol kg^{-1} of positive surface charge at pH 3. The dominant clay minerals of the Rhodic Ferralsol are kaolinite, gibbsite, and hematite. The free Fe_2O_3 content of the clay fraction is as high as 15.9%. The clay fraction does not carry a noticeable amount of permanent negative surface charge. The quantity of variable negative surface charge is smaller than that of the above–mentioned two types of soils. The quantity of positive surface charge at pH 3 is 5 cmol kg^{-1}. For this type of soil the ZPNC of the clay fraction is at pH 3.9, below which the quantity of positive surface charge exceeds that of negative surface charge.

The magnitudes of variation in net surface charge with the change in pH for the three types of soils are also different. For the clay fraction of Rhodic Ferralsol, Ferrali–Haplic Acrisol, and Ali–Haplic Acrisol, the variation occurs within the pH range of 3–7 are 8.8, 6.5 and 9.5 cmol kg^{-1},

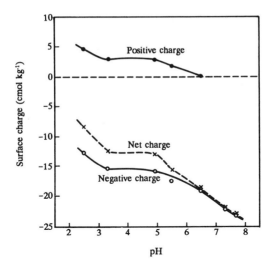

Fig. 2.26. Change in surface charge with pH for the clay fraction of Ferrali–Haplic Acrisol (Zhang and Jiang, 1964).

respectively. If calculated on the basis of relative value, it can be shown that the variability is the largest for the Rhodic Ferralsol and the smallest for the Ali–Haplic Acrisol, with the Ferrali–Haplic Acrisol intermediate in position.

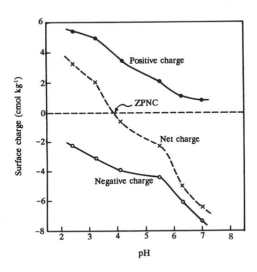

Fig. 2.27. Change in surface charge with pH for the clay fraction of Rhodic Ferralsol (Zhang and Jiang, 1964).

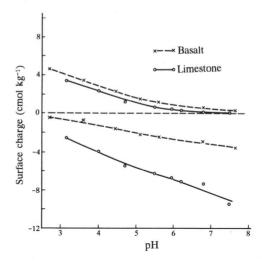

Fig. 2.28. Effect of parent material on surface charge of variable charge soils (Kunming).

The difference in surface charge properties among various types of soils is caused by the difference in chemical composition as a result of soil genesis. On the other hand, it is also frequently observed that within the same region and under the same climatic conditions, parent material can affect the composition and thus the surface charge properties of the soil. For example, in the Kunming region, the Ferric-Haplic Acrisol derived from limestone and the Hyper-Rhodic Ferralsol derived from basalt contain 15.4% and 21.2% of free Fe_2O_3, respectively. As seen in Fig. 2.28, the latter soil carries more positive surface charge than does the former soil. The former soil carries a much larger quantity of negative surface charge, due to the presence of a certain amount of vermiculite.

The quantities of surface charge of three Ferrali-Haplic Acrisols derived from different parent materials in the Xishuangbanna region are shown in Fig. 2.29. The three soils contain a similar amount of free iron oxides. Hence, the quantities of positive surface charge are close to one another. However, because soils derived from phyllite and from sandstone-shale contain a considerable amount of hydrous mica, they carry a larger quantity of negative surface charge than does the soil derived from a Quaternary deposit containing chiefly kaolinite.

The quantity of surface charge among different horizons within the same soil profile may be different. For variable charge soils, the profile is generally quite thick and within a certain depth the mineral composition generally does not differ much. However, if there is a large amount of organic matter accumulated in the surface layer, this organic matter may

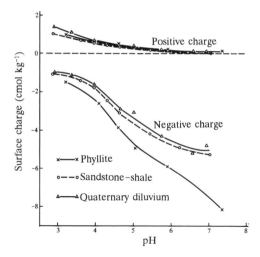

Fig. 2.29. Surface charges of Ferrali–Haplic Acrisols derived from different parent materials (Xishuangbanna).

affect the surface charge properties of the soil. It is frequently observed that this layer carries more negative surface charge and less positive surface charge than do lower horizons. It is for this reason that the pH_0 of the B horizon of tropical soils in Brazil is 4.0–6.1 while that of the A horizon is in the range of 2.2–3.4 (Morais et al., 1976).

The quantity of surface charge carried by various size fractions of a soil is also different.

Generally speaking, most of the surface charges of a soil concentrate in the clay fraction. Other fractions also carry a certain amount of surface charge. Therefore, the quantity of surface charge carried by a soil is related to the mechanical composition. A comparison of the quantities of positive, negative, and net surface charges between the clay fraction and the whole soil for two variable charge soils is presented in Table 2.5. For the whole soil of Ali–Haplic Acrisol, the quantity of both negative surface charge and net surface charge approximately corresponds to one–half of that of the clay fraction, and the quantity of positive surface charge is larger than this ratio. This soil contains about 40% of clay. Thus, it appears that the contribution of coarse fractions to negative surface charge of this soil is not large. The Rhodic Ferralsol contains about 75% of clay. At pH 7 the whole soil carries less negative and net surface charge but more positive surface charge than does the clay fraction. Chemical analyses have shown that the contents of iron oxides and titanium oxides of the coarser size fraction are higher than those of the clay fraction. This is the reason why the whole soil of this Rhodic Ferralsol carries more positive surface charge as compared

Table 2.5 Surface Charges (Unit: cmol kg^{-1}) of Whole Soil and Clay
Fraction (Zhang and Jiang, 1964; Zhang et al., 1979)

Soil	Fraction	pH 3			pH 7		
		Positive	Negative	Net	Positive	Negative	Net
Ali–Haplic	Soil	2.0	7.0	−5.0	0.5	9.5	−9.0
Acrisol	Clay	3.5	14.3	−10.8	0	20.6	−20.6
Rhodic	Soil	5.5	3.0	2.5	1.5	6.0	−4.5
Ferralsol	Clay	5.0	2.8	2.2	0.9	7.3	−6.4

to the clay fraction.

2.4.2 pH$_0$

According to the definition, pH$_0$ is the zero point of charge of the variable
charge component of a soil.

The relevant data for some soils are given in Table 2.6. It can be seen
that the pH$_0$ of various soils is related to the content of free Fe$_2$O$_3$ to some
extent and is also affected by the amount of 2:1–type clay minerals. The
pH$_0$ of two Hyper–Rhodic Ferralsols is the highest, because these soils
contain a high amount of iron oxides and contain kaolinite as the dominant
clay mineral. By contrast, the pH$_0$ of the Ferric–Haplic Acrisol is not very
high, although the soil contains 15.44% of free Fe$_2$O$_3$, owing to the presence
of a certain amount of vermiculite.

The pH$_0$ of a soil is also closely related to the content of organic matter.
For soils containing large amounts of iron and aluminum oxides in
Queensland, Australia, the pH$_0$ decreases markedly with the increase in
organic carbon content (Gillman and Yu, 1986).

2.4.3 Surface Charge Density

Surface charge density of a soil is the quantity of electric charge per unit
surface area. This surface charge density determines the field strength
around soil particles, and therefore it is directly related to the structure of
the electric double layer, which in turn is the basis for soil colloids to
possess a series of physicochemical properties. Owing to the heterogeneity
in mineralogical composition and in the composition and form of organic
and inorganic coatings, the distribution of surface charge density on the

Table 2.6 Zero Point of Charge (pH_0) of Variable Charge Soils

Soil	Parent Material	$Fe_2O_3{}^a$ (%)	pH_0	Dominant Minerals[b]
Hyper-Rhodic Ferralsol	Paleo-crust	22.31	6.05	Kl., Gb.
Hyper-Rhodic Ferralsol	Basalt	21.16	6.05	Kl., Gb.
Ferric Acrisol	Limestone	15.44	3.6	Kl., Vm.
Terra Rossa	Limestone	7.15	3.3	Kl., Vm.
Ferrali-Haplic Acrisol	Quaternary deposit	6.72	3.3	Kl.
Ferrali-Haplic Acrisol	Phyllite	7.29	3.0	Kl., Il.
Ferrali-Haplic Acrisol	Granite	5.86	2.6	Kl.
Ferrali-Haplic Acrisol	Sandstone-shale	6.58	2.95	Kl., Il.
Ali-Haplic Acrisol	Quaternary clay	4.52	3.2	Kl., Il.
Xanthic-Haplic Acrisol	Quaternary clay	9.58	3.2	Kl., Vm., Il.

[a]Free oxide
[b]Kl., kaolinite; Gb, gibbsite; Vm., vermiculite; Il., hydrous mica.

surface of soil particles is uneven. At present, it is still not possible to distinguish the location of individual charge sites and the surface area in detail. Therefore, the surface charge density calculated from the ratio of the quantity of surface charge to total surface area is only an average value.

Soil colloidal surfaces can be classified as inner surface and outer surface. Surfaces between crystal layers of swelling clay minerals belong to the inner surface, while the outer surface of clay minerals and surfaces of humus and amorphous materials such as free iron and aluminum oxides belong to the outer surface.

Generally, clay minerals contribute to a large part of surface area of a soil. The approximate ranges of surface area per gram of various clay minerals are: montmorillonite, 766–810 m², including 15–100 m² of outer surface; illite, 67–100 m², mostly outer surface; kaolinite, 7–30 m², exclusively outer surface; vermiculite, close to that of montmorillonite in total surface area for particles less than 2 μm in size, while it is only 100–200 m² for coarser particles; halloysite, about 45 m²; hydrohalloysite, 430 m², including 30 m² of outer surface; allophane, 260–300 m², exclusively outer surface.

The surface area of variable charge soils is consistent with the composition of the principal clay minerals. Some examples in this respect are shown in Table 2.7.

It can be seen from the table that for the colloid fraction of Rhodic Ferralsol dominated by kaolinite and gibbsite and containing a large amount

Table 2.7 Surface Area of Soil Colloids (<1 µm)[a] (Zhang and Jiang, 1964)

Soil	Treatment	Surface Area (m² g⁻¹)			Dominant Minerals[b]
		Inner	Outer	Total	
Rhodic	Original	13.8	43.1	56.9	Kl., Gb.
Ferralsol	Fe-removed	18.4	19.2	37.6	
Ali-Haplic	Original	38.5	53.5	92.0	Kl., Il., Vm.
Acrisol	Fe-removed	42.6	40.2	82.8	
Vertisol	Original	198	47.0	245	Kl., Mt.

[a]Organic matter removed.
[b]Kl., kaolinite; Gb., gibbsite; Il., hydrous mica; Vm., vermiculite; Mt., montmorillonite.

of iron oxides, the surface area is only 56.9 m² per gram, chiefly of outer surface. The colloid fraction of Ali-Haplic Acrisol derived from Quaternary red clay with kaolinite, illite, and vermiculite as the principal clay minerals has a surface area of 92 m² g⁻¹ in which the outer surface is slightly larger than the inner surface. By contrast, the young tropical soil, Vertisol, containing principally kaolinite and montmorillonite has the largest surface area which is composed mostly of inner surface. It can also be seen that free iron oxides exert a remarkable influence on the surface area of variable charge soils. The total surface area of the soil colloid decreased after the removal of free iron oxides, especially for the colloid of Rhodic Ferralsol with a high content of free iron oxides. Since free iron oxides only possess outer surface, the decrease in outer surface after removal of these oxides was more pronounced. The "apparent specific surface areas" calculated from the decrease in outer surface area and the amount of iron oxides for the Rhodic Ferralsol and the Ali-Haplic Acrisol are 170 m² and 187 m² per gram of free iron oxides, respectively.

Free iron oxides may affect the surface area of soil colloid in two ways. These iron oxides themselves possess a quite large surface area. On the other hand, they may also mask a part of the surface area of crystal minerals in the form of coatings. From the data presented in Table 2.7 it may be suggested that the first effect exceeds the second effect in magnitude, because the total surface area of soil colloids decreased after the removal of free iron oxides. Since the calculated apparent specific surface area of free iron oxides is larger than that of crystalline minerals by

Table 2.8 Surface Charge Density (C) of Colloids of Variable Charge Soils at Different pH (Zhang and Jiang, 1964)

Treatment	Ali–Haplic Acrisol		Ferrali–Haplic Acrisol		Rhodic Ferralsol	
	pH	μC cm^{-2}	pH	μC cm^{-2}	pH	μC cm^{-2}
Original	3.3	-12.9	3.0	-8.5	3.3	+3.6
	4.9	-13.3	4.2	-11.8	4.1	+0.7
	5.5	-16.3	6.0	-16.0	5.4	-1.4
	6.5	-19.6	6.9	-19.3	6.3	-6.1
	7.3	-22.9	7.4	-21.0	7.0	-9.0
	7.7	-24.0	7.7	-22.5	–	–
Fe_2O_3– removed	5.1	-21.3	4.2	-28.0	3.2	-1.5
	5.7	-24.2	6.4	-35.3	4.2	-6.7
	7.3	-27.4	7.0	-38.0	5.8	-12.3
	7.7	-28.8	7.8	-42.7	7.1	-17.7

several times, it may be supposed that these oxides are coated on a part of the crystalline minerals unevenly in the form of porous finely–dispersed aggregates.

The surface charge density of soils or soil colloids is mostly of the order of $1-3.5 \times 10^{-7}$ mol cm^{-2} within the pH range of 4–8, corresponding to 10–37 μC cm^{-2} (Yu, 1976). It is about 10 μC cm^{-2} for the Rhodic Ferralsol and is 15–20 μC cm^{-2} for the Ferrali–Haplic Acrisol and the Ali–Haplic Acrisol at pH 7. The corresponding value for a constant charge soil in central China is 23 μC m^{-2} (Zhang and Jiang, 1964).

Since the quantity of net negative surface charge of a soil increases with the rise in pH while the surface area is practically independent of pH, the surface charge density will increase with the increase in pH. The surface charge densities of three representative variable charge soils at different pH are given in Table 2.8, showing such a tendency clearly. Among the three soils the surface charge density is the lowest for the Rhodic Ferralsol and the highest for the Ali–Haplic Acrisol. The surface charge density increased markedly after the removal of free iron oxides, caused by the increase in net negative surface charge and the decrease in surface area of the soil.

It appears from the data given above that the surface charge density of the colloid fraction of variable charge soils is lower than that of constant charge soils. Actually the difference is more remarkable than in the figure given above, because under field conditions the pH of most variable charge soils is lower than 7. This point is of practical significance. As shall be seen in later chapters, the adsorptive strength of variable charge soils for cations, particularly Rhodic Ferralsols, is weaker than that of constant charge soils. The difference in surface charge density between these two categories of soils may be the principal cause of this phenomenon.

BIBLIOGRAPHY

Anderson, S. J. and Sposito, G. (1992) *Soil Sci. Soc. Am. J.*, 56:1437–1443.

Arnold, P. W. (1977) *J. Soil Sci.*, 28:393–402.

Arnold, P. W. (1978) in *The Chemistry of Soil Constituents* (D. J. Greenland and M. H. B. Hayes, eds.). John Wiley & Sons, Chichester, pp. 355–404.

Barber, R. G. and Rowell, D. L. (1972) *J. Soil Sci.*, 23:135–146.

Barrow, N. J. (1985) *Adv. Agron.*, 38:183–230.

Barrow, N. J., Bowden, J. W., Posner, A. M., and Quirk, J. P. (1980) *Aust. J. Soil Res.*, 18:37–48.

Basu, P. K., Nayak, D. C., Berman, A. K., Vanadachod, C., and Ghosh,K. (1986) *J. Indian Soc. Soil Sci.*, 34:24–28.

Bolland, M. D. A., Posner, A. M., and Quirk, J. P. (1976) *Aust. J. Soil Res.*, 14:197–216.

Bowden, J. W., Posner, A. M., and Quirk, J. P. (1977) *Aust. J. Soil Res.*, 15:121–136.

Breeuwsma, A. and Lyklema, J. (1973) *J. Colloid Interface. Sci.*, 43:437–448.

Chan, K. Y., Davey, B. G., and Greening, H. R. ((1979) *Soil Sci. Soc. Am. J.*, 43:301–303.

Duquette, M. and Hendershot, W. (1993a) *Soil Sci. Soc. Am. J.*, 57:1222–1227.

Duquette, M. and Hendershot, W. (1993b) *Soil Sci. Soc. Am. J.*, 57:1228–1234.

El-Swaify, S. A. and Sayegh, A. H. (1975) *Soil Sci.*, 120:49–56.

Espinoza, W., Gast, R. G., and Adams, R. S. Jr. (1975) *Soil Sci. Soc. Am. Proc.*, 39:842–846.

Fey, M. V. and Le Roux, J. (1976) *Soil Sci. Soc. Am. J.*, 40:359–364.

Fieldes, M. and Schofield, R. K. (1960) *N. Z. J. Sci.*, 3:563–579.

Galindo, G. G. and Bingham, F. T. (1977)Soil Sci. Soc. Am. J.,40:883–886.

Gallez, A., Juo, A. S. R. and Herbillon, A. J. (1976) *Soil Sci. Soc. Am. J.*, 40:601–608.

Gast, R. G. (1977) in *Minerals in Soil Environments* (J. B. Dixon and S. B. Weed, eds.). Soil Science Socciety of America, Madison, WI, pp. 27–73.

Gillman, G. P. and Bell, L. C. (1976) *Aust. J. Soil Res.*, 14:351–360.

Gillman, G. P. and Uehara, G. (1980) *Soil Sci. Soc. Am. J.*, 44:252–255.

Gillman, G. P. and Yu, T. R. (1986) in *Proceedings of the International Symposium on Red Soils* (Institute of Soil Science, ed.). Science

Press/Elsevier, Beijing/Amsterdam, pp. 251–261.

Gonzales–Batista, A., Moreno, J. M. H., Galdas, E. F., and Herbillon, A. J. (1982) *Clays Clay Miner.*, 30:103–110.

Greenland, D. J. (1986) in *Proceedings of the International Symposium on Red Soils* (Institute of Soil Science, ed.). Science Press/Elsevier, Beijing/Amsterdam, pp. 262–273.

Greenland, D. J. (1974) *Trans. 10th Intern. Congr. Soil Sci.*, II:278–285.

Greenland, D. J. and Mott, C. J. B. (1978) in *The Chemistry of Soil Constituents* (D. J. Greenland and M. H. B. Hayes, eds.). John Wiley & Sons, Chichester, pp. 321–353.

Hendershot, W. H. (1978) Can. *J. Soil Sci.*, 58:439–442.

Hendershot, W. H. and Lavkulich, L. M. (1978) *Soil Sci. Soc. Am. J.*, 42:468–472.

Hendershot, W. H. and Lavkulich, L. M. (1979) *Soil Sci.*, 128:136–141.

Hendershot, W. H. and Lavkulich, L. M. (1983) *Soil Sci. Soc. Am. J.*, 47:1252–1260.

Hendershot, W. H., Singleton, G. A., and Lavkulich, L. M. (1979) *Soil Sci. Soc. Am. J.*, 43:387–389.

Hingston, F. J., Atkinson, R. J., Posner, A. M., and Quirk, J. P. (1968) Trans. 9th Intern. Congr. Soil Sci., I:669–678.

Hingston, F. J., Posner, A. M., and Quirk, J. P. (1972) *J. Soil Sci.*, 23: 177–192.

Keng, J. C. W. and Uehara, G. (1974) *Soil Crop Sci. Soc. Florida, Proc.*, 33:119–126.

Laverdier, M. R. and Weaver, R. M. (1977) *Soil Sci. Soc. Am. J.*, 41: 505–510.

Li, X. Y. (1985) *Acta Pedol. Sinica*, 22:120–126.

Madrid, L. and Arambarri, P. De (1978) *Geoderma*, 21:199–208.

Madrid, L., Diaz, E., and Cabrera, F. (1984) *J. Soil Sci.*, 35:373–380.

Mattson, S. (1931) *Soil Sci.*, 32:343–365.

Mattson, S. (1932) *Soil Sci.*, 34:209–240.

Morais, E. I., Page, A. L., and Lund, L. J. (1976) *Soil Sci. Soc. Am. J.*, 40:521–527.

Okamura, Y. and Wada, K. (1983) *J. Soil Sci.*, 34:287–295.

Parfitt, R. L. (1978) *Adv. Agron.*, 30:1–50.

Parker, J. C., Zelazny, L. W., Sampath, S., and Harris, W. G. (1979) *Soil Sci. Soc. Am. J.*, 43:668–674.

Parks, G. A. (1965) *Chem. Rev.*, 65:177–198.

Parks, G. A. (1967) *Advances in Chemistry Series*, no. 67. American Chemical Society, Washington D.C.

Parks, G. A. and de Bruyn, P. L. (1962) *J. Phys. Chem.*, 66:967–973.

Perrot, K. W. (1977) *Clays Clay Miner.*, 25:417–421.

Pyman, M. A. F., Bowden, J. W. and Posner, A. M. (1979a) *Clay Miner.*,

14:87-92.

Pyman, M. A. F., Bowden, J. W., and Posner, A. M. (1979b) *Aust. J. Soil Res.*, 17:191-195.

Rajan, S. S. S. (1978) *Soil Sci. Soc. Am. J.*, 42:39-44.

Rengasamy, P. and Oades, J. M. (1977) *Aust. J. Soil Res.*, 15:235-242.

Sakurai, K., Ohdate, Y., and Kyuma, K. (1988) *Soil Sci. Plant Nutr.*, 34:171-182.

Sakurai, K., Nakayama, A., Watanabe, T., and Kyuma, K. (1989) *Soil Sci. Plant Nutr.*, 35:623-633.

Sakurai, K., Ohdate, Y., and Kyuma, K. (1989a) *Soil Sci. Plant Nutr.*, 35:21--31.

Sakurai, K., Ohdate, Y., and Kyuma, K. (1989b) *Soil Sci. Plant Nutr.*, 35:89-100.

Sakurai, K., Teshima, A., and Kyuma, K. (1990) *Soil Sci. Plant Nutr.*, 36:73-81.

Schofield, R. K. (1949) *J. Soil Sci.*, 1:1-8.

Schulthess, C. P. and Sparks, D. L. (1986) *Soil Sci. Soc. Am. J.*, 50:1406-1411.

Schulthess, C. P. and Sparks, D. L. (1987) *Soil Sci. Soc. Am. J.*, 51:1136-1144.

Schulthess, C. P. and Sparks, D. L. (1988) *Soil Sci. Soc. Am. J.*, 52:92-97.

Shao, Z. C. (1990) *Acta Pedol. Sinica*, 27:159-165.

Shao, Z. C. and Wang, W. J. (1991) *Pedosphere*, 1:29-39.

Singh, U. and Uehara, G. (1986) in *Soil Physical Chemistry* (D. L. Sparks, ed.). CRC Press, Boca Raton, FL, pp. 1-38.

Sposito, G. (1981) *Soil Sci. Soc. Am. J.*, 45:292-297.

Sposito, G. (1983) *Soil Sci. Soc. Am. J.*, 47:1058-1059.

Sposito, G. (1992) in *Environmental Particles* (J. Buffle and H. P. van Leeuwen, eds.). Lewis Pubishing Co., Boca Raton, FL, pp. 292-314.

Sprycha, R. (1989) *J. Colloid Interface Sci.*, 127:12-25.

Stoop, W. A. (1980) *Geoderma*, 23:303-314.

Sumner, M. E. (1963) *Clay Miner. Bull.*, 5:218-226.

Sumner, M. E. and Davidtz, J. C. (1965) *S. Afr. J. Agric. Sci.*, 8:1045-1050.

Tan, K. H. and Dowling, P. S. (1984) *Geoderma*, 32:89-101.

Tang, L. Y. and Chen, J. F. (1987) *Acta Pedol. Sinica*, 24:306-312.

Tessens, E. and Shamshuddin, J. (1982) *Pedologie*, 30:85-106.

Tessens, E. and Zauyah, S. (1982) *Soil Sci. Soc. Am. J.*, 46:1103-1106.

Theng, B. K. G. (ed.) (1980) *Soils with Variable Charge*. New Zealand Soil Science Society, Parmerston North.

Uehara, G. and Gillman, G. P. (1980) *Soil Sci. Soc. Am. J.*, 44:250-252.

Uehara, G. and Gillman, G. P. (1981) *The Mineralogy, Chemistry and Physics of Tropical Soils with Variable Charge Clays*. Westview Press, Boulder, CO.

van Olphen, H. (1977) *Introduction to Clay Colloid Chemistry*. Interscience, New York.

Van Raij, B. and Peech, M. (1972) *Soil Sci. Soc. Am. Proc.*, 36:587–593.

Wada, K. and Harward, M. E. (1974) *Adv. Agron.*, 26:211–260.

Wada, K. and Okamura, Y. (1983) *Soil Sci. Soc. Am. J.*, 47:902–905.

Wada, S. I. and Wada, K. (1985) *J. Soil Sci.*, 36:21–29.

Wann, S. S. and Uehara, G. (1978a) *Soil Sci. Soc. Am. J.*, 42:565–570.

Wann, S. S. and Uehara, G. (1978b) *Soil Sci. Soc. Am. J.*, 42:886–888.

White, G. N. and Zelazny, L. W. (1986) in *Soil Physical Chemistry* (D. L. Sparks, ed.). CRC Press, Baca Raton, FL, pp. 39–81.

Yu, T. R. and Zhang, X. N. (1983) in *Red Soils of China* (C. K. Li, ed.). Science Press, Beijing, pp. 74–90.

Yu, T. R. (1976) *Electrochemical Properties of Soils and Their Research Methods*. Science Press, Beijing, pp. 9–48.

Yu, T. R. and Zhang, X. N. (1986) in *Proceedings of the International Symposium on Red Soils* (Institute of Soil Science, ed.). Science Press/Elsevier, Beijing/Amsterdam, pp. 409–441,

Yu, T. R. and Zhang, X. N. (1987) in *Soils of China* (Institute of Soil Science, ed.). Science Press, Beijing, pp. 418–432.

Yu, T. R. (ed.) (1987) *Principles of Soil Chemistry*. Science Press, Beijing.

Yuan, C. L. (1981) *Acta Pedol. Sinica*, 18:345–352.

Zhang, G. Y., Zhang, X. N., and Yu, T. R. (1987) *J. Soil Sci.*, 38:29–38.

Zhang, G. Y., Zhang, X. N., and Yu, T. R. (1991) *Pedosphere*, 1:17–28.

Zhang, X. N. and Jiang, N. H. (1962) *Science Cir.*, no. 9, 52–53.

Zhang, X. N. and Jiang, N. H. (1964) *Acta Pedol. Sinica*, 12:120–131.

Zhang, X. N., Jiang, N. H., Shao, Z. C., Pan, S. Z., and Zhang, W. G. (1979) *Acta Pedol. Sinica*, 16:145–156.

Zhang, X. N., Zhang, G. Y., Zhao, A. Z., and Yu, t. R. (1989) *Geoderma*, 44:275–286.

Zhang, X. N. (1985) in *Physical Chemistry of Paddy Soils* (T. R. Yu, ed.). Science Press/Springer Verlag, Beijing/Berlin, pp. 111–156.

Zhang, X. N. and Zhao, A. Z. (1984) *Acta Pedol. Sinica*, 21:358–367.

Zhang, X. N. and Zhao, A. Z. (1988) *Acta Pedol. Sinica*, 25:164–174.

Zhao, A. Z. and Zhang, X. N. (1991) *Soils*, 23:231–235.

Zhao, A. Z. and Zhang, X. N. (1992) *Acta Pedol. Sinica*, 29:392–400.

3

ELECTROSTATIC ADSORPTION OF CATIONS

G. L. Ji and H. Y. Li

Adsorption of ions is a direct consequence of the carrying of surface charge for soils. Owing to the characteristics of variable charge soils in chemical and mineralogical compositions, these soils possess distinct amphoteric properties. Therefore, they can adsorb cations as well as anions. Under field conditions, most of the variable charge soils carry more negative surface charge than positive surface charge, hence they adsorb more cations than anions. Under certain conditions the quantities of adsorbed cations and anions are equal to each other. In this case the soil is said to be at its iso-ionic point.

Generally, for most cations commonly found in soils, the interaction force between them and the surface of soil particles during adsorption is electrostatic in nature. However, owing to the characteristics of variable charge soils, a specific force may also be involved in the adsorption of some cations. This latter topic shall be discussed in Chapter 5. In this chapter, only electrostatic adsorption is dealt with.

In the present chapter, the mechanism of electrostatic adsorption of cations by variable charge soils and the factors that may affect this type of adsorption are presented first. Then, the dissociation of adsorbed cations is discussed. Finally, the competitive adsorptions of potassium ions with sodium ions and of potassium ions with calcium ions are examined.

3.1 MECHANISM OF ELECTROSTATIC ADSORPTION OF CATIONS

3.1.1 Concept of Ion Adsorption

According to the definition in physical chemistry, the concentration of solute in the surface layer of the solution is different from that in the interior of the bulk solution. If the concentration of solute in the surface layer is higher than that in the interior, the phenomenon is called *adsorption*. Conversely, it is called *negative adsorption*. In soil science, on the other hand, the heterogeneity in distribution of ions in soil colloidal systems is interpreted mainly in terms of electrostatic interactions occurring at the interface between soil colloidal particles and the liquid phase (Bear, 1964). Owing to adsorption or negative adsorption, the concentration of ions at the surface of soil colloidal particles or adjacent to the surface is higher or lower than that in the diffuse layer or the free solution. Hence, from a microscopical viewpoint, adsorption should be the difference in concentration of ions between the surface of soil colloid and the diffuse layer. In practice, however, adsorption is generally examined as the difference in ion concentration between the whole electric double layer and the free solution.

Soils invariably carry surface charges. According to the principle of electroneutrality, there should always be an equivalent amount of counter-ions adsorbed on the surface of soil colloids. Therefore, ion adsorption is generally observed through ion–exchange with another ion species. Ion adsorption and ion–exchange differ in meaning; Ion adsorption denotes interactions between ions and the surface of soil colloid, while ion–exchange denotes interactions between ion species. In terms of free energy, the former reaction refers to adsorption energy or bonding energy, whereas the latter reaction refers to energy of ion–exchange.

The free energy of adsorption at the surface of adsorbents may be distinguished as three portions:

$$\Delta G_{ads.} = \Delta G_{coul.} + \Delta G_{chem.} + \Delta G_{reac.}$$

(3-1)

In the equation, $\Delta G_{coul.}$ refers to the change in free energy caused by electrostatic interaction between the point charge of ions and the electric field, and it is of most concern in this chapter. This energy change is only related to the electric charge of ions and is independent of other properties of the ions. Hence, it is the electrostatic portion of the whole free energy of adsorption.

$\Delta G_{chem.}$ denotes the change in free energy caused by specific adsorption of ions on the surface of adsorbent. The numerical value of this energy may be positive, negative, or zero in sign. The bonding forces related to $\Delta G_{chem.}$ are determined by the nature of the adsorbent and that of the ion species. They include coordination force, van der Waals force, and polarization force. Ions with an adsorption free energy including one or more of these forces can be adsorbed on surfaces free of electric charge. If this $\Delta G_{chem.}$ is so high that it can overcome the electrostatic repulsive force, ions can be adsorbed on surfaces with the same sign of charge as the ions.

$\Delta G_{reac.}$ is related to the size and polarizability of adsorbed ions as well as the structure of the solution adjacent to the surface of adsorbent. Its numerical value may be either positive or negative in sign. It differs from $\Delta G_{chem.}$ in that it may change with the change in surface potential. When the surface does not carry an electric charge, this term becomes zero.

Despite the arbitrariness in distinguishing adsorption free energy described above, the concept may frequently agree with experimental results. If the ions can only be adsorbed on surfaces with opposite charge, the $\Delta G_{chem.}$ term in $\Delta G_{ads.}$ is very small. Such ions are called *inert ions* or *nonspecific ions*. If selective adsorption occurs among more than one ion species with the same sign of electric charge, that is, if the adsorbed amount of one ion species is not proportional to the relative activity of that ion species in solution, and at the same time that ion species can only be adsorbed on oppositely charged surfaces, it means that $\Delta G_{react.}$ is operative. When the ions can be adsorbed on surfaces with the same sign of charge as the ions or with zero charge, the $\Delta G_{chem.}$ term would be important. In the last case, the ions are called *charging ions*.

Ions that can transfer their electric charges to the surface of adsorbents are of two types. If these ions are the component of both the solid phase and the solvent, such as H^+ or OH^-, they are called *potential–determining ions*. Some other ions, although unable to enter the crystal lattice, can also affect the surface charge of adsorbents. Such ions are called *specific ions*.

3.1.2 Properties of Cations Relating to Electrostatic Adsorption

For a given negatively–charged adsorbent, the fundamental factor in affecting the electrostatic adsorption of a cation is the charge/size ratio of that cation, because, according to the Coulomb's law, the attracting force is proportional to the number of electric charge carried by that cation and inversely proportional to the square of the distance between the surface of the adsorbent and the center of that cation. Therefore, a divalent cation is always adsorbed more strongly than a monovalent cation of the same size.

A complicating factor in aqueous systems is that charged ions are hydrated. Because of the presence of ion–dipole force, some water molecules would orient around the ion. A part of these water molecules adjacent to the ion are bound so tightly that they can move with the moving ion. Therefore, this ion together with this part of water may be regarded as an entity. Water molecules within this region are said to be in the "primary region." Another part of water, located within the outer hydration shell or the "secondary region," are affected by both the orientation force exerted by the ion and the tendency of forming network structure among water molecules themselves. Thus, they are only partly oriented. The total volume of the primary region and the secondary region can be regarded as an effective volume within which water molecules are affected by the center ion. The number of water molecules within this volume is called *hydration number of the ion.* Depending on the method of determination, that is, on how many layers of water molecules are regarded as within the effective volume, the hydration number as determined by various authors differs greatly. For example, for Na^+ ions, it may be 1, 2, 2.5, 4.5, 6–7, 16.9, 44.5, and 71 (Bockris, 1970). For the same reason, the effective radius of a hydrated ion cited by various authors differs by a factor of two, three, or even more.

The free energy of hydration of ion species i is proportional to the square of z_ie_0 and inversely proportional to $2r_i$ (Bockris, 1970). Therefore, as can be seen in Table 3.1, the hydration energy of divalent cations and especially of trivalent cations is much higher than that of monovalent cations. For cations of the same valency, those with a smaller radius are more hydrated and hence possess a larger hydrated radius.

When cations are electrostatically adsorbed by negatively–charged soils, it can be expected that the affinity of soil colloid surface for cations of different valencies should generally be of the order $M^+ < M^{2+} < M^{3+}$ (Table 3.1). For variable charge soils Ali-Haplic Acrisol and Rhodic Ferralsol, it has been observed that the order of affinity was $Na^+ < Ca^{2+} < Mn^{2+} < Al^{3+}$ (Yu and Zhang, 1990). This is consistent with their order of charge–size functions z^2/r, which are 2.2, 10.3, 12.5, and 43.6 $C^2 m^{-1} \times 10^{28}$, respectively (Thomas and Hargrove, 1984). Based on the electric double layer theory, Helmy (1967) suggested that when the solution contains equal concentrations of mono–, di–, and trivalent cations the soil should chiefly adsorb trivalent cations. For cations of the same valency, the adsorption strength should be determined principally by the hydrated radius of the ions (Table 3.1).

However, in the literature the orders of adsorption strength among various cations obtained by different authors with different materials are quite inconsistent.

For example, with cation–exchange resins, clay minerals, and many hydrated metal oxides, the selectivity in adsorption to alkali metal ions is generally of the order

$$Li^+ < Na^+ < K^+ < Rb^+ < Cs^+$$

On the other hand, with some hydrated metal oxides, the reversed order in selectivity may be observed (Kinniburgh and Jackson, 1981; McBride, 1989; Milnes and Fitzpatrick, 1989).

If cations of different valencies are compared, the situation may be more complicated.

In order to interpret these inconsistent results, various suggestions have been made. Some authors proposed the concept of binding between ions and adsorption sites. Some authors applied the ion–pair theory or the surface complexation theory to adsorption in soils. Some raised the question

Table 3.1 Radius (r), Hydrated Radius (r_H), Polarizability, and Hydration Energy of Cations in Relation to Relative Exchange Ability in NH_4–Smectite (Nightingale, 1959; Scheffer and Schachtschabel, 1992)

Ion Species	r (nm)	r_H (nm)	Polarizability (cm^3)	Hydration energy ($kJ\ mol^{-1}$)	Relative Exchange (%)
Li^+	0.076	0.382	0.03	503	32
Na^+	0.102	0.358	0.17	419	33
K^+	0.138	0.331	0.80	356	51
Rb^+	0.152	0.329	1.42	335	63
Cs^+	0.167	0.329	2.35	314	69
Mg^{2+}	0.072	0.428	0.10	1802	69
Ca^{2+}	0.100	0.412	0.54	1571	71
Sr^{2+}	0.118	0.412	0.87	1425	74
Ba^{2+}	0.135	0.404	1.68	1341	73
Al^{3+}	0.054	0.475	–	4649	85
La^{3+}	0.103	0.452	–	3268	86

of fitness of ions in size. Some authors considered that soils may carry different adsorptive sites, with some sites preferring one ion species while others preferring another ion species. Another possibility is that the theory of selectivity of glass electrodes to monovalent cations proposed by Eisenman (1967) may be applied to soils (Juo and Barber, 1969; Yu, 1976).

According to this theory, the selectivity of adsorbents to cations is determined by the difference in reaction free energy of these ions with water molecules and with adsorption sites of the adsorbent. Therefore, the fundamental factors in influencing selectivity are effective field strength of adsorption sites and hydration energy of ions. Under ordinary conditions, hydration energy of ions plays the dominant role, and therefore the order of adsorption strength among various ions of the same valency is consistent with the lyotropic series. However, if the field strength near the adsorption site is so high that the reaction force with ions exceeds the hydration energy of ions, these ions may be dehydrated or partially dehydrated. In this case, the dominant factor in determining the affinity between ions and the surface of adsorbents would be the atomic radius rather than the hydrated radius of ions. As a result, the order of relative adsorption strength may be at variance with the lyotropic series. According to this theory, the adsorbability of various cations is affected by the nature of both the ions and the soil, even though no specific force, which shall be treated in Chapter 5, is involved.

3.1.3 Selectivity Coefficient Involving Two Ion Species

When there are two kinds of cations present in the system, the soil will adsorb them in different proportions.

We take the adsorption of sodium ions by a soil originally adsorbed with potassium ions as an example:

$$SK + Na^+ \rightleftharpoons SNa + K^+ \tag{3-2}$$

where SK and SNa are adsorbed cations, and Na^+ and K^+ are cations in solution, respectively. At equilibrium, the following relation should hold:

$$\frac{[SNa]}{[SK]} = k_s \frac{a_{Na}}{a_K} \tag{3-3}$$

In this equation, k_s is called the *selectivity coefficient* or *selectivity constant*. In principle, this index can characterize the relative affinity of a soil with respect to the two cation species. In practice, however, the numerical value of this index should not be considered too seriously, although large amounts of experimental data regarding cations in soil systems have accumulated (Bolt, 1979). The main reason lies in the fact that the numerical value of this calculated selectivity coefficient can remain constant only within a limited concentration range under a specified condition, even though many attempts have been made to widen the applicable range by modifying equation (3-3).

Here two difficulties are encountered. In equation (3-3), a_{Na} and a_K are activities of the two ion species in the equilibrium solution. The commonly adopted practice is to compute the activity coefficient of ions from total ionic strength. As is well known, single ion activities cannot be calculated accurately in this way, especially at a high ionic strength. The situation may be even more complex when two or more ion species with different valencies, such as K^+ and Ca^{2+}, are treated.

Another difficult problem is the computation of the activity of adsorbed ions, which is necessary in the application of equation (3-3). What is the meaning and the actual value of the activity of adsorbed ions? No one knows exactly, although some assumptions have been made in the literature.

A prerequisite for the application of equation (3-3) or any modified form of it is that the surface properties of the adsorbent, such as a soil, must not be affected by adsorption. This requirement cannot be fulfilled fully even for clay minerals and constant charge soils in many cases. For variable charge soils, as has been seen in Chapter 2, the surface charge properties may be affected by a variety of factors, including the adsorption of ions. Therefore, it may be expected that the applicability of equation

(3–3) would be rather limited in both theory and practice.

Based on the above reasoning, in this chapter the index selectivity coefficient is not used in discussions. Instead, the activity ratio between two ion species remaining in solution after adsorption, which can be determined directly with two ion–selective electrodes, is employed when the adsorbabilities of two ion species are compared.

3.1.4 Models for Describing Ion Adsorption on a Variable Charge Surface

According to the chemical nature, the surface of soil colloids can be classified into two basic types: surface with variable surface charge (constant surface potential) and surface with constant surface charge.

In order to describe the adsorption of ions by a variable charge surface, various models have been proposed (Barrow, 1985; Bowden et al., 1980; McBride, 1989; Sparks, 1986; Sposito, 1983, 1984; Uehara and Gillman, 1981; Westall and Hohl, 1980). Bowden et al. (1977) and Barrow et al. (1980) proposed a "Stern variable surface charge–variable surface potential (VSC–VSP) adsorption model." In this model, the adsorption layer at the solid–liquid interface may be divided into three sublayers. The first sublayer consists of potential–determining ions, such as H^+ and OH^-. The second sublayer consists of specifically adsorbed ions. The third sublayer is the diffuse layer, composed of both counterions and co-ions. With the proposed model, the quantities of electric charge of the three sublayers can be calculated. Barrow et al. (1980), when studying the adsorption of NaCl, Na_2SO_4 and $CaCl_2$ by goethite with the proposed model, obtained results in good agreement with the experimental data. The model has also been tested for the adsorption of other ions (Barrow, 1985), including phosphate ions (Bowden et al., 1980).

At present, however, no model can be regarded as universally applicable to all soil conditions.

3.2 FACTORS AFFECTING ADSORPTION

3.2.1 Surface Charge of Soil

The adsorption of cations by soils is caused by negative surface charge but not necessarily by net negative surface charge. Most of the soils carry permanent negative surface charge. For variable charge soils, there is also a large quantity of variable negative surface charge. Therefore, all soils can adsorb cations. However, owing to differences in chemical composition and mineralogical composition, the adsorption capacity of various soils differs greatly. For example, paddy soils derived from Rhodic Ferralsols carry large quantities of variable positive and negative surface charges and only a small

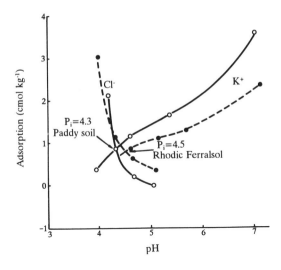

Fig. 3.1. Adsorption of potassium ions and chloride ions by Rhodic Ferralsol and paddy soil derived from it (P_i is iso-ionic point) (Zhang and Zhao, 1984).

amount of permanent negative surface charge. By contrast, for paddy soils derived from Cambisols the quantity of permanent negative surface charge is large while that of positive surface charge is small. As a consequence, the former type of soils adsorbs less potassium ions and more chloride ions than does the latter type of soils. As is seen from Fig. 3.1, for the Rhodic Ferralsol and a paddy soil derived from it there is an iso-ionic point at pH 4.5 and pH 4.3, respectively. For Cambisols, this iso-ionic point generally cannot be observed.

When adsorbed cations are in a state of adsorption-dissociation equilibrium, the activity of these ions in the system is closely related to the surface charge properties of the soil. For example, Fig. 3.2 shows that for the paddy soil derived from Cambisol the activity of potassium ions is quite small when the total amount of these ions is less than the quantity of negative surface charge of the soil. On the other hand, for the paddy soil derived from Rhodic Ferralsol the activity of potassium ions in the soil suspension is comparatively large, and it increases nearly proportionally with the increase in potassium ions present in the system. The behavior of paddy soil derived from Ali-Haplic Acrisol lies intermediate between that of the above two soils.

The adsorption of cations is directly controlled by negative surface charge but not necessarily by net surface charge of the soil. For instance, Rhodic Ferralsol can still adsorb cations when the pH is below its isoelectric point

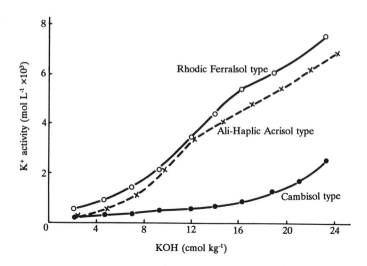

Fig. 3.2. Activities of potassium ions in three paddy soils (Zhang and Zhao, 1984).

4.7, although the soil carries net positive surface charge at this time. Of course, the quantity of adsorbed cations may also be affected by positive surface charge to some extent.

Despite the close relationship between the quantity of adsorbed cations and surface charge of variable charge soils, for a given ion species the adsorbed amount is not proportional to the quantity of surface charge in a simple way. The reason for this is that when the quantity of surface charge is changed, the composition of the solution would inevitably change, and thus the competition of other ions for adsorption sites on the surface of soil colloids and the interactions of these ions with the adsorbed cations make the situation complicated.

It has been seen in Chapter 2 that iron oxides are extremely important materials in affecting the surface charge properties of variable charge soils. Therefore, the presence of free iron oxides makes the relationship between ion adsorption and soil type even more complicated. On one hand, iron oxides cause the decrease in exposed negative surface charge sites and thus the decrease in amount of adsorbed cations. On the other hand, under ordinary pH conditions the positive surface charge carried by these oxides may exert a repulsive action on cations. If the electrolyte concentration of the soil suspension is sufficiently high so that the diffuse layer of the soil colloid is very thin, negative and positive surface charge sites would adsorb or repulse cations of the solution independently. In this case, the measured amount of adsorbed cations is the result of a geometrical mean of these two effects. If the electrolyte concentration of the soil suspension is very low so

Table 3.2 Change in Adsorption of Potassium Ions and Sodium Ions in Variable Charge Soils After Removal of Iron Oxides[a]

Soil	K–Adsorption (mmol kg⁻¹)		Na–Adsorption (mmol kg⁻¹)	
	Original	Fe–removed	Original	Fe–removed
Ali–Haplic Acrisol	3.46	4.24	0.32	0.86
Ferrali–Haplic Acrisol	2.18	3.49	0.13	0.80
Rhodic Ferralsol	3.37	3.80	0.46	0.67
Hyper–Rhodic Ferralsol	3.06	3.97	0.44	0.71

[a]Initial electrolyte concentration, 2×10^{-3} mol L⁻¹; $C_K/C_{Na} = 1$; same pH for original soil and Fe_2O_3-removed soil.

that the diffuse layers near positive and negative surface charge sites overlap partially, the quantity and strength of adsorption of cations would also be decreased.

The changes in adsorption of potassium ions and sodium ions by four variable charge soils after the partial removal of iron oxides are shown in Table 3.2. Clearly, the adsorption of both ion species increases after treatment. Since in the experiment only a part of free iron oxides can be removed by exposure to daylight in an oxalic acid–ammonium oxalate solution, it can be expected that the increments in adsorption of potassium ions and sodium ions would be much larger than those given in Table 3.2, if all the iron oxides are removed, especially for the Rhodic Ferralsol and the Hyper–Rhodic Ferralsol.

3.2.2 pH

Soil pH can affect the quantity of surface charge as well as the form of ions. At low pH there may be the involvement of competitive adsorption of hydrogen and aluminum ions. Therefore, pH may exert important influences on electrostatic adsorption of cations by the soil.

As seen in Fig. 3.3, the adsorption of potassium ions by three variable charge soils decreases with the decrease in pH, especially within the higher pH range. Within the lower pH range the change in adsorption is not so distinct. When the pH is very low, variable negative surface charge may disappear nearly entirely. However, at this time the soil still carries a certain quantity of permanent negative surface charge, which is unaffected by pH. This is the reason why the amount of adsorbed potassium ions is nearly

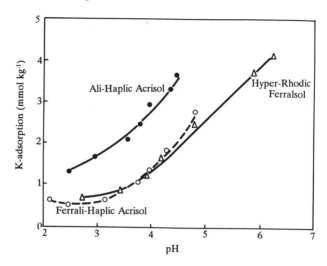

Fig. 3.3. Effect of pH on adsorption of potassium ions by variable charge soils (initial concentration, 1 mmol L⁻¹) (Li and Ji, 1992b).

independent of pH when the pH is lower than about 3.5. Similarly, the adsorption of sodium ions by variable charge soils also decreases with the decrease in pH (Fig. 3.4), especially within the pH range of 4.5–6.5, although the extent of decrease is smaller than that of potassium ions.

In addition to the effect on quantity of surface charge, pH may also affect the form of some cations in solution. Chaussidon (1963) and Chen et al. (1979) found that at pH 10, magnesium ions in $MgCl_2$ solution were not in the form of Mg^{2+} but were in the form of $Mg(OH)^+$. Thus, magnesium ions can be adsorbed by soils as monovalent ions. They suspected that when the pH is sufficiently high all the alkaline earth metal ions would be in the form of $M(OH)^+$. The effect of pH on the form of cations is more pronounced for trivalent cations. Some authors suggested that at low pH, calcium ions may be adsorbed by kaolinite in the form of ion–pair Ca^+Cl, thus behaving similar to monovalent cations (Marcano–Martinez and McBride, 1989).

When the pH is lower than a certain value, the competition of hydrogen ions and especially aluminum ions for ion–exchange sites cannot be overlooked. For instance, in a 0.001 M KCl solution of pH 3 the hydrogen ion concentration would be 0.001 mol L⁻¹, corresponding to the concentration of potassium ions. If the solution is in equilibrium with a soil, hydrogen ions and aluminum ions with a higher bonding energy would compete for adsorption sites on the surface of soil colloids with potassium ions, resulting in the decrease in adsorption of potassium ions. This kind of effect is more remarkable in dilute electrolyte solutions.

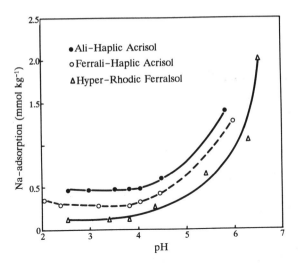

Fig. 3.4. Effect of pH on adsorption of sodium ions by variable charge soils (initial concentration, 1 mmol L^{-1}) (Li and Ji, 1992b).

In order to examine the relationship between ion adsorption and surface charge of the soil at different pH, one can plot the ratio (amount of cations adsorbed)/(quantity of negative surface charge) against pH. As can be seen from Fig. 3.5, the ratio is not a constant, but instead it increases with the rise in pH. Thus, it may be assumed that, for ammonium ions, the effect of

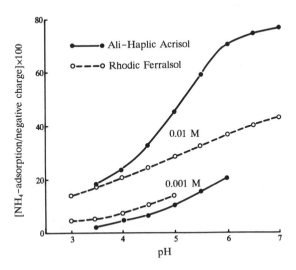

Fig. 3.5. Relationship between adsorption of ammonium ions and pH at different ammonium concentrations (Zhang et al., 1979).

pH on adsorption is caused chiefly by the direct effect of pH change, while the effect caused by the change in surface charge is secondary in importance.

3.2.3 Concentration of Ions

The amount of cations adsorbed by variable charge soils is also related to the concentration of that ion species. Van Raij and Peech (1972) found that the amounts of sodium ions adsorbed by an Oxisol in a 0.2 M NaCl solution and in a 0.01 M NaCl solution with the same pH were 1.7 cmol kg^{-1} and 0.6 cmol kg^{-1}, respectively.

The relationship between adsorption and concentration of potassium ions is shown in Fig. 3.6. The adsorption increases with the increase in potassium ion concentration of the equilibrium solution, especially within the lower concentration range. For the Ferrali–Haplic Acrisol, adsorption approaches a maximum value at which all the negative exchange sites that can adsorb potassium ions are occupied by these ions.

Figure 3.7 shows the adsorption of sodium ions in relation to the equilibrium sodium ion concentration in solution for four variable charge soils.

In the two examples given above, the order of amount of potassium ions or sodium ions adsorbed by the four soils seems to be inconsistent with the order of quantity of negative surface charge. Actually, this phenomenon is related to the difference in pH of the electrodialyzed soils. The pH of the

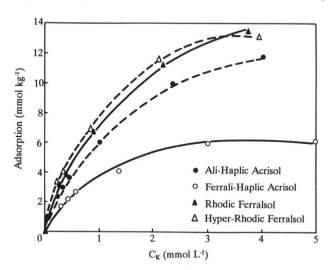

Fig. 3.6. Adsorption of potassium ions in relation to equilibrium potassium ion concentration.

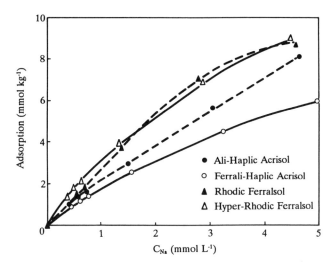

Fig. 3.7. Adsorption of sodium ions in relation to equilibrium sodium ion concentration

electrodialyzed Ali–Haplic Acrisol, Ferrali–Haplic Acrisol, Rhodic Ferralsol, and Hyper–Rhodic Ferralsol in 0.001 M KCl solution are 4.48, 4.80, 5.01, and 5.38, respectively. The effect of pH on negative surface charge of different soils causes the phenomenon mentioned above.

It can be seen from a comparison of the two figures that, at low ion concentrations, the effect of concentration on ion adsorption is much greater for potassium ions than for sodium ions.

The amount of cations adsorbed by a soil is not proportional to the amount of that ion species present in the system in a simple manner. The proportion of adsorption generally decreases with the increase in ion concentration, and the higher the pH, the more pronounced this tendency. For example, at pH 4 the proportion of potassium ions adsorbed by Rhodic Ferralsol in 0.01 M and 0.001 M KCl solutions are 5% and 13%, respectively, whereas at pH 5 the corresponding figures increases to 7% and 26%. For the adsorption of ammonium ions by Ali–Haplic Acrisol, the proportions in 0.01 M and 0.001 M NH$_4$Cl solutions are 9% and 18% at pH 4 and 25% and 65% at pH 5.5, respectively (Fig. 3.8).

At present, the mechanism of the effect of ion concentration on ion adsorption is not known. According to the theory of diffuse double layer, for variable charge soils containing large amounts of iron and aluminum oxides, the surface charge density on the surface of soil colloid would increase after adsorption of ions (van Olphen, 1977). This would result in an accompanied increase in adsorption of ions. On the other hand, because the thickness of the diffuse double layer is inversely proportional to the square root of ion

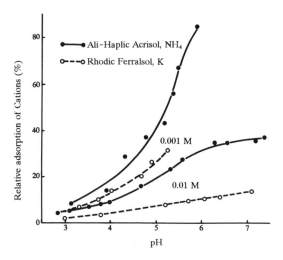

Fig. 3.8. Relative adsorption of cations at different cation concentrations by Ali-Haplic Acrisol and Rhodic Ferralsol (Zhang et al., 1979).

concentration, when the thickness extends to a certain extent following the decrease in ion concentration, the diffuse layers of adjacent negative and positive surface charge sites on the surface of soil colloids may overlap partially with one another, resulting in a mutual compensation of a part of positive and negative surface charges. This would result in a decrease in adsorption of cations and anions. Based on their experimental results, Barber and Rowell (1972) suggested that both of these mechanisms may operate.

3.2.4 Accompanying Anions

The adsorption of cations by variable charge soils is also affected by the kind of accompanying anions.

When Ali-Haplic Acrisol, Ferrali-Haplic Acrisol, and two Rhodic Ferralsols were in adsorption-dissociation equilibrium in mixed solutions containing potassium ions and calcium ions with chloride, nitrate, or sulfate as the accompanying anions, it was found that in the case of chloride ions and nitrate ions the equilibrium pH and the amount of adsorbed potassium ions or adsorbed calcium ions were similar. By contrast, in the case of sulfate ions, the equilibrium pH rose and the adsorbed amounts of both cation species increased (Table 3.3). This means that the presence of sulfate ions caused the increase in adsorption of cations, whereas chloride ions and nitrate ions did not show such an effect. It can also be seen from Table 3.3 that in the former case both the pK-0.5pCa value and the adsorption

Table 3.3 Effect of Accompanying Anions on Adsorption of Cations (Li and Ji, 1992b)

Soil	Accompanying Anions[a]	pH	pK-0.5pCa	Adsorption[b] K	Adsorption[b] Ca	Ads. ratio (K/2Ca)
Ali–Haplic	Cl	4.27	1.40	5.55	2.50	1.11
Acrisol	NO$_3$	4.32	1.40	5.55	2.50	1.11
	SO$_4$	4.86	1.37	7.92	4.53	0.88
Ferrali–Hap.	Cl	4.64	1.30	3.85	1.99	0.97
Acrisol	NO$_3$	4.57	1.29	3.99	2.25	0.89
	SO$_4$	5.11	1.24	6.21	4.13	0.75
Rhodic	Cl	4.81	1.31	5.34	3.19	0.84
Ferralsol	NO$_3$	4.81	1.31	5.22	3.10	0.84
	SO$_4$	5.54	1.18	8.35	4.88	0.86
Hyper–Rhodic	Cl	5.17	1.36	5.34	2.72	0.98
Ferralsol	NO$_3$	5.17	1.36	5.34	2.72	0.98
	SO$_4$	6.08	1.30	8.10	4.71	0.86

[a]Initial potassium to calcium equivalent ratio, 1:1.
[b]mmol kg^{-1}.

equivalent ratio K/2Ca were smaller than in the latter case, implying an increase in selectivity of variable charge soils to calcium ions caused by sulfate ions.

Anions can affect the adsorption of cations in several ways: (1) effect on ionic strength of the solution; (2) formation of ion–pairs [according to the ion–pair theory, the formation constants K of a same cation species with different anion species may be different; this would affect the adsorption of cations (Karmarkar et al., 1991; Rao et al., 1968; Shainberg and Kemper, 1967; Sposito, 1991)]; and (3) specific adsorption of anions. Owing to this specific adsorption, the surface charge properties of soil colloids and thus the capacity for adsorbing cations may change. The effect of sulfate ions on adsorption of potassium ions and calcium ions shown in Table 3.3 should be caused mainly by the change in surface charge properties of these soils after the specific adsorption of sulfate ions. Some authors suggested that Ca^{2+} ions can form neutral ion–pair $CaSO_4^0$ with SO_4^{2-} ions and are adsorbed on the surface of soils in this form (Bolan et al., 1993; Marcano--

Martinez and McBride, 1989).

3.3 DISSOCIATION OF ADSORBED CATIONS

3.3.1 Degree of Dissociation and Fraction Active

Under ordinary conditions, soil colloid carries negative surface charges, and therefore it behaves like a large-sized anion in many respects. When adsorbing cations, soil colloid becomes a colloidal electrolyte. Its properties are similar to that of a weak electrolyte, and the adsorbed cations may undergo dissociation in water. According to the theory of electrolyte solution, the degree of dissociation of a weak electrolyte is related to the nature and concentration of that electrolyte. This degree of dissociation can reflect itself in ion activities. According to the concept raised by Marshall (1964), this degree of dissociation, called fraction active f, can be expressed by the ratio of the activity a of a given ion species dissociated from the soil colloid measured with an ion-selective electrode to the concentration C of these ions in the system:

$$f = a/C \tag{3-4}$$

According to the theory of weak electrolytes, the degree of dissociation decreases with the increase in concentration. As is shown in Fig. 3.9, the degree of dissociation of potassium ions adsorbed by several variable charge

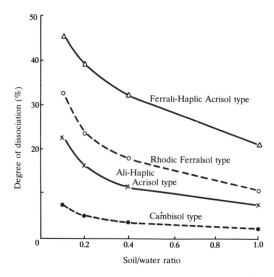

Fig. 3.9. Degree of dissociation of potassium ions in variable charge paddy soils in relation to suspension concentration (Zhang and Zhao, 1984).

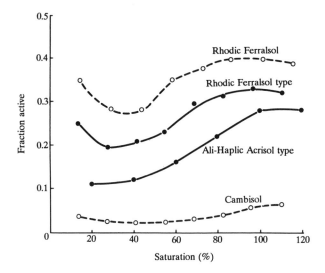

Fig. 3.10. Fraction active of potassium ions in variable charge upland soils and paddy soils in relation to K-saturation (Zhang and Zhao, 1984).

paddy soils decreases with the increase in suspension concentration in a nearly exponential manner, following the general behavior of weak electrolytes. Figure 3.10 shows the relationship between fraction active of potassium ions and degree of K-saturation. The results indicate that the degree of dissociation of adsorbed cations is the largest for the Rhodic Ferralsol and the smallest for the constant charge soil Cambisol. This reflects the characteristics of variable charge soils.

The degree of base-saturation of a soil is equivalent to the ratio of the salt of a weak acid to the acid. A 100% base-saturation is equivalent to a ratio of 0% of weak acid to 100% of its salt. As seen in Fig. 3.10, within the K-saturation range of 30-90%, the fraction active of potassium ions increases with the increase in K-saturation for all of the soils, especially for variable charge soils. Thus, the situation is similar to that of weak electrolytes.

3.3.2 Bonding Energy

When adsorbed cations are in adsorption-dissociation equilibrium with the soil, a parameter called *mean bonding free energy*, or *bonding energy*, may be used to characterize the interaction force between these cations and the soil. According to a concept suggested by Marshall (1964), this bonding energy ΔF can be calculated from the equation

$$\Delta F = RT \ln(C/a) = RT \ln(1/f) \qquad (3-5)$$

where C and a are the concentration and activity of the cations in the soil suspension, respectively.

Bonding energy is in reality the electrostatic attraction energy between the surface of soil colloid and ions. Therefore, this energy should be closely related to the surface charge properties of the soil. The quantity of negative surface charge, the amount of cations, and environmental factors all affect the magnitude of this bonding energy. The quantity of negative surface charge of soil colloid is determined by the nature of the soil and the suspension concentration. The amount of cations is related to the degree of cation–saturation or the concentration of that cation species in the system. Among environmental factors, pH plays the outstanding role. These will be examined in the following section.

3.3.3 Factors Affecting Bonding Energy

As can be seen from Tables 3.4 and 3.5, at the same pH the bonding energy with sodium ions and potassium ions is the highest for the constant charge soil Cambisol carrying a large quantity of permanent negative charge and is the lowest for the Rhodic Ferralsol carrying only a small quantity of

Table 3.4 Bonding Energy of Soils with Na Ions at Different pH (Xuan et al.,1965)

Soil	Apparent CEC (cmol kg^{-1})	pH	pNa	a_{Na}/C_{Na} (%)	$(\Delta F)_{Na}$ (J mol^{-1})
Rhodic Ferralsol	5.4	4.79	1.98	87.1	340
		5.24	2.02	79.4	550
		6.23	2.16	57.5	1340
		7.36	2.36	36.3	2430
Ali–Haplic Acrisol	7.8	4.88	2.02	79.4	550
		5.72	2.24	47.9	1760
		6.58	2.60	20.9	3770
		7.19	2.92	9.9	5530
Cambisol	13.8	4.71	2.02	79.4	550
		5.65	2.24	47.9	1760
		6.61	2.64	19.1	3980
		7.33	3.12	6.3	6660

Table 3.5 Bonding Energy of Soils with K Ions at Different pH (Xuan et al., 1965)

Soil	Apparent CEC (cmol kg^{-1})	pH	pK	a_K/C_K (%)	$(\Delta F)_K$ (J mol^{-1})
Rhodic Ferralsol	5.4	4.38	2.15	58.9	1300
		5.83	2.26	45.7	1890
		6.51	2.35	37.2	2390
		7.03	2.50	26.3	3230
Ali–Haplic Acrisol	7.8	3.99	2.17	56.2	1380
		4.64	2.30	41.7	2100
		5.07	2.42	31.6	2770
		6.32	3.07	7.1	6370
Cambisol	13.8	3.93	2.18	55.0	1470
		4.62	2.34	38.0	2350
		5.79	2.70	16.6	4320
		6.66	3.33	3.9	7920

negative surface charge, while the Ali–Haplic Acrisol is intermediate. This means that the bonding energy of various soils is related to the soil type,

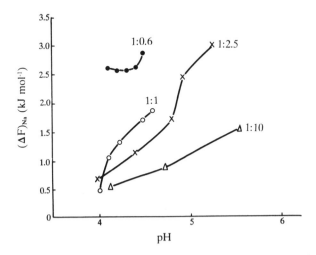

Fig. 3.11. Effect of soil-to-water ratio on bonding energy of sodium ions with Rhodic Ferralsol (numerals in the figure are soil-to-water ratio) (Xuan et al., 1965).

owing to the difference in quantity of negative surface charge.

When the concentration of sodium ions was kept constant while the soil-to-water ratio was changed in the system, it was observed that within the ratio range of 1:0.6 to 1:10, the larger the ratio the higher the bonding energy with sodium ions (Fig. 3.11).

Since iron oxides carry positive charges under ordinary pH conditions, it would be expected that if a soil is coated with iron oxides so that the quantity of positive surface charge is increased and a part of the original negative surface charge is masked, the bonding energy with sodium ions would decrease; the higher the pH, the more noticeable this effect (Fig. 3.12).

Figure 3.13 shows that, for the Ali–Haplic Acrisol and Rhodic Ferralsol, both the quantity of net negative surface charge and the bonding energy with sodium ions increase when the pH is raised. This implies again that the quantity of negative surface charge on the surface of soil colloids is an important factor in determining the bonding energy with cations. However, it can also be seen from the figure that the bonding energy–pH curve does not parallel the net negative surface charge–pH curve. This means that the bonding energy is not related to the quantity of net negative surface charge of soils in a simple manner.

When the concentration of soil suspension was kept constant while the concentration of potassium ions was changed, it was observed that at pH 5.4 the bonding energy of Rhodic Ferralsol was 710 J mol^{-1} in a 0.01 M KCl solution while it was 3990 J mol^{-1} in a 0.001 M KCl solution (Table 3.6).

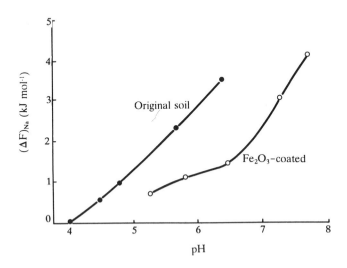

Fig. 3.12. Change in bonding energy of Rhodic Ferralsol with sodium ions after coating with iron oxides (Xuan et al., 1965).

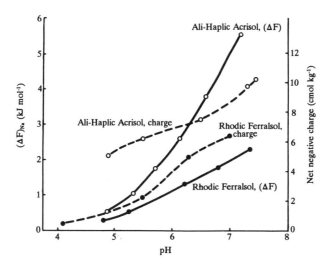

Fig. 3.13. Bonding energy of two variable charge soils with sodium ions in relation to quantity of net negative surface charge (Xuan et al., 1965).

Table 3.6 Bonding Energy of Fe_2O_3-Coated Rhodic Ferralsol with Cations at Different Cation Concentrations (Xuan et al., 1965)

Cation Species	C (M)	pH	pX	a_x/C_x (%)	$(\Delta F)_x$ (J mol^{-1})
Na	0.01	5.26	2.20	63.1	1170
		5.75	2.25	56.2	1430
		6.75	2.33	46.8	1890
		7.52	2.63	23.4	3650
	0.001	5.33	3.07	85.1	420
		5.50	3.30	50.1	1720
		5.90	3.47	33.9	2720
K	0.01	4.96	2.12	75.9	590
		5.83	2.15	70.8	880
		6.80	2.40	39.8	2310
		7.24	2.55	28.2	3180
	0.001	5.06	3.40	39.8	2310
		5.23	3.58	26.3	3350
		5.37	3.69	20.4	3990

This means that the lower the cation concentration, the higher the bonding energy. Research showed that the activity of adsorption sites of soils was not constant. When the percentage of coverage by potassium ions on the surface of soil colloid was low, the adsorption strength with potassium ions was comparatively strong (Jardine and Sparks, 1984b). Such a finding is consistent with the variation in bonding energy presented above.

When the degree of cation–saturation is changed, both the amount of cations present in the system and the pH are changed. Thus, the situation is a little complicated, especially for variable charge soils. For example, for the Rhodic Ferralsol with a great variability in surface charge, owing to the effect of increased negative surface charge following the rise in pH, the bonding energy may increase markedly (Table 3.7).

Despite the change in bonding energy of variable charge soils with the change in pH, one cannot interpret this phenomenon merely through the change in surface charge carried by soil colloids. This is because at low pH the competition of hydrogen ions and particularly aluminum ions for exchange sites would lead to the decrease in bonding energy of alkali metal ions. Besides, pH may also affect the distribution of ions within the electric double layer and the activity of different adsorption sites.

3.4 COMPETITIVE ADSORPTION OF POTASSIUM IONS WITH SODIUM IONS

For actual soil systems there is always the presence of two or more cation

Table 3.7 Bonding Energy of Soils with Sodium Ions at Different Na--Saturations (Xuan et al., 1965)

Soil	Na–Saturation (%)	pH	pNa	a_{Na}/C_{Na} (%)	$(\Delta F)_{Na}$ (J mol^{-1})
Rhodic Ferralsol	16.7	4.90	2.66	110.0	–210
	33.3	5.47	2.62	60.2	1200
	66.7	6.31	2.50	39.8	2220
	83.3	6.80	2.43	37.2	2390
	100.0	7.23	2.36	36.3	2430
Cambisol	16.7	5.06	3.56	13.8	4780
	33.3	6.38	3.32	7.9	6120
	66.7	6.56	3.27	6.8	6540
	83.3	7.07	3.04	9.1	5780
	100.0	7.37	2.86	11.5	5240

species. These cations would be adsorbed by soils simultaneously in a competitive manner. The relative magnitudes of adsorption of these ion species are determined mainly by the nature of the relevant ions. In this section, the competitive adsorption of two monovalent cation species, taking potassium ions and sodium ions as examples, will be examined. The competitive adsorption of monovalent cations with divalent cations will be dealt with in Section 3.5.

It is generally considered that the adsorption of sodium ions and potassium ions by soils is electrostatic in mechanism. However, Udo (1978) and Deist and Talibudeen (1967) thought that kaolinite may adsorb potassium ions specifically to a certain extent. Nir et al. (1986) also suggested that montmorillonite can adsorb monovalent cations Li, Na, and K specifically. It appears that for most cations it is difficult to make a clear-cut distinction between electrostatic adsorption and specific adsorption. Therefore, results of studies on the competitive adsorption of potassium ions with sodium ions in variable charge soils under identical conditions would be helpful in elucidating the mechanism of interactions between soils and cations.

3.4.1 Comparison of Adsorption of Two Ion Species

When variable charge soils were in equilibrium with mixed solutions containing equal amounts of potassium ions and sodium ions, it was found that the activity ratio a_K/a_{Na} in soil suspensions as directly determined with a potassium ion–selective electrode and a sodium ion–selective electrode was smaller than unity in all the concentration ranges studied (Fig. 3.14). This implies that under identical experimental conditions the adsorption of potassium ions is always stronger than that of sodium ions.

If the amount of adsorbed potassium ions is compared with that of adsorbed sodium ions (Table 3.8), it can be found that the presence of sodium ions did not affect the adsorption of potassium ions, because the adsorbed amount of potassium ions in mixed systems was nearly the same as that in single ion systems. By contrast, the presence of potassium ions caused a drastic decrease in adsorption of sodium ions. When the pH is sufficiently low there may be the occurrence of negative adsorption of sodium ions. Thus, the selectivities of variable charge soils to potassium ions and to sodium ions differ markedly. The ratio in the selectivity coefficient may be much larger than the value 1.37 suggested by Heald et al. (1964).

Variable charge soils of tropical and subtropical regions contain large amounts of iron and aluminum oxides and their hydrates. When the pH is lower than the zero point of charge of these oxides, they would carry positive charges, exerting a repulsive action on cations. When the repulsive force of positive surface charge sites of soil colloids exceeds the attractive

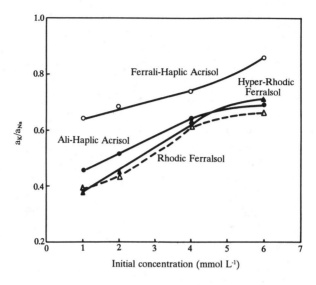

Fig. 3.14. Effect of initial solution concentration on activity ratio a_K/a_{Na} in soil suspension (initial C_K/C_{Na} in solution, 1).

force of negative surface charge sites, negative adsorption of cations would occur. Under ordinary conditions, because variable charge soils carry a

Table 3.8 Comparison of K–Adsorption and Na–Adsorption Between Single–Ion Systems and Mixed–Ion Systems[a]

| Soil | pH | Adsorption (mmol kg⁻¹) | | | |
| | | K | | Na | |
		Single	Mixed	Single	Mixed
Ali–Haplic Acrisol	3.0	1.7	1.5	0.35	0.00
	4.0	2.7	2.5	0.45	0.05
Ferrali–Haplic	3.0	0.5	0.5	0.25	−0.10
Acrisol	4.0	1.3	1.2	0.30	−0.08
Rhodic Ferralsol	3.0	0.7	0.6	0.10	−0.28
	4.0	1.2	1.2	0.25	−0.20

[a]Potassium and sodium ion concentration in mixed systems, 1 mmol L⁻¹ each.

certain quantity of negative surface charge, they can adsorb sodium ions when these ions are present, even when the pH is quite low. On the other hand, when there is also the presence of potassium ions, because these ions are adsorbed preferentially and occupy most of the negative surface charge sites, the attractive force of soil colloidal surface with sodium ions is weakened. In this case, the repulsive effect of positive surface charge sites on sodium ions may play the dominant role, resulting in the negative adsorption of these ions. This phenomenon can occur only in variable charge soils, but not in constant charge soils. Apparently, the lower the pH is from the zero point of charge of the soil, the more remarkable the negative adsorption. This is the reason why the negative adsorption of sodium ions in Rhodic Ferralsol is more remarkable than in Ferrali–Haplic Acrisol and why for a given soil the negative adsorption is more remarkable at pH 3 than at pH 4 as shown in Table 3.8. For the Ali–Haplic Acrisol, no negative adsorption occurs, owing to the surplus of negative surface charge over positive surface charge even at low pH.

If examined in a quantitative way, it can be seen from Fig. 3.15 that in mixed solutions the adsorbed amount of sodium ions is inversely related to that of potassium ions. For Ali–Haplic Acrisol, Hyper–Rhodic Ferralsol, and Ferrali–Haplic Acrisol, the slopes of the correlation line are -0.56, -0.65, and -0.76, respectively. This means that the affinity of variable charge soils for potassium ions is greater than that for sodium ions, especially in the Ali–Haplic Acrisol. Negative adsorption of sodium ions occurred when the amount of adsorbed potassium ions exceeded a certain

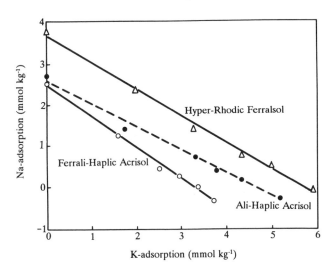

Fig. 3.15. Adsorption of sodium ions in relation to adsorption of potassium ions in mixed solutions (Li and Ji, 1992a).

value. For Ali–Haplic Acrisol, Ferrali–Haplic Acrisol, and Hyper–Rhodic Ferralsol in a mixed solution with a total ion concentration of 2 mmol L^{-1}, the concentrations of potassium ions of the equilibrium solution at which negative adsorption of sodium ions occurred were 0.55, 0.65, and 0.60 mmol L^{-1}, respectively.

It has been observed that, within the initial molar ratio C_K/C_{Na} range of 0:5 to 5:0 with the same initial total K+Na concentration, when the ratio was increased, the pH of the equilibrium suspension decreased. For the Ali–Haplic Acrisol, Ferrali–Haplic Acrisol and Rhodic Ferralsol, the variation ranges in pH were 5.34–4.35, 5.27–4.56 and 5.98–5.46, respectively. This should be an indication of the replacement of more hydrogen ions and aluminum ions by potassium ions when the ratio was increased.

3.4.2 Effect of pH

In Section 3.2.2, the effect of pH on the adsorption of potassium ions and sodium ions by variable charge soils in single–ion systems was discussed. If there are two cation species present together in the system, a combined effect should also exist, and the situation may be more complex than that in single–ion systems.

It can be seen from Fig. 3.16 that in the equilibrium solutions the activity ratios a_K/a_{Na} are always smaller than unity for the four variable charge soils. This means that the change in pH does not affect the general tendency, that

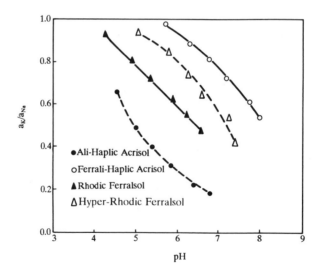

Fig. 3.16. Effect of pH on activity ratio a_K/a_{Na} in soil suspension (initial concentrations of sodium ions and potassium ions, 1 mmol L^{-1}) (Li and Ji, 1992a).

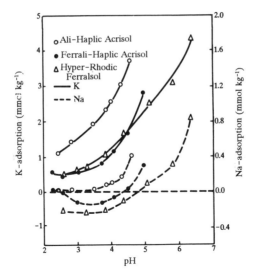

Fig. 3.17. Effect of pH on adsorption of potassium ions and sodium ions (initial concentrations of potassium ions and sodium ions, 1 mmol L^{-1}) (Li and Ji, 1992a).

is, the affinity of these soils for potassium ions is greater than that for sodium ions. This difference in affinity could only be caused by the difference in nature between the two ion species, because the adsorption occurs under identical experimental conditions. On the other hand, the activity ratio a_K/a_{Na} decreased with the rise in pH. Since the rise in pH also means an increase in quantity of negative surface charge carried by the soil, it appears that the higher the negative surface charge density of a soil, the stronger the competition of potassium ions with sodium ions.

The effect of pH on competitive adsorption of potassium ions with sodium ions can also be examined from the quantities of ions adsorbed. As seen in Fig. 3.17, in a mixed solution the effect of pH on the adsorption of potassium was nearly identical to that when there were only potassium ions alone as shown in Fig. 3.3. This implies that the presence of sodium ions practically does not affect the adsorption of potassium ions by variable charge soil, irrespective of changes in pH. On the other hand, the presence of potassium ions greatly affected the adsorption of sodium ions. The adsorption of sodium ions decreased with the decrease in pH. For the Ferrali–Haplic Acrisol and the Hyper–Rhodic Ferralsol, negative adsorption of sodium ions may occur when the pH is sufficiently low. The threshold pH below which negative adsorption occurred for the two soils were 4.4 and 4.8, respectively.

Table 3.9 Effect of Iron Oxides on a_K/a_{Na} Ratio in Soil Suspensions[a]

	a_K/a_{Na}	
Soil	Original	Fe_2O_3-Removed
Ali–Haplic Acrisol	0.45	0.34
Ferrali–Haplic Acrisol	0.65	0.48
Rhodic Ferralsol	0.45	0.41
Hyper–Rhodic Ferralsol	0.56	0.38

[a]Original soil and Fe_2O_3-removed soil were adjusted to the same pH.

3.4.3 Effect of Iron Oxides

The presence of iron oxides in variable charge soils causes the increase in positive surface charge and the decrease in negative surface charge. This would affect the competitive adsorption of potassium ions with sodium ions. As seen in Table 3.9, while the removal of iron oxides did not affect the general tendency that the affinity of variable charge soils for potassium ions

Table 3.10 Effect of Dielectric Constant of Solvent on Adsorption of Potassium Ions and Sodium Ions (Li and Ji, 1992a)

Soil	Ethanol (%)	K–Adsorption mmol kg^{-1}	Rel.[a]	Na–Adsorption mmol kg^{-1}	Rel.
Ali–Haplic Acrisol	0	10.83	1	0.40	1
	20	13.77	1.27	0.78	1.95
Ferrali–Haplic Acrisol	0	7.59	1	0.51	1
	20	8.76	1.15	0.82	1.61
Rhodic Ferralsol	0	11.09	1	1.59	1
	20	12.68	1.14	1.81	1.14
Hyper–Rhodic Ferralsol	0	13.16	1	1.95	1
	20	16.75	1.27	2.09	1.07

[a]Relative value.

was stronger than that for sodium ions, the difference in affinity became greater when iron oxides were removed. This should be a consequence of the increase in negative surface charge density, similar to the effect of increased pH shown in Fig. 3.16.

3.4.4 Effect of Dielectric Constant of Solvent

If the dielectric constant of the solvent was decreased by the addition of ethanol, it can be found that the amounts of both adsorbed potassium ions and adsorbed sodium ions increased (Table 3.10). According to the principle of electrostatics, the electrostatic force exerted on ions by an electric field is inversely proportional to the dielectric constant of the medium. When the dielectric constant of the solvent is decreased, the electrostatic force between ions and the surface of soil colloids increases. This would result in an increase in adsorption of cations. This implies that the adsorption of potassium ions and sodium ions is caused mainly by electrostatic attraction force. It can be calculated that the dielectric constant of water is about 1.16 times that of the aqueous solution containing 20% of ethanol. For the adsorption of potassium ions, the ratio of adsorbed amount in the two media was in the range of 1.14–1.27. In the case of sodium ions, the ratios for two Rhodic Ferralsols were also within this range. For the Ali–Haplic Acrisol and the Ferrali–Haplic Acrisol, however, the ratio was much larger. The reason is not known.

3.4.5 Effect of Accompanying Anions

If a variable charge soil was allowed to react with mixed solutions containing potassium salts and sodium salts composed of different anions, it can be observed that the activity ratios a_K/a_{Na} in the equilibrium suspension were close to each other for salts composed of nitrate or perchlorate as the accompanying anions (Table 3.11). The ratio was slightly smaller for chloride. By contrast, the ratio was much smaller if sulfate was present as the accompanying anions.

Apparently, this situation is chiefly caused by the difference of various anions on surface charge properties of the soils. In the literature, it is generally considered that such anions as chloride, nitrate and perchlorate are adsorbed by soils solely through electrostatic force. In recent years, the results of a series of studies conducted in the Institute of Soil Science, which shall be presented in later chapters of this book, showed that, at least for chloride ions, in addition to electrostatic force, a specific force is involved in the adsorption by variable charge soils. However, generally speaking, the effect of these three anion species on surface properties of soils is not large when compared with sulfate ions. By contrast, there is a pronounced

Table 3.11 Effect of Accompanying Anions on a_K/a_{Na} Ratio in Soil Suspensions (Li and Ji, 1992a)[a]

Soil	Anions	pH	a_K/a_{Na}
Ali–Haplic Acrisol	Cl	4.59	0.24
	NO_3	4.53	0.27
	ClO_4	4.55	0.27
	SO_4	5.82	0.13
Ferrali–Haplic Acrisol	Cl	5.04	0.39
	NO_3	4.81	0.40
	ClO_4	4.84	0.42
	SO_4	5.61	0.17
Rhodic Ferralsol	Cl	5.01	0.25
	NO_3	5.00	0.30
	ClO_4	4.98	0.31
	SO_4	6.26	0.15
Hyper–Rhodic Ferralsol	Cl	5.38	0.27
	NO_3	5.35	0.29
	ClO_4	5.30	0.30
	SO_4	6.42	0.10

[a]Initial total concentration, 1 mmol L^{-1}; $C_K = C_{Na}$.

specific adsorption of sulfate ions by variable charge soils. This kind of adsorption would result in the release of OH^- ions (Table 3.11) and the increase in negative surface charge density of soil colloids (cf. Chapter 6). This should be the principal cause of the increase in selectivity to potassium ions when the accompanying anion species is sulfate. Shainberg et al. (1987) also found that the increase in surface charge density of montmorillonite caused an increase in selectivity coefficient of potassium ions relative to sodium ions in adsorption.

From the materials presented in this section it may be concluded that the adsorption of potassium ions and sodium ions by variable charge soils is controlled principally by electrostatic attraction force. The selectivity to potassium ions by various variable charge soils is always greater than that to sodium ions. Among these soils, the Ali–Haplic Acrisol carries the largest quantity of negative surface charge and adsorbs the largest amount of cations. For a given soil, such factors as pH, iron oxides, and accompanying anions affect the adsorption of cations by affecting the surface charge properties of the soil. The decrease in dielectric constant of the solvent causes the increase in adsorbed amount. All these phenomena can be explained by electrostatic attraction between the surface of soil colloids and cations. The phenomenon that in mixed systems composed of two cation species the attraction force of soils to potassium ions is greater than that to sodium ions is understandable, because for cations of the same valence the smaller the hydrated radius, the stronger the electrostatic force between these ions and the surface of soil colloids, and thus the easier the adsorption. The hydrated radius of potassium ions is 0.532 nm, corresponding to only 67% of that of sodium ions with a hydrated radius of 0.790 nm. Thus, the difference in behavior with respect to adsorption between these two ion species is similar to that in the adsorption by constant charge soils and clay minerals.

However, in the competitive adsorption of potassium ions with sodium ions there are also some complicated situations. The adsorption ratio between the two ion species is always larger, and sometimes much larger, than that predicated from the ratio of hydrated radius. The adsorption ratio differs with the kind of soil. For a given soil the ratio varies with the change in negative surface charge density. The concentration of relevant cations in solution can affect the relative amount of adsorption of the two ion species. These phenomena lead to the suggestion that, in addition to electrostatic force, a certain specific force may affect the adsorption of these cations, at least of potassium ions, by variable charge soils.

In the adsorption of monovalent cations by variable charge soils, a characteristic is that a negative adsorption of sodium ions may occur when the pH is sufficiently low. In single sodium ion systems for the Rhodic Ferralsols, the numerical value of bonding energy may be negative in sign when the degree of Na–saturation is lower than about 20% (Yu and Zhang, 1990). In this case it should be the competition of aluminum ions for negative adsorption sites that induces the negative adsorption of sodium ions. In mixed systems in which potassium ions and sodium ions coexist, the former ion species with a stronger affinity for soil colloid can also play a role similar to that of aluminum ions. Of course, for the occurrence of negative adsorption of sodium ions there must be a prerequisite, that is, the positive surface charge density of soil colloid is sufficiently high. For constant charge soils, such negative adsorption would never occur.

3.5 COMPETITIVE ADSORPTION OF POTASSIUM IONS WITH CALCIUM IONS

Potassium and calcium are two important elements for plant nutrition in soils. Their interactions with the surface of soil colloids are one of the focuses in soil chemical studies (Bruggenwert and Kamphorst, 1979). Marshall (1964) discussed the factors affecting potassium to calcium activity ratio. Beckett (1964), based on Schofield's "ratio law," proposed to use activity ratio $a_K/(a_{Ca})^{1/2}$ as an index for evaluating the availability of potassium to plants in soils. Ulrich (1961) called the value pK-0.5pCa as potassium–calcium potential. However, because of the complexities in chemical composition of soils and the difference in nature of the two ion species, the relative selectivity in adsorption between these two kinds of ions for constant charge soils and clay minerals obtained by various authors differs greatly, and it frequently varies with experimental conditions (Ferris and Japson, 1975; Helmke, 1980; Hoover, 1944; Jensen, 1973; Levy et al., 1988; Lim et al., 1980; Raney and Hoover, 1946; Rhue and Mansell, 1988; Schwertmann, 1962; Sparks and Huang, 1985; Talibudeen and Goulding, 1983; Udo, 1978). Variable charge soils contain both constant charge components and variable charge components, and therefore a situation more complex than that in the case of competitive adsorption of potassium ions with sodium ions may be expected when potassium ions and calcium ions interact with these soils in mixed systems. In this section, based on the activity ratio of potassium ions to calcium ions as determined directly with two ion–selective electrodes (Li and Ji, 1991; Yu et al., 1989b; Yu et al., 1990), the competitive adsorption of potassium ions with calcium ions will be discussed. In this case, because the adsorption of the two cation species proceeds under identical experimental conditions, any difference in relative selectivity should be directly related to the nature of the ions.

3.5.1 Competitive Adsorption in Relation to Quantity of Ions

When a soil was in equilibrium with mixed solutions of different concentrations but contained equivalent amounts of potassium ions and calcium ions, it was observed that the value of pK-0.5pCa in the equilibrium suspension decreased with the increase in total concentration of the ions (Table 3.12). This means that the selectivity to calcium ions relative to potassium ions increased with the increase in total ion concentration. Among the four soils the value of pK-0.5pCa for Ali-Haplic Acrisol was the largest, especially when the total concentration of ions was low, indicating the smallest selectivity to calcium ions as compared to the other three types of soils. At high total ion concentrations the difference became less remarkable. Nevertheless, in this case the relative selectivity of the four soils to the two ion species can still be evaluated from the pK-0.5pCa values.

If the equivalent ratio of potassium ions to calcium ions in adsorption is calculated, it can be clearly seen that this ratio decreased with the increase in total ion concentration, indicating a tendency of increased selectivity to

Table 3.12 Effect of Ion Concentration on Adsorption of Potassium Ions and Calcium Ions (Li and Ji, 1992b)

Soil	K+Ca (mol L^{-1})^a	pH	pK-0.5pCa	Adsorption^b K	Ca	Ratio (K/2Ca)
Ali-Haplic Acrisol	1.5×10^{-3}	4.42	1.65	3.31	1.36	1.22
	3.0×10^{-3}	4.28	1.37	5.12	2.32	1.10
	6.0×10^{-3}	4.07	1.15	8.01	4.26	0.94
	1.2×10^{-2}	4.03	0.94	11.23	7.73	0.73
	2.4×10^{-2}	3.97	0.72	15.59	17.15	0.46
Ferrali-Haplic Acrisol	1.5×10^{-3}	4.67	1.48	2.44	1.30	0.94
	3.0×10^{-3}	4.48	1.24	3.26	2.26	0.72
	6.0×10^{-3}	4.28	1.06	4.90	3.99	0.61
	1.2×10^{-2}	4.22	0.87	6.20	7.44	0.42
	2.4×10^{-2}	4.14	0.68	7.73	16.62	0.23
Rhodic Ferralsol	1.5×10^{-3}	4.82	1.49	3.14	1.84	0.85
	3.0×10^{--}	4.66	1.27	5.00	3.27	0.77
	^{3}6.0×10^{-3}	4.50	1.09	7.73	5.44	0.71
	1.2×10^{-2}	4.33	0.91	9.87	8.28	0.60
	2.4×10^{-2}	4.36	0.70	12.55	16.62	0.38
Hyper-Rhodic Ferralsol	1.5×10^{-3}	5.04	1.57	3.23	1.65	0.98
	3.0×10^{-3}	5.09	1.33	4.77	2.55	0.93
	6.0×10^{-3}	5.10	1.12	7.15	4.26	0.84
	1.2×10^{-2}	4.96	0.93	9.87	6.55	0.75
	2.4×10^{-2}	4.82	0.71	15.59	14.36	0.54

^{a}quivalent ratio, 1
^{b}(mmol kg^{-1})

calcium ions. The adsorption equivalent ratio was generally smaller than unity. However, for the Ali-Haplic Acrisol the ratio may be larger than unity when the ion concentration is low. This means that in this case the selectivity of this soil to potassium ions was greater than that to calcium ions. This may be caused by the presence of a substantial amount of 2:1-type clay minerals such as hydrous mica.

The phenomenon mentioned above indicates that the activities of ion-exchange sites on the surface of variable charge soils are different. Some sites possess a strong ability to adsorb potassium ions. When the concentration of potassium ions in the solution is low, these sites would adsorb potassium ions preferentially. After the saturation of these sites by potassium ions, other sites with a smaller selectivity to potassium ions begin to operate. This is the reason why the relative selectivity of soils to potassium ions and calcium ions varies with the concentration of ions. This phenomenon has also been observed for other kaolinitic soils (Levy *et al.*, 1988; Rhue and Mansell, 1988).

The relative selectivity of variable charge soils to potassium ions and calcium ions also varies with the initial equivalent ratio between the two ion species. It was observed that, under conditions of constant total concentration, the amount of adsorbed potassium ions decreased while that of adsorbed calcium ions increased with the decrease in equivalent ratio, resulting in a gradual increase in $pK-0.5pCa$ (Table 3.13). At an equivalent ratio of 1 the value $pK-0.5pCa$ was the largest for the Ali–Haplic Acrisol, indicating again the smallest selectivity to calcium ions as compared to other types of soils.

For a given soil, when the initial equivalent ratio of potassium ions to calcium ions varied from 2.5:1 to 1:2, it was found that, despite the decrease in adsorption of potassium ions and the increase in adsorption of calcium ions with the decrease in this ratio, the total quantity of the two adsorbed ion species was approximately the same in all the systems. This implies that the increment in adsorption of calcium ions was equivalent to the decrement in adsorption of potassium ions if compared on a molar basis, that is, the ratio was 1:1 but not 1:2. It should also be noted in Table 3.13 that the pH of the equilibrium solution remained practically constant, irrespective of the variation of the initial equivalent ratio. This means that one mole of calcium ions and one mole of potassium ions replaced the same quantity of hydrogen ions. Thus, whether calcium ions can be adsorbed in the form of CaX^+ (X^- may be Cl^-) by variable charge soils remains one interesting question deserving further elucidation.

3.5.2 Effect of Iron Oxides

Since the presence of iron oxides causes the increase in positive surface charge and the decrease in negative surface charge of variable charge soils, it can be expected that these soils would adsorb more potassium ions and calcium ions after the removal of iron oxides. This is just the case. It has been found that the amounts of potassium ions and calcium ions adsorbed from a mixed solution with an equivalent ratio of 1 increased from 8.01 mmol kg^{-1} to 10.69 mmol kg^{-1} and from 4.26 mmol kg^{-1} to 7.92 mmol kg^{-1}, respectively, for an Ali–Haplic Acrisol after the removal of iron oxides. For the Ferrali–Haplic Acrisol, Rhodic Ferralsol and Hyper–Rhodic Ferralsol, the corresponding figures were as follows: potassium, 4.90 → 7.15,

Table 3.13 Effect of Potassium to Calcium Equivalent Ratio on Adsorption of Ions (Li and Ji, 1992b)

Soil	Equivalent Ratio (K/2Ca)	pH	pK-0.5pCa	Adsorption (mmol kg^{-1}) K	Ca	Ratio (K/2Ca)
Ali–Haplic	2.5:1	4.35	1.18	7.02	1.13	3.11
Acrisol	1:1	4.37	1.41	5.55	2.38	1.17
	1:2	4.32	1.63	4.27	3.21	0.67
	1:4	4.35	1.91	2.92	4.26	0.34
	1:10	4.37	2.32	1.52	4.93	0.31
Ferrali–Haplic	2.5:1	4.54	1.06	5.11	1.06	2.41
Acrisol	1:1	4.51	1.27	3.85	2.38	0.81
	1:2	4.52	1.50	2.94	3.34	0.44
	1:4	4.55	1.75	1.85	3.71	0.25
	1:10	4.56	2.12	0.96	4.20	0.11
Rhodic Ferralsol	2.5:1	4.78	1.05	7.27	1.81	2.01
	1:1	4.77	1.30	5.65	3.49	0.81
	1:2	4.76	1.54	4.12	4.75	0.43
	1:4	4.76	1.80	2.77	6.03	0.46
	1:10	4.75	2.19	1.46	7.27	0.10
Hyper–Rhodic	2.5:1	5.16	1.12	7.02	1.46	2.40
Ferralsol	1:1	5.10	1.35	5.44	2.92	0.93
	1:2	5.11	1.57	3.96	4.04	0.49
	1:4	5.16	1.85	2.72	4.77	0.29
	1:10	5.19	2.26	1.46	5.28	0.14

7.73 → 8.81, 7.15 → 8.28; calcium, 3.99 → 6.99, 5.44 → 8.01, 4.26 → 6.99. All these data indicate a suppressing effect of iron oxides on adsorption of cations in variable charge soils.

One more interesting point is the effect of the decrease in negative surface charge density caused by the presence of iron oxides on the relative selectivity to the two ion species in variable charge soils. The changes in adsorption equivalent ratio of potassium ions to calcium ions for several variable charge soils after the removal of iron oxides are shown in Figs. 3.18 and 3.19. As seen in the figures, the ratio decreases for all the soils. This means an increase in selectivity to calcium ions relative to potassium ions. For the four soils the increase is of the order Hyper–Rhodic Ferralsol >

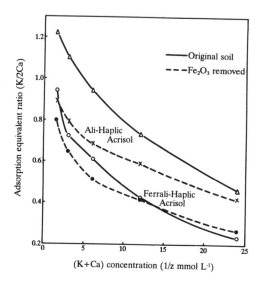

Fig. 3.18. Adsorption equivalent ratio K/2Ca in Ali–Haplic Acrisol and Ferrali–Haplic Acrisol before and after removal of free iron oxides (initial equivalent ratio, 1).

Rhodic Ferralsol > Ali–Haplic Acrisol > Ferrali–Haplic Acrisol. This order is consistent with the order of content of iron oxides in the soils. Thus,

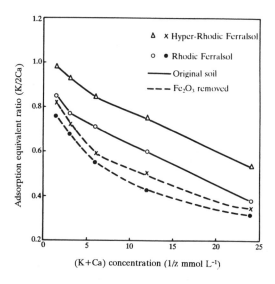

Fig 3.19. Adsorption equivalent ratio K/2Ca in Rhodic Ferralsols before and after removal of free iron oxides (initial equivalent ratio, 1).

similar to the case in the competitive adsorption of potassium ions with sodium ions, the increase in negative surface charge density of soil colloids is favorable for the adsorption of calcium ions with a stronger competitive power as compared to potassium ions.

3.5.3 Effect of pH

The situation for the effect of pH on competitive adsorption of potassium ions with calcium ions is more complex than that of potassium ions with sodium ions (Pleysier et al., 1979; Pratt et al., 1962; Rhue and Mansell, 1988). Here at least three important factors are involved. (1) The quantity of negative surface charge of soils is affected. According to the Donnan theory, the increase in negative surface charge density caused by the increase in pH would be favorable to the competition of divalent calcium ions against potassium ions. (2) The possibility of specific adsorption of the two ion species. It appears that the specific adsorption of potassium ions is affected very little by pH. On the other hand, the proportion of calcium ions adsorbed in the form of $CaOH^+$ increases with the rise in pH (Chan et al., 1979; Yu, 1976). (3) The competitive adsorption of hydrogen ions at low pH and particularly the competitive adsorption of aluminum ions when the pH is lower than about 5.5 makes the competition of potassium ions with calcium ions even more unfavorable. These factors together would cause the adsorption equivalent ratio of potassium ions to calcium ions to have a maximum at a certain pH. This hypothesis has been verified (Yu et al., 1989b). For the Ali–Haplic Acrisol and Rhodic Ferralsol, the values pH–0.5pCa of the equilibrium solution after adsorption were the highest within the pH range of 5.5–6.5 when the initial equivalent concentrations of the two ion species were the same.

On the other hand, the competitive adsorption of two ion species at different pH is also related to the relative amounts of these ions. In Figs. 3.20 and 3.21, the effects of pH on pK–0.5pCa of variable charge soils when the initial molar ratios C_K/C_{Ca} are 2:1 and 1:2, respectively, are shown. Within the lower pH range (<6), the pK–0.5pCa decreases with the decrease in pH, especially when the potassium to calcium ratio is small. This perhaps chiefly reflects the effect of the competitive adsorption of hydrogen ions and aluminum ions, and this will be discussed in the next section. Within the higher (6–7) pH range, the pK–0.5pCa decreases slightly with the rise in pH when the molar ratio is 2:1 and increases when the molar ratio is 1:2.

The value pK–0.5pCa is an overall reflection of the activities of potassium ions and calcium ions in the equilibrium solution of soils. When the pH is changed, the activity of the two ion species may increase or decrease to a same direction, or it may also increase for one ion species but

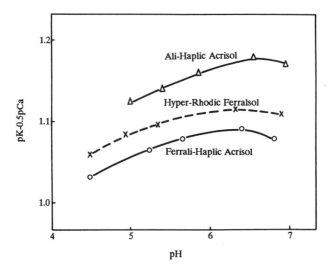

Fig. 3.20. Effect of pH on pK–0.5pCa in soil suspension at an initial K:Ca ratio of 2:1 (Li and Ji, 1992b).

decrease for another ion species. Besides, when the activity is expressed in the form of pK or pCa the variation is in a logarithmic manner but not in a proportional manner. Hence, this would add complexities when interpreting the change in pK–0.5pCa with the change in pH. Therefore, sometimes it is necessary to examine the parameters pK and pCa separately.

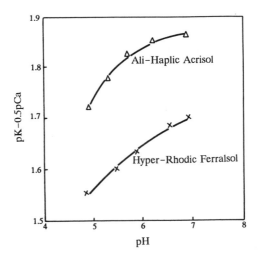

Fig. 3.21. Effect of pH on pK–0.5pCa in soil suspension at an initial K:Ca ratio of 1:2 (Li and Ji, 1992b).

It was found (Yu et al., 1989b) that at all the potassium-to-calcium equivalent ratios both pK and pCa of the equilibrium suspension of Ali-Haplic Acrisol and Rhodic Ferralsol increased with the rise in pH, indicating an increase in adsorption of both of the two cation species. On the other hand, the increment in pK was larger when the initial ratio was small, whereas the increment in pCa was larger when the ratio was great. Thus, within the pH range of higher than 6 the increase in pK-0.5pCa with the rise in pH shown in Fig. 3.21 chiefly reflects a larger relative variation range in adsorption of potassium ions when the amount of these ions present in the system is small, while the slight decrease in pK-0.5pCa shown in Fig. 3.20 perhaps chiefly reflects a larger relative variation range in adsorption of calcium ions when the amount of these ions is small. In the latter case, both the increase in negative surface charge density of soil colloids and the occurrence of specific adsorption of calcium ions following the rise in pH may cause the marked decrease in calcium concentration of the soil suspension.

It follows from the above discussions that factors involved in the effect of pH on selectivity of soils to various cations in adsorption are quite complex. This is perhaps the main reason why the selectivity coefficients to various cations obtained by different authors under different experimental conditions varied so widely.

3.5.4 Effect of Accompanying Ions

Other ions present in variable charge soils may affect the competitive adsorption of potassium ions with calcium ions through a variety of ways. Cations would compete for negative surface charge sites of soil colloids with the two ion species, with the extent of competition determined by the affinities of these cations relative to potassium and calcium ions for the exchange sites. Anions can affect the adsorption of these two ion species through affecting the existing form of calcium ions and the surface properties of soils.

It has been mentioned several times in this chapter that hydrogen ions and aluminum ions present in acid soils may affect the adsorption of such cations as sodium, potassium, calcium, and so on. Experimental results have shown that the competition of aluminum ions for adsorption sites at low pH was the main cause for the lowering of bonding energy of variable charge soils with alkali metal cations (Yu and Zhang, 1990). This competition is also the cause for the large difference in bonding energy between potassium ions and sodium ions at low pH. The effect of aluminum ions on the competitive adsorption of potassium ions with calcium ions can be noticed from the data given in Table 3.14. When originally adsorbed cations are hydrogen and aluminum ions, the pK value of the soil suspension may be

Table 3.14

Effect of Accompanying Cations on Competitive Adsorption of Potassium Ions with Calcium Ions[a] (Yu et al., 1989a)

Soil	Soil:Water	pK		pK–0.5pCa	
		Na–sat.[b]	H,Al–sat.	Na–sat.	H,Al–sat.
Ali–Haplic Acrisol	1:1	3.30	2.49	1.20	1.14
	1:2	3.54	2.71	1.21	1.25
	1:4	3.79	2.98	1.20	1.35
	1:8	4.05	3.23	1.31	1.48
Rhodic Ferralsol	1:1	2.92	2.56	1.11	1.12
	1:2	3.12	2.78	1.16	1.24
	1:4	3.30	3.01	1.23	1.34
	1:8	3.53	3.33	1.27	1.46

[a]Total K + Ca, 0.5 symmetry value with an equivalent ratio of 1:1.
[b]Saturated cations.

smaller by 0.8 unit than that when originally adsorbed cations are sodium ions, indicating a much stronger competitive power of hydrogen and aluminum ions with potassium ions as compared to sodium ions. Within the soil–to–water ratio range of 1:2 to 1:8, the pK–0.5pCa value in H,Al–saturated systems was larger than that in Na–saturated systems. This means that the adsorption of calcium ions is also suppressed by hydrogen and aluminum ions.

The difference in the effect on the competitive adsorption of potassium ions with calcium ions exerted by aluminum ions and sodium ions can be seen more clearly from Fig. 3.22. The adsorbent is manganese oxide. Manganese oxide is a commonly found mineral in variable charge soils, although the amount is not large. The presence of sodium ions can cause a decrease in adsorption of potassium ions while only cause a slight decrease in adsorption of calcium ions when the amount of calcium ions is high. Hence, the pK–0.5pCa value of the soil suspension is smaller than that of the check treatment when the amount of calcium ions is low. By contrast, the strong affinity of aluminum ions for the adsorbent also causes a pronounced decrease in adsorption of calcium ions. This would result in the largest value of pK–0.5pCa in the suspension.

Besides aluminum ions, there may be the presence of ferrous ions and manganese ions in variable charge soils. The amounts of these ions are

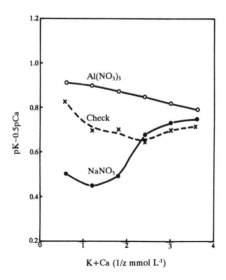

Fig. 3.22. Effect of aluminum ions and sodium ions on competitive adsorption of potassium ions with calcium ions in Na-saturated manganese oxide [quantities of $Al(NO_3)_3$ and $NaNO_3$, 1.2 1/z mol g^{-1})] (Data of T. R. Yu, F. L. Wang, and P. M. Huang).

particularly large when the soil is submerged, especially if the soil contains a large amount of organic matter, as shall be seen in Chapters 13 and 14. These ions would compete for exchange sites with potassium ions and calcium ions. Likewise, when these ions are added to a soil, they would replace originally adsorbed potassium and calcium ions (Table 3.15). In the table, because the soil contains much more calcium ions than potassium ions, the amount of calcium ions released is invariably much larger than that of potassium ions. It can also be noted that a small amount of manganese ions is sufficient to replace most of the adsorbed potassium ions. If the amount of manganese ions is further increased, more calcium ions may be replaced until nearly all the adsorbed calcium ions together with the potassium ions are replaced, so that the pK-0.5pCa value attains a near constant.

Similarly, ferrous ions can affect the activity ratio of calcium ions to potassium ions in Ali-Haplic Acrisol and kaolinite (Fig. 3.23). The activity ratio increased markedly after the addition of a small amount of ferrous ions. With the gradual increase in amount of ferrous ions, the ratio tended to attain a constant value at which all the adsorbed potassium ions and calcium ions were replaced.

From the materials presented above, it is apparent that the effect of accompanying cations on the competitive adsorption of potassium ions with

Table 3.15

Effect of Manganese Ions on Release of Adsorbed Potassium Ions and Calcium Ions in Ali–Haplic Acrisol and Kaolinite (Yu et al., 1990)

Material	Treatment	MnSO$_4^a$	pK-0.5pCa	K^b	Ca^b	Ca/K	Release^c K	Release^c Ca
Kaolinite	Check	0	2.376	9.0	9.2	1.02	–	–
		1	2.719	11.9	78	6.56	2.9	69
		2	2.726	12.5	87	6.96	3.5	79
		3	2.715	13.2	93	7.05	4.2	84
	Submerged	0	2.722	6.3	22.4	3.56	–	–
		2	2.990	7.9	118	14.9	1.6	96
		3	2.956	9.2	139	15.1	2.9	117
Ali–Haplic Acrisol	Check	0	2.065	87	206	2.39	–	–
		2	2.187	102	489	4.79	15	283
		4	2.205	105	569	5.42	18	263
		6	2.207	109	615	5.64	22	409
	Submerged	0	2.108	79	214	2.70	–	–
		2	2.203	113	650	5.75	34	436
		4	2.222	121	815	6.73	42	601
		6	2.226	127	906	7.13	48	692

^{a}4½cmol kg^{-1}; ^{b}1/z μmol L^{-1}; ^{c}1/z μmol 100g^{-1}.

calcium ions is determined by the nature of this foreign cation species. If the affinity of this cation species for the soil is smaller than that of both potassium ions and calcium ions, such as in the case of sodium ions, this ion species mainly affects the adsorption of potassium ions. If the affinity is larger than that of both potassium ions and calcium ions, such as in the case of aluminum ions, the adsorption of calcium ions may also be affected. Since in acid variable charge soils there is invariably the presence of a large amount of aluminum ions, the effect of this ion species on competitive adsorption of potassium ions with calcium ions is of important practical significance.

With regard to the effect of accompanying anions, the effect of sulfate ions on the competitive adsorption of potassium ions with sodium ions has been presented in Section 3.4.5. A similar effect can be observed for the

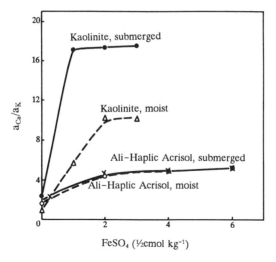

Fig. 3.23. Effect of ferrous ions on activity ratio a_{Ca}/a_K in suspensions of Ali–Haplic Acrisol and kaolinite (Yu, et al., 1990).

competitive adsorption of potassium ions with calcium ions. As seen in Table 3.16, for soils saturated with either sodium ions, calcium ions, or hydrogen and aluminum ions, when the potassium to calcium equivalent ratio is either 3:1 or 1:3, the pK of the equilibrium suspension is invariably larger when the accompanying anions are sulfate than when they are chloride. This is mainly caused by the increase in negative surface charge as a result of the rise in pH in the sulfate system. The difference in pCa between the two anion systems is even much larger than the difference in pK. This implies that the increased adsorption of calcium ions is also related to the coordinative adsorption between sulfate ions and calcium ions (Bolan et al., 1993). Actually, as shall be seen in Chapter 4, the surface charge properties of variable charge soil may even be affected by such anions as chloride. Thus, when the effects of accompanying cations and accompanying anions are combined together, the situation in the competitive adsorption of potassium ions with calcium ions is rather complex.

3.6 CONCLUDING REMARKS

In this chapter, the adsorption of cations by variable charge soils is discussed mainly from the standpoint of electrostatic adsorption. No doubt, for principal nutrient cations in soils and under ordinary conditions, the electrostatic attraction force between negative surface charges of soils and cations plays the determinative role in the interactions between them. However, under certain circumstances, specific forces may also exert their

Table 3.16 Effect of Accompanying Anions on pK and pCa in Soil
Suspensions (Yu et al., 1989b)

Soil	K:Ca[a]	Anions	pK			pCa		
			Na[b]	Ca	H,Al	Na	Ca	H,Al
Ali–Haplic	3:1	Cl	2.48	2.39	2.32	3.70	2.76	3.26
Acrisol		SO₄	2.53	2.43	2.39	3.92	2.90	3.52
	1:3	Cl	3.01	2.93	2.81	3.06	2.60	2.76
		SO₄	3.08	2.97	2.90	3.28	2.72	3.00
Rhodic	3:1	Cl	2.63	2.41	2.25	3.94	2.71	3.18
Ferralsol		SO₄	2.71	2.48	2.33	4.16	2.92	3.40
	1:3	Cl	3.21	3.01	2.77	3.16	2.52	2.70
		SO₄	3.30	3.09	2.86	3.40	2.70	2.90

[a]Equivalent ratio; total amount, 1 symmetry value.
[b]Exchangeable cation species.

effects. Unfortunately, these two kinds of forces cannot be easily distinguished from each other in ordinary experiments.

In this chapter, discussions are focused on ion adsorption but not on ion–exchange. Ion adsorption is concerned with the relationship between ions and the surface of soils while ion–exchange referrs to the replacement of one adsorbed ion species by another ion species. Actually, during the genesis of soils, at the moment when a negative surface charge is produced there would always be an equivalent of countercation adsorbed from the medium to maintain electroneutrality. Therefore, an ion–exchange reaction would always be involved in all the experiments in ion adsorption.

From the discussions made in this chapter it is apparent that in the electrostatic adsorption of cations by variable charge soils there are a variety of characteristics that are different from constant charge soils. In particular, both the quantity and strength of adsorption vary with environmental conditions. Negative adsorption may also occur. Such a phenomenon can never occur in constant charge soils. These features are determined by the particular nature of the surface of variable charge soils. Under field conditions these environmental factors are invariably in a state of constant change. Therefore, the interrelations between these soils and various cations are constantly changing.

BIBLIOGRAPHY

Barber, R. G. and Rowell, D. L. (1972) *J. Soil Sci.*, 23:135-146.

Barrow, N. J. (1985) *Adv. Agron.*, 38:183-230.

Barrow, N. J., Bowden, J. W., Posner, A. M., and Quirk, J. P. (1980) *Aust. J. Soil Res.*, 18:37-47.

Bear, F. E. (ed.) (1964) *Chemistry of the Soil.* Reinhold, New York.

Beckett, P. H. T. (1964) *J. Soil Sci.*, 15:9-23.

Bockris, J. O'M (1970) *Modern Electrochemistry.* vol. 1. MacDonald, New York.

Bolan, N. S., Syers, J. K., and Sumner, M. E. (1993) *Soil Sci. Soc. Am. J.*, 57:691-696.

Bolt, G. H. (1979) *Soil Chemistry. B. Physico-chemical Models.* Elsevier, Amsterdam.

Bowden, J. W., Posner, A. M. and Quirk, J. P. (1977) *Aust. J. Soil Res.*, 15:121-136.

Bowden, J. W., Posner, A. M. and Quirk, J. P. (1980) in *Soils with Variable Charge* (B. K. G. Theng, ed.). New Zealand Socciety of Soil Science, pp. 147-166.

Bruggenwert, M. G. M. and Kamphorst, A. (1979) in *Soil Chemistry. B. Physicochemical Models* (G. H. Bolt, ed.). Elsevier, Amsterdam. pp.141-203.

Chan, K. Y., Davey, B. G., and Geering, H. R. (1979) *Soil Sci. Soc. Am. J.*, 43:301-304.

Chaussidon, J. (1963) *Soil Sci.*, 95:131-133.

Deist, J. and Talibudeen, O. (1967) *J. Soil Sci.*, 18:138-148.

Eisenman, G. (ed.) (1967) *Glass Electrodes for Hydrogen and Other Cations.*Marcel Dekker, New York.

Ferris, A. P. and Jepson, W. B. (1975) *J. Colloid Interface. Sci.*, 51:245-249.

Fischer, W. R. (1990) *Z. Pflanzenernähr. Bodenk.*, 153:93-95.

Heald, W. R., Frere, M. H. and De Wit, C. T. (1964) *Soil Sci. Soc. Am. Proc.*, 28:622-627.

Helmy, A. K. (1967) *J. Soil Sci.*, 18:35-38.

Hensen, H. E. (1973) *Agrochimica*, 17:181-190.

Hoover, C. D. (1944) *Soil Sci. Soc. Am. Proc.*, 9:66-71.

Jardine, P. M. and Sparks, D. L. (1984a) *Soil Sci. Soc. Am. J.*, 48:39-45.

Jardine, P. M. and Sparks, D. L. (1984b) *Soil Sci. Soc. Am. J.*, 48:45-50.

Juo, A. S. R. and Barber, S. A. (1969) *Soil Sci. Soc. Am. Proc.*, 33:360-363.

Karmarkar, S. V., Dudley, L. M., Jurinak, J. J., and James, D. W. (1991) *Soil Sci. Soc. Am. J.*, 55:1268-1274.

Kinniburgh, D. G. and Jackson, M. L. (1981) in *Adsorption of Inorganics at Solid-liquid Interfaces* (M. A. Anderson and A. J. Rubin, eds.). Ann Arbor Science, Ann Arbor, MI, pp. 91-160.

Levy, G. J., van der Watt, H. V. R., Shainberg, I. and du Plessis, H. M. (1988) *Soil Sci. Soc. Am. J.*, 52:1259-1264.

Li, C. K. (ed.) (1983) Red Soils of China. Science Press, Beijing.

Li, H. Y. and Ji, G. L. (1991) *Pedosphere*, 1:363-369.

Li, H. Y. and Ji, G. L. (1992a) *Pedosphere*, 2:245-254.

Li, H. Y. and Ji, G. L. (1992b) *Pedosphere*, 2:255-264.

Li, H. Y. and Ji, G. L. (1992c) *Chem. Sensors*, 12:45-50.

Lim, C. H., Jackson, M. L., Koons, R. D., and Helmke, P. A. (1980) *Clays Clay Miner.*, 28:223-229.

Marcano-Martinez, E. and McBride, M. B. (1989) *Soil Sci. Soc. Am. J.*, 53:63-69.

Marshall, C. E. (1964) *The Physical Chemistry and Mineralogy of Soils.* vol. I. John Wiley & Sons, New York.

McBride, M. B. (1989) in *Minerals in Soil Environments* (J. B. Dixon and S. B. Weed, eds.). Soil Science Society of America, Madison, WI, pp. 35-88.

Milnes, A. R. and Fitzpatrick, R.W. (1989) in *Minerals in Soil Environments* (J. B. Dixon and S. B. Weed, eds.). Soil Science Society of America, Madison, WI, pp. 1131-1205.

Nightingale, E. R. (1959) *J. Phys. Chem.*, 63:1381-1387.

Nir, S., Hirsch, D., Navrot, J., and Banin, A. (1986) *Soil Sci. Soc. Am. J.*, 50:40-45.

Odahara, K. and Wada, S. I. (1992) *Jn. J. Soil Sci. Plant Nutr.*, 63:64-71.

Pleysier, J. L., Juo, A. S. R., and Herbillon, A. J. (1979) *Soil Sci. Soc. Am Am. J.*, 43:875-880.

Pratt, P. F., Whittig, L. D., and Grover, B. L. (1962) *Soil Sci. Soc. Am. Proc.*, 26:227-230.

Raney, W. A. and Hoover, C. D. (1946) *Soil Sci. Soc. Am. Proc.*, 10: 231-237.

Rao, T. S., Page, A. L., and Coleman, N. T. (1968) *Soil Sci. Soc. Am. Proc.*, 32:639-643.

Rhue, R. D. and Mansell, R. S. (1988) *Soil Sci. Soc. Am. J.*, 52:641-647.

Scheffer, F. and Schachtschabel, P. (1992) *Lehrbuch der Bodenkunde.* Ferdinand Enke, Stuttgart, pp. 90-107.

Schwertmann, U. (1962) *Z. Pflanzenernähr. Dung. Bodenk.*, 97:9-25.

Shainberg, I., Alperovitch, N. I., and Keren, R. (1987) *Clays Clay Miner.*, 35:68-73.

Shainberg, I. and Kemper, W. D. (1967) *Soil Sci.*, 103:4-9.

Sparks, D. L. (ed.) (1986) *Soil Physical Chemistry.* CRC Press, Boca Raton, FL.

Sparks, D. L. and Huang, P. M. (1985) in *Potassium in Agriculture* (R. D. Munson, ed.). American Society of Agronomy, Madison, WI, pp. 201-276.

Sposito, G. (1983) *J. Colloid Interface Sci.*, 91:329–340.

Sposito, G. (1984) *The Surface Chemistry of Soils.* Oxford University Press, New York.

Sposito, G. (1991) *Soil Sci. Soc. Am. J.*, 55:965–967.

Talibudeen, O. and Goulding, K. W. T. (1983) *Clays Clay Miner.*, 31: 137–142.

Theng, B. K. G. (ed.) (1980) *Soils with Variable Charge.* New Zealand Society of Soil Science. pp. 147–166.

Thomas, G. W. and Hargrove, W. L. (1984) in *Soil Acidity and Liming* (F. Adams, ed.). American Society of Agronomy, Madison, WI.

Udo, E. J. (1978) *Soil Sci. Soc. Am. J.*, 42:556–560.

Uehara, G. and Gillman, G. P. (1981) The Mineralogy, Chemistry and *Physics of Tropical Soils with Variable Charge Clays.* Westview Press, Boulder, CO.

Ulrich, B. (1961) *Landw. Forschung*, 14:225–228.

van Olphen, H. (1977) *An Introduction to Clay Colloid Chemistry.* Interscience, New York.

Van Raij, B. and Peech, M. (1972) *Soil Sci. Soc. Am. Proc.*, 36:587–593.

Westall, J. and Hohl, H. (1980) *Adv. Colloid Interface. Sci.*, 12:265–294.

Xuan, J. X., Zhang, W. G., and Yu, T. R. (1965) *Acta Pedol. Sinica*, 13:427–436.

Yu, T. R. (1976) *Electrochemical Properties of Soils and Their Research Methods.* Science Press, Beijing.

Yu, T. R. (1985) *Physical Chemistry of Paddy Soils.* Science Press/Springer Verlag, Beijing/Berlin.

Yu, T. R., Beyme, B., and Richter, J. (1989a) *Z. Pflanzenernähr. Bodenk.*, 152:353–358.

Yu, T. R., Beyme, B., and Richter, J. (1989b) *Z. Pflanzenernähr. Bodenk.*, 153:359–365.

Yu, T. R. and Zhang, X. N. (1990) in *Soils of China* (Institute of Soil Science, ed.). Science Press, Beijing, pp. 494–513.

Yu, T. R., Wang, F. L., and Huang, P. M. (1990) *Trans. 14th Intern. Congr. Soil Sci.*, II:68–73.

Zhang, X. N. and Zhao, A. Z. (1984) *Acta Pedol. Sinica*, 21:358–367.

Zhang, X. N. and Zhao, A. Z. (1988) *Acta Pedol. Sinica*, 25:164–174.

Zhang, X. N., Jiang, N. H., Shao, Z. C., Pan, S. Z., and Zhang, W. G. (1979) *Acta Pedol. Sinica*, 16:145–156.

4

ELECTROSTATIC ADSORPTION OF ANIONS

G. L. Ji

Electrostatic adsorption of anions is one of the important characteristics of variable charge soils. This is caused by the fundamental feature that these soils carry a large quantity of positive surface charge. However, because these soils carry positive as well as negative surface charges, they may exert both attractive and repulsive forces on anions. Therefore, the situation in the adsorption of anions by these soils may be quite complex. There may also be the occurrence of negative adsorption of anions. Besides, for some anion species both electrostatic force and specific force may be involved during their interactions with variable charge soils. As shall be seen in this chapter, such specific force may be operative even for some anion species such as chloride that are generally considered as solely electrostatic in nature during adsorption.

Because of historical reasons, the literature on electrostatic adsorption of anions by soils is very limited. Nevertheless, as shall be seen in this chapter, the topic is of interest in both theory and practice.

In the present chapter, adsorption of anions shall be discussed mainly from the viewpoint of electrostatic adsorption. The other type of adsorption, specific adsorption or coordination adsorption, shall be dealt with in Chapter 6.

4.1 PROPERTIES OF ANIONS RELATING TO ELECTROSTATIC ADSORPTION

4.1.1 Anions versus Cations

The radius of anions is generally much larger than that of cations (Table 4.1). Thus, the charge density on anions would be low. When hydrated, because of the smaller ion–dipole force exerted on water molecules, anions are less hydrated than cations. This can be seen in Table 4.1. The r_H/r_c ratio for cations ranges from 2.22 to 6.37, while that for anions is smaller

Table 4.1 Crystal Radii r_c, Hydrated Radii r_H, and Acid Dissociation Constants of Some Cations and Anions (Bowden et al., 1980b; Nightingale, 1959; Stumm and Morgan, 1970)

Ion	r_c (nm)	r_H (nm)	r_H/r_c	pK_a^a
Li^+	0.060	0.382	6.37	13.8
Na^+	0.095	0.358	3.77	15
K^+	0.133	0.331	2.49	16
Rb^+	0.148	0.329	2.22	–
Ca^{2+}	0.099	0.412	4.16	12.6
F^-	0.136	0.352	2.59	3.3
Cl^-	0.181	0.332	1.83	–6
NO_3^-	0.264	0.335	1.27	–1.2
ClO_4^-	0.292	0.338	1.16	–7
SO_4^{2-}	0.290	0.379	1.31	–2

$^a pK_a$ of the acid for anions.

than 2 except for F^-. The orientation of water molecules around anions, especially in the primary hydration region, is also different from that around cations (Conway, 1981).

Because of the small r_H/r_c ratio, hydration does not induce the change in order of size when anions of the same valency are compared. For example, the crystal radii of Cl^-, NO_3^-, and ClO_4^- are 0.181, 0.264, and 0.292 nm, respectively, while the hydrated radii of these ions are 0.332, 0.335, and 0.338, respectively. This is in contrast to cations for which the smaller the radius, the larger the hydrated radius. This point is of significance in the adsorption of anions by soils, as shall be seen in later sections of this chapter.

Besides the sign of electric charge, another fundamental difference in nature between anions and cations is the acidity strength. For anions such as Cl^-, NO_3^- and ClO_4^-, they have a very weak affinity for protons. Therefore, they are the base of strong acids. On the contrary, monovalent cations commonly found in soils are very weak acids, because it is extremely difficult for them to dissociate a proton from their enveloping sheath of water molecules, as is indicated by a very large pK_a value (Table 4.1).

4.1.2 Negative Adsorption

Negative adsorption can be viewed microscopically or macroscopically.

Microscopically, the closer to the surface of negatively-charged soil colloid, the stronger would be the repulsive force against anions and thus the lower the anion concentration as compared to free solution. However, negative adsorption of anions is generally viewed in a macroscopical way, that is, it is expressed as the increment in anion concentration after the addition of a soil to an electrolyte solution as compared to the original solution.

For clay minerals and constant charge soils, Wiklander (Bear, 1955), based on the Donnan theory, derived the following equations, in which ions originated from the colloid and from the free solution are considered separately:

$$[Z_M + Y_M]^n \cdot Y_A^m = X_M^n \cdot X_A^m \tag{4-1}$$

$$\left[\frac{Z_M + Y_M}{X_M}\right]^{1/m} = \left[\frac{X_A}{Y_A}\right]^{1/n} \tag{4-2}$$

namely

$$\frac{X_A}{Y_A} = \left[\frac{Z_M + Y_M}{X_M}\right]^{n/m} \tag{4-3}$$

or

$$\frac{X_A}{Y_A} = \left[\frac{f_{M_i}}{f_{M_o}} \cdot \frac{[M]_i}{[M]_o}\right]^{n/m} \tag{4-4}$$

In the equations, Z represents the activity of cations dissociated from the colloid, Y the activity of ions of the free electrolyte in solution within the electric double layer, X the activity of ions in free solution, M the cation species, A the anion species, and m and n the valences of the two ion species. Because $(Z_M + Y_M)/X_M$ is larger than 1, the numerical value of X_A/Y_A (i.e., the magnitude of negative adsorption) will increase with the increase in valence n of the anion species, but will decrease with the increase in valence m of the cation species. For example, Mattson (cf. Bear, 1955) found that, in Na-saturated bentonite, the order of negative adsorption among various anions was

$$Cl^- = NO_3^- < SO_4^{2-} < Fe(CN)_6^{4-}$$

For clays saturated with different cations in equilibrium with solutions containing the chloride of the relevant cations, the negative adsorption of chloride ions was of the following order:

$$Na^+ > K^+ > Ca^{2+} > Ba^{2+}$$

For symmetrical electrolytes, negative adsorption is affected principally by the ratio of activity coefficient of ions between the inner solution and the outer solution, f_{M_i}/f_{M_o}. The larger the ratio, the more pronounced the negative adsorption.

Negative adsorption is also related to the nature of the soil. Wiklander used $m{\cdot}C_7/V$ to replace the activity Z_M of adsorbed cations in equation (4-3), in which m is the weight of soil in gram, C_7 is the cation–exchange capacity at pH 7, and V is the volume of the inner solution. For MA–type electrolytes,

$$\frac{X_A}{Y_A} = \frac{\dfrac{m \cdot C_7}{V} + Y_M}{X_M} \tag{4-5}$$

The above equation indicates that, when other factors are kept constant, the magnitude of negative adsorption will increase with the increase in quantity and cation–exchange capacity of the soil. Therefore, generally, the magnitude of negative adsorption of anions for various clay minerals may have the following decreasing order: montmorillonite > hydrous mica > kaolinite.

Equation (4–5) also shows that the concentration of electrolytes may have an effect on negative adsorption. When diluted, because X_M will decrease while m and C_7 remain unchanged, the magnitude of negative adsorption would increase.

4.2 FACTORS AFFECTING ELECTROSTATIC ADSORPTION OF ANIONS

The electrostatic adsorption of anions by variable charge soils is caused by positive surface charges of the soil. This kind of adsorption is controlled solely by electrostatic force between ions and the surface of soil colloid. Therefore, all factors that can affect such force would affect the adsorption. These factors include the nature of the ions, the properties of the soil, and environmental conditions, such as pH of the solution, electrolyte concentration, dielectric constant of the solvent and accompanying ions. These will be discussed in the following sections.

4.2.1 Nature and Quantity of Ions

Electrostatic adsorption of anions by variable charge soils is the result of electrostatic interactions between the surface of soil colloids and anions. The magnitude of such electrostatic forces obeys Coulomb's law. Therefore, the electric charge and hydration radius of the ions directly affect the interaction force. For the same soil and under identical environmental conditions, the higher the valency of the ions, the stronger the attractive force.

The radius of ions can affect the thickness of their hydration shell. For ions with a large hydration radius the proximal distance with the surface of soil colloids would be large and therefore according to Coulomb's law, the attractive force between them would be weak.

Besides electric charge and hydration radius, the structure and electron distribution of ions can also affect the adsorption property. For example, according to an experiment, when Ali–Haplic Acrisol, Ferrali–Haplic Acrisol, and Rhodic Ferralsol were in equilibrium with mixed solutions containing identical initial concentrations of chloride ions and nitrate ions, within the pH range under study, the ion activity ratios a_{Cl}/a_{NO_3} in the suspension were always smaller than unity (Fig. 4.1). This means that the amount of adsorbed chloride ions was larger than that of adsorbed nitrate ions. Since the adsorption proceeded in mixed solutions with identical initial concentrations of the two ion species, the difference in affinity of soils with the two ion species can only be attributed to the difference in properties between these ions themselves. However, both chloride and nitrate are monovalent ions with a nearly identical hydration radius. Then, the

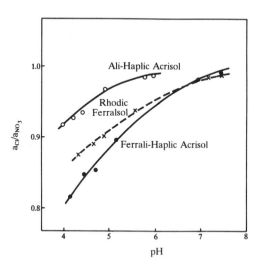

Fig. 4.1. Activity ratio of chloride ions to nitrate ions in suspensions of variable charge soils (initial concentration of chloride ions and nitrate ions, 1 mmol L^{-1}, each) (Wang et al., 1987).

difference in adsorption must be related to the difference in structure or electron distribution of the ions.

The quantity of adsorption is also related to the concentration of anions. For example, it was found that, at pH 4, Rhodic Ferralsol adsorbed chloride by an amount of 1.4 cmol kg^{-1} in a 0.01 M KCl solution and of 3.1 cmol kg^{-1} in a 0.1 M KCl solution. For the Ali–Haplic Acrisol, at pH 4, the amounts of adsorbed chloride in 0.001 M and 0.01 M NH$_4$Cl solutions were 0.1 cmol kg^{-1} and 0.4 cmol kg^{-1}, respectively (Zhang et al., 1979). Kinjo and Pratt (1971a) found that within the concentration range 0.03–0.2 M of nitrate ions the adsorption by two soils could be described with the Langmuir equation. For the other two soils, the adsorption was also in conformity with the Langmuir equation within the concentration range of 0.08–0.22 M, while it was more consistent with the Freundlich equation when the concentration of nitrate ions in the equilibrium solution was lower than 0.08 M.

When there are two anion species present together, ion concentration can also affect the relative adsorption of the two kinds of ions. As can be seen in Table 4.2, following the increase in total ion concentration, the increment in adsorption of chloride ions becomes greater than that of nitrate ions in the three variable charge soils, so that the adsorption ratio of nitrate ions to chloride ions decreases. This once again reflects the difference in nature between the two ion species, which shall be discussed

Table 4.2   Effect of Ion Concentration on Adsorption of Anions (Wang et al., 1987)

Soil	Total Concentration $(M)^a$	pH	Adsorptiton (cmol kg^{-1}) Cl	NO$_3$	Ratio (NO$_3$/Cl)
Ali–Haplic	2×10^{-3}	4.0	0.182	0.136	0.746
Acrisol	2×10^{-2}	4.1	0.642	0.462	0.720
Ferrali–Haplic	2×10^{-3}	4.1	0.283	0.197	0.695
Acrisol	2×10^{-2}	4.3	0.858	0.528	0.615
Rhodic Ferralsol	2×10^{-3}	4.3	0.245	0.191	0.780
	2×10^{-2}	4.7	1.120	0.804	0.718

$^a C_{Cl} = C_{NO_3}$.

in Section 4.4.

At present, the mechanism for the effect of ion concentration on electrostatic adsorption of anions is not exactly known. Some authors attempted to interpret this phenomenon in terms of the electric double layer of colloids. Most of the soils possess both polarized surface with constant surface charge and unpolarized reversible surface with constant surface potential (Arnold, 1977). Highly weathered tropical soils contain large amounts of oxides of iron and aluminum and are therefore dominated by constant potential surfaces. The charge density σ of such surfaces is determined by both the pH of the solution and the concentration of adsorbed ions in the medium. According to the diffuse double layer theory, surface charge density is proportional to the square root of the concentration n of adsorbed ions, that is, $\sigma = K{\cdot}n^{1/2}$, where K is a constant. Therefore, the higher the concentration of adsorbed ions in solution, the higher the surface charge density. This should be an important cause for the effect of ion concentration on adsorption. Besides, according to the diffuse double layer theory, the thickness of the diffuse layer is inversely proportional to the square root of ion concentration in solution. Thus, the thickness of the diffuse layer will increase when the electrolyte concentration is reduced. When the diffuse layer extends to a certain extent, the diffuse layers around adjacent positive and negative charge sites at the surface of soil colloids may overlap partly with one another, resulting in the mutual compensation of a part of positive surface charge and of negative surface charge. This would lead to the decrease in adsorption of both cations and anions. This may be another cause of the effect of electrolyte concentration on ion adsorption. Barber and Rowell (1972) suggested that both of these mechanisms may operate.

4.2.2 Surface Properties of Soil

Since electrostatic adsorption of anions by variable charge soils is caused by electrostatic attraction force between positive surface charge of the soil and anions, it would be expected that this kind of adsorption is closely related to the quantity of positive surface charge and the corresponding charge density. Free iron and aluminum oxides in soils are the principal carriers of positive charge. Hydroxyl aluminum at the edges of kaolinite may also carry positive charge under certain conditions. Extensive research about the relationship between anion adsorption and iron oxides have been carried out. In Fig. 4.2, the effect of iron oxides on sulfate adsorption in a Rhodic Ferralsol is shown. It can be seen that the adsorption decreases markedly after the removal of iron oxides from the soil, especially at low pH. Despite the involvement of specific adsorption of sulfate, the decrease in quantity of positive surface charge after the removal of iron oxides should be

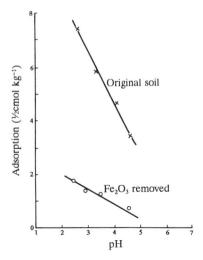

Fig. 4.2. Effect of iron oxides on adsorption of sulfate ions by Rhodic Feralsol (Zhang et al., 1979).

the principal cause for the difference in sulfate adsorption shown in the figure. The situation is even more pronounced in the adsorption of chloride ions. For example, it was observed that the amount of chloride adsorbed by a Rhodic Ferralsol was 1.5 cmol kg^{-1} at pH 4, and it reduced to 0.05 cmol kg^{-1} after the removal of free iron oxides. The adsorbed amount in the iron-free soil was still as low as 0.2 cmol kg^{-1} at pH 3. Since the adsorption is affected by pH, anions may show negative adsorption when the pH is sufficiently high. The pH at which anion adsorption becomes negative adsorption is called *zero point of adsorption* for the anion species. For the iron-free Rhodic Ferralsol, the zero point of adsorption was 4.6, while it was 6.6 for the untreated soil.

The increase in positive surface charge caused by the presence of iron oxides can also affect the relative adsorption of two anion species when present together. Such an effect for three variable charge soils is shown in Table 4.3. The amount of both adsorbed chloride ions and adsorbed nitrate ions increased after the coating with iron oxides, especially for nitrate ions. The increments in anion adsorption for the Rhodic Ferralsol were relatively small among the three soils. This is related to the large amount of iron oxides already contained in this soil. It should be noticed that, because the pH of electrodialyzed soils increased after the coating with iron oxides, the actual effect of iron oxides on adsorption of anions by variable charge soils should be more pronounced than that shown in Table 4.3, if the effect of pH on anion adsorption is taken into account.

In addition to the quantity factor, the crystalline form of iron oxides can

Table 4.3 Effect of Iron Oxides on Adsorption of Anions (Wang et al., 1987)

| Soil | Treatment | pH | Adsorption[a] | | Ratio (NO$_3$/Cl) |
			Cl	NO$_3$	
Ali–Haplic Acrisol	Original soil	4.0	0.182	0.136	0.746
	Fe$_2$O$_3$–coated	4.7	0.249	0.202	0.810
Ferrali–Haplic Acrisol	Original soil	4.1	0.283	0.197	0.695
	Fe$_2$O$_3$–coated	4.9	0.387	0.304	0.786
Rhodic Ferralsol	Original soil	4.3	0.245	0.191	0.780
	Fe$_2$O$_3$–coated	5.6	0.263	0.203	0.772

[a](cmol kg^{-1}).

also affect electrostatic adsorption of anions by soils. Borggaard (1984) used EDTA and dithionite + EDTA to remove amorphous iron oxides and crystalline iron oxides from soils, respectively. He ascribed the difference in amount of chloride adsorbed at pH 5 and at pH 7 (ΔCl) to the adsorption of chloride ions by variable charge groups. He found that the adsorption was related to both forms of iron oxides. Of course, the magnitudes of the effect exerted by the two forms were different.

The adsorption of anions by variable charge soils is also affected by negative surface charges. This is because these negative charges can exert a repulsive effect on anions. Black and Waring (1979) showed that adsorption of nitrate ions may not occur when the quantity of net negative surface charge is sufficiently large. On the other hand, the soil could adsorb nitrate ions even when it carries a small amount of net negative surface charge, implying that the positive and negative surface charge sites are discrete but are not neutralized by each other. This is the basic reason why many variable charge soils can adsorb anions under ordinary pH conditions, although they carry net negative surface charge.

Besides clay minerals carrying permanent negative surface charges, organic matter is also one kind of important carriers of negative charge in variable charge soils. Therefore, the presence of organic matter may induce the decrease in amount of anions adsorbed. In Fig. 4.3, the increase in chloride adsorption after the removal of organic matter from Rhodic Ferralsol is shown. As seen in the figure, the effect of organic matter is more remarkable at high pH. This is because organic matter carries more

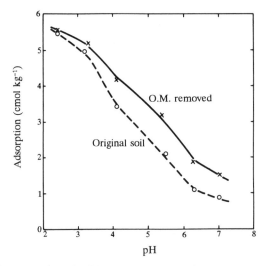

Fig. 4.3. Increase in chloride adsorption by Rhodic Ferralsol after removal of organic matter. (Drawn from data of Zhang and Jiang, 1964.)

negative charge when the pH is high. Owing to the effect of organic matter, it may be observed (Black and Waring, 1979) that the amount of nitrate adsorbed by variable charge soils increased with the increase in depth within the profile, being 0.16 cmol kg^{-1} at a depth of 45–90 cm and 0.45 cmol kg^{-1} at a depth of 360–600 cm. When the organic carbon content was higher than 10 g kg^{-1}, the soils did not adsorb nitrate ions or showed a negative adsorption.

It can be concluded that both positive and negative surface charges can affect electrostatic adsorption of anions by soils. Variable charge soils generally carry positive as well as negative surface charges. Therefore, the relationship between their surface properties and electrostatic adsorption of anions may be rather complex.

4.2.3 pH

The pH of the medium can affect electrostatic adsorption of ions by affecting either the surface charge properties of the soil or the form of ions.

It was shown in Chapter 2 that, for a given soil, pH is the determinating factor in affecting its surface charge properties. This of course would affect the adsorption of anions.

The effects of pH on the adsorption of chloride ions and nitrate ions by the Rhodic Ferralsol, Ferrali–Haplic Acrisol, and Ali–Haplic Acrisol are shown in Figs. 4.4 and 4.5, respectively. The amount of adsorbed anions increases with the decrease in pH for all three soils. This is closely related

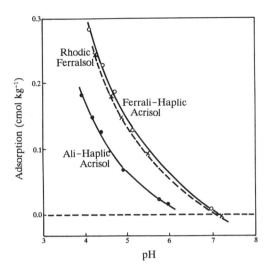

Fig. 4.4. Effect of pH on chloride adsorption in soils (Wang et al., 1987).

to the increase in positive surface charge of the soils.

If the amount of adsorbed anions is calculated as the percentage of the quantity of positive surface charge at different pH based on the surface charge–pH curve and the corresponding adsorption curve, it can be found that this percentage is not a constant; rather, it increases with the decrease in pH. For example, when chloride ions were adsorbed by a Rhodic Ferralsol in a 0.01 M KCl solution, the percentages were 22%, 33%, and 46% at pH 5, 4, and 3, respectively (Zhang et al., 1979). This situation is similar to the case presented in chapter 3, in which the variations in cation adsorption and in bonding energy at different pH do not parallel the variation in negative surface charge of the soil.

Figures 4.4 and 4.5 show that Rhodic Ferralsol and Ferrali–Haplic Acrisol can still adsorb a small amount of anions, particularly chloride ions, at pH 6.5. Singh and Kanehiro (1969) also observed that some soils could adsorb nitrate ions when the pH was higher than 7. These results indicate that these soils can carry anion-adsorbing sites even under neutral and alkaline conditions.

4.2.4 Dielectric Constant of Solvent

According to Coulomb's law, in vacuum, the electrostatic force between two charged particles is proportional to the product of electric charges (q) and is inversely proportional to the square of distance (r). In the presence of a medium, the reaction force (P) is

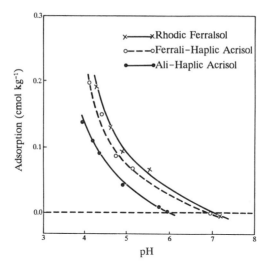

Fig. 4.5. Effect of pH on nitrate adsorption in soils (Wang et al., 1987).

$$P = \frac{q_1 q_2}{e r^2} \qquad (4\text{-}6)$$

where e is the dielectric constant of the medium. If q_1 and q_2 are of the same sign, P would be a repulsive force, while P would be an attractive force if the signs of q_1 and q_2 are opposite.

Since electrostatic adsorption of anions by soils is determined by electrostatic attractive force between the surface of soil colloid and anions, the dielectric constant of the solvent in soil suspension should affect this adsorption. Such an effect on the adsorption of chloride ions and nitrate ions for three variable charge soils is shown in Table 4.4. When the solvent contains 20% of ethanol, the adsorbed amounts of both anion species are larger than those in water, especially for nitrate ions. The cause for this increase in adsorption may be that the dielectric constant of the solvent is reduced after the addition of ethanol, resulting in an increase in electrostatic attractive force between the surface of soil colloids and ions. It can be calculated that the ratio of dielectric constant between the two solvents is about 1.16. The ratio of adsorbed chloride or nitrate in the two cases for the three soils ranges from 1.08 to 1.28, with 1.16 on average.

4.2.5 Accompanying Cations

Owing to the difference in chemical properties among various cations, when the accompanying cation species of the anions are different, the pH of the solution may be different. This would affect the adsorption of anions by

Table 4.4 Effect of Dielectric Constant of Solvent on Adsorption of
Anions[a] (Wang et al., 1987)

Soil	Water:Ethanol	Adsorption[b]		Ratio
		Cl	NO_3	(NO_3/Cl)
Ali–Haplic	100:0	0.249	0.202	0.810
Acrisol	80:20	0.281	0.237	0.840
Ferrali–Haplic	100:0	0.387	0.304	0.786
Acrisol	80:20	0.417	0.334	0.801
Rhodic	100:0	0.263	0.203	0.772
Ferralsol	80:20	0.315	0.260	0.825

[a]Dielectric constants of water and ethanol at 25°C are
78.54 and 24.3, respectively.
[b](cmol kg^{-1}), initial concentration, $C_{Cl} = C_{NO_3}$ = 0.001 M.

affecting surface charge properties of soils. Another more important factor
is that, according to the ion-pair theory (Heald et al., 1964; Rao et al.,
1968; Shainberg and Kemper, 1967), the ability of various cations in forming
ion–pairs with the same anion species may be different. As a result,
accompanying cations may affect the adsorption of anions.

The effect of accompanying cation species on adsorption of chloride ions
and nitrate ions in three variable charge soils is shown in Table 4.5. When
the accompanying cation species was calcium, Ferrali–Haplic Acrisol and
Rhodic Ferralsol adsorbed more anions than when the accompanying cation
species was potassium. The ratio of adsorption between the two cases was
consistent with the ratio of adsorption of cations. For the Ali–Haplic
Acrisol, the effect was not distinct.

4.3 NEGATIVE ADSORPTION

Variable charge soils carry negative as well as positive surface charges. The
combined effect of repulsive force and attractive force on anions exerted by
the two kinds of surface charges makes the situation in negative adsorption
complicated. However, it can be expected that, when the pH of the medium
is sufficiently high so that the negative charge density exceeds the positive
charge density on the surface of soil colloids, negative adsorption of anions
may occur. In Figs. 4.4 and 4.5, such a tendency can be observed for both

Table 4.5 Effect of Accompanying Cations on Adsorption of Anions
(Wang t al., 1987)

| Soil | Cation | pH | Adsorption[a] | | | Ratio |
			Cation	Cl	NO$_3$	(NO$_3$/Cl)
Ali–Haplic	K	4.2	0.656	0.193	0.146	0.756
Acrisol	Ca	4.2	0.653	0.189	0.132	0.698
Ferrali–Haplic	K	4.4	0.664	0.284	0.206	0.725
Acrisol	Ca	4.3	0.743	0.320	0.233	0.728
Rhodic Ferralsol	K	4.4	0.689	0.256	0.202	0.789
	Ca	4.4	0.809	0.302	0.245	0.811

[a](cmol kg^{-1}).

chloride ions and nitrate ions in three variable charge soils. The negative adsorption of chloride ions shown in Fig. 4.6 is even clearer. The magnitude of negative adsorption of chloride ions increased with the rise in pH. When the pH was higher than 7, the amount of negative adsorption in the two soils attained 0.1 cmol kg^{-1}.

Thus, like the case for cations presented in Chapter 3, both adsorption and negative adsorption of anions may occur in variable charge soils. This is one of the important differences between variable charge soils and constant charge soils.

The pH at which there is neither adsorption nor negative adsorption of anions may be called *zero point of adsorption* for the soil. Apparently, this zero point of adsorption is related to both the properties of the soil and the nature of the anion species. As can be seen in Figs. 4.4 and 4.5, the zero points of adsorption of chloride ions in Rhodic Ferralsol, Ferrali–Haplic Acrisol, and Ali–Haplic Acrisol are about 7.2, 7.1, and 6.3, respectively. For nitrate ions, the corresponding values are 7.0, 6.9, and 6.0, respectively. This means that the zero point of adsorption for chloride ions is a little higher than that for nitrate ions.

When Fig. 4.4 and Fig. 4.6 are compared, it can be noted that, for the same soil, the zero points of adsorption of chloride ions shown in the two figures are remarkably different. For the Ali–Haplic Acrisol, they are 4.8 and 6.3 and for the Rhodic Ferralsol, they are 6.2 and 7.2. The cause of this difference is that, besides the difference in experimental conditions, the concentrations of KCl in the two cases are different. The variation in zero

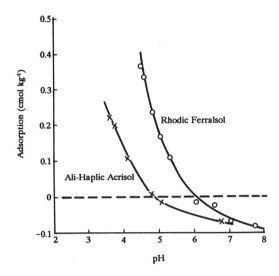

Fig. 4.6. Adsorption and negative adsorption of chloride ions by variable charge soil. (Drawn from data of Zhang et al., 1979.)

point of adsorption of chloride ions with the difference in KCl concentration is shown in Fig. 4.7. When the concentrations were 0.001 M and 0.01 M, the

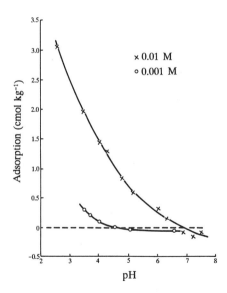

Fig. 4.7. Zero point of chloride adsorption of Ali–Haplic Acrisol in KCl solutions of different concentrations.

zero points of adsorption were 4.6 and 6.8, respectively. Another experiment showed that the zero points of adsorption of chloride ions in Rhodic Ferralsol were 6.0 and 6.6 and in Ali–Haplic Acrisol were 5.1 and 6.2 when the NH_4Cl concentrations were 0.001 M and 0.01 M, respectively (Zhang et al., 1979). Thus, the zero point of adsorption of anions in dilute solution is lower than that in concentrated solution, and such a difference is more remarkable in Ali–Haplic Acrisol than in Rhodic Ferralsol. One possible explanation for this phenomenon may be as follows. The relative magnitudes of effective positive surface charge density and effective negative surface charge density of these two types of soils in electrolyte solutions of different concentrations are different. When the electrolyte concentration is decreased, the thickness of the diffuse layer around positive and negative surface charge sites would increase. This may cause a partial overlap among these diffuse layers, resulting in decreases in effective positive and effective negative surface charges in equivalent quantities. For the Ali–Haplic Acrisol, the quantity of negative surface charge is much larger than that of positive surface charge. Hence, the ratio of effective negative surface charge density to effective positive surface charge density at low electrolyte concentration is distinctly larger than the ratio at high electrolyte concentration. This means that in the former case the repulsive force of the soil to chloride ions is stronger while the attractive force is weaker as compared to the latter case. This is the reason why the zero point of adsorption is relatively lower in the former case. On the other hand, for the Rhodic Ferralsol, because the quantities of positive and negative surface charges differ very little, the change in the ratio of effective negative surface charge density to effective positive surface charge density caused by the partial overlap among diffuse layers is small. Hence, the zero point of adsorption of anions was less affected by electrolyte concentration.

The zero point of adsorption of anions for the Rhodic Ferralsol decreased markedly after the removal of free iron oxides, caused by the increase in the ratio of negative surface charge density to positive surface charge density.

4.4 MECHANISMS OF ADSORPTION OF ANIONS

It is generally assumed that chloride, nitrate, and perchlorate are typical electrostatically adsorbed ions and that in adsorption only electrostatic attractive force between anions and the surface of soil particles is involved. Then, in the absence of other specific ions in the solution, the surface potential of the colloid should be solely controlled by the concentration of potential–determining ions H^+ and OH^-, that is, by pH of the solution.

Arnold (1977) doubted the existence of any real inert electrolyte. Hiroki and Hideo (1980) used $NaNO_3$, $NaCl$, or Na_2SO_4 as the supporting

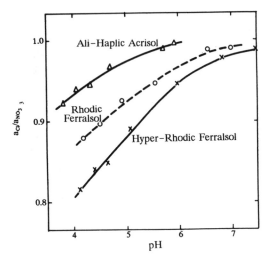

Fig. 4.8. Relationship between activity ratio a_{Cl}/a_{NO_3} and pH in soil suspensions (Ji and Kong, 1992).

electrolyte to determine the zero point of charge of the soil and found that this point increased in the order $NO_3^- < Cl^- < SO_4^{2-}$. Barrow et al. (1980) and Davis et al. (1978), when calculating the amount of adsorbed chloride ions based on a theoretical model, found that only under the assumption that these ions were adsorbed specifically to a certain extent the calculated

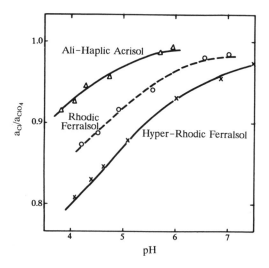

Fig. 4.9. Relationship between activity ratio a_{Cl}/a_{ClO_4} and pH in soil suspensions (Ji and Kong, 1992).

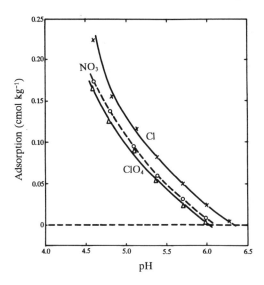

Fig. 4.10. Effect of pH on adsorption of anions by Ali–Haplic Acrisol (Ji and Kong, 1992).

results were consistent with experimental data. Parfitt and Russell (1977) observed that, when HCl and HNO_3 were evaporated on the surface of goethite separately, chloride ions underwent interactions with the solid

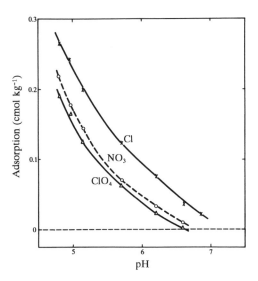

Fig. 4.11. Effect of pH on adsorption of anions by Rhodic Ferralsol (Ji and Kong, 1992).

surface to some extent, while nitrate ions did not. Thus, the real mechanisms in the adsorption of these supposedly electrostatic anions deserve close examination.

One desirable way in this respect should be to directly determine the activity ratio between two anion species in the suspension with the respective ion–selective electrodes after interactions of a soil with a mixed solution containing two anion species with identical concentrations. Since ion activity represents the effective ion concentration actually present in the suspension, any deviation in activity ratio from the original value 1 must be caused by the difference in adsorption between the two ion species.

The variations in activity ratios a_{Cl}/a_{NO_3} and a_{Cl}/a_{ClO_4} with the change in pH in three variable charge soil suspensions after reacting with mixed solutions containing identical concentrations of two anion species are shown in Figs. 4.8 and 4.9, respectively. Within the pH range under study, the activity ratios in the soil suspension were invariably smaller than 1. The lower the pH of the medium, the more remarkable the difference in the adsorption. This means that these soils adsorbed more chloride ions than nitrate ions or perchlorate ions. Since the adsorption proceeded under identical conditions, the difference in affinity should only be caused by the difference in the nature of the anions.

Figures 4.10 and 4.11 show that the quantities of adsorbed chloride, nitrate, and perchlorate ions decrease with the increase in pH for the three variable charge soils. When the pH is higher than a certain value, negative adsorption may occur. This should be related to the change in surface charge of the soil. On the other hand, the figures also show that, in mixed solutions with identical concentrations of the anion species, the quantities of adsorption are of the order $Cl^- > NO_3^- > ClO_4^-$.

For adsorptions of chloride ions and nitrate ions (Table 4.6) or of chloride ions and perchlorate ions (Table 4.7), the sequence of adsorption does not affect the adsorbed amounts. This is different from adsorptions of phosphate ions and fluoride ions, in which a hysteresis may occur (Ji, 1986).

Table 4.8 shows that the amount of adsorbed chloride, nitrate, or perchlorate ions increased after the coating with iron oxides on variable charge soils, especially for the Ali–Haplic Acrisol with a relatively low content of iron oxides. If comparisons are made under the same pH, the effect of iron oxides would be more remarkable. However, the coating cannot change the order of affinity for soils among the three anion species.

If the dielectric constant of the medium is reduced by the addition of 20% of ethanol, the adsorbed amounts of chloride, nitrate, and perchlorate ions would increase (Table 4.9). On the other hand, the change in dielectric constant cannot induce the change in the order of affinity for the surface of soil colloids among the three anion species.

Table 4.6 Effect of Sequence of Adsorption on Adsorption of Chloride
Ions and Nitrate Ions (Wang et al., 1987)

Soil	Adsorption Sequence	Adsorption (cmol kg^{-1})	
		Cl	NO$_3$
Ali–Haplic	Simultaneous	0.182	0.136
Acrisol	Separate	0.170	0.127
Ferrali–Haplic	Simultaneous	0.283	0.197
Acrisol	Separate	0.256	0.183
Rhodic Ferralsol	Simultaneous	0.245	0.191
	Separate	0.250	0.205

Table 4.7 Effect of Sequence of Adsorption on Adsorption of Chloride
Ions and Perchlorate Ions (Ji and Kong, 1992)

Soil	Adsorption Sequence	Adsorption (cmol kg^{-1})	
		Cl	ClO$_4$
Ali–Haplic	Simultaneous	0.176	0.113
Acrisol	Separate	0.181	0.121
Rhodic Ferralsol	Simultaneous	0.275	0.173
	Separate	0.281	0.176
Hyper–Rhodic	Simultaneous	0.301	0.203
Ferralsol	Separate	0.295	0.211

Table 4.8 Effect of Iron Oxides on Adsorption of Anions[a] (Ji and Kong, 1992)

Soil	Treatment	pH	Adsorption[b] Cl	NO$_3$	ClO$_4$	Adsorption Ratio (NO$_3$/Cl)	(ClO$_4$/Cl)
Ali–Haplic	Original	4.1	0.150	0.118	0.091	0.787	0.607
Acrisol	Fe-coated	4.8	0.219	0.185	0.151	0.845	0.690
Rhodic	Original	4.5	0.225	0.160	0.145	0.711	0.644
Ferralsol	Fe-coated	5.6	0.266	0.191	0.174	0.718	0.654
Hyper–Rhodic	Original	4.8	0.265	0.217	0.190	0.819	0.717
Ferralsol	Fe-coated	5.8	0.296	0.251	0.227	0.848	0.767

[a] $C_{Cl} = C_{NO_3} = C_{ClO_4} = 0.001\ M$.

[b] (cmol kg^{-1}).

Table 4.9 Effect of Dielectric Constant of Solvent on Adsorption of Anions[a]

Soil	Water:Ethanol	Adsorption[b] Cl	NO$_3$	ClO$_4$	Adsorption Ratio (NO$_3$/Cl)	(ClO$_4$/Cl)
Ali–Haplic	100:0	0.154	0.123	0.105	0.799	0.682
Acrisol	80:20	0.190	0.153	0.133	0.805	0.701
Rhodic Ferralsol	100:0	0.215	0.168	0.139	0.781	0.647
	80:20	0.253	0.196	0.172	0.775	0.680
Hyper–Rhodic	100:0	0.271	0.209	0.185	0.771	0.683
Ferralsol	80:20	0.320	0.253	0.228	0.790	0.713

[a] Initial concentration, $C_{Cl} = C_{NO_3} = C_{ClO_4} = 0.001\ M$.

[b] (cmol kg^{-1}).

From the phenomena mentioned above it can be concluded that, while the adsorption of chloride, nitrate, and perchlorate ions is principally controlled by electrostatic attractive force, the affinity of the three anion species for soils is invariably of the order chloride > nitrate > perchlorate. The pH of the medium, the iron oxide content of the soil, the ion concentration, the dielectric constant of the solvent, and the sequence of adsorption cannot cause the change in this order. Since the results presented above were obtained under identical experimental conditions, this difference in affinity can only be attributed to the difference in the nature of the three ion species.

Except for the structure and electron distribution, other chemical properties of chloride, nitrate, and perchlorate ions in aqueous solution are similar, such as hydration radius. If the adsorption of these ions by variable charge soils is caused solely by electrostatic force, according to the electric double layer theory, these ions would only function as compensating ions of positive charges on the surface of soil colloids, and the adsorbed amounts from mixed solutions with identical initial concentrations of relevant ion species should be the same. In reality, however, this is not the case. Thus, it is not possible to interpret the adsorption of these anions only in terms of electrostatic adsorption.

It is noticeable that the surface properties of variable charge soils not only affect the adsorbed amount of anions, but also affect the relative affinity for different anion species. It has been calculated that, at pH 4.2 ± 0.1, the selectivity coefficients with respect to nitrate and chloride ions for Ali–Haplic Acrisol, Ferrali–Haplic Acrisol, and Rhodic Ferralsol were 0.68, 0.55, and 0.68, respectively (Wang et al., 1987). The selectivity coefficient for a cation–exchange resin was calculated to be 5.0. This is understandable because the lipophilicity of nitrate ions is greater than that of chloride ions. Thus the nature of the adsorbent plays an important role in determining adsorption. But, according to the principle of electrostatic adsorption, an adsorbent only provides charged surface for adsorption, and therefore the difference in adsorption of chloride, nitrate, and perchlorate ions should be independent of the nature of the adsorbent. This is at odds with experimental results.

If chloride, nitrate, and perchlorate are typical electrostatically adsorbed ions, then, following the decrease and vanishing in positive surface charge with the rise in pH, the difference in adsorption among the three ion species should vanish eventually, and there should be the occurrence of adsorption or negative adsorption at the same pH. On the contrary, actually, as is shown in Figs. 4.12 and 4.13, following the rise in pH, the difference in adsorption among the three ion species does not decrease but increases, especially for the difference between chloride ions and the other two ion species. This means that, following the decrease in positive surface

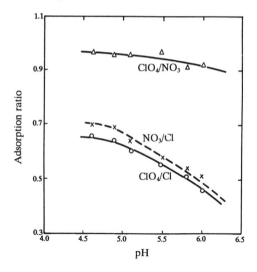

Fig. 4.12. Adsorption ratio of anions in Ali–Haplic Acrisol in relation to pH (Ji and Kong, 1992).

charge of the soil, the role of electrostatic force in inducing adsorption of the three ion species by variable charge soils becomes smaller, whereas the role of one or more other forces becomes greater.

According to the principle of electrostatic adsorption, if a variable charge soil is treated with chloride, nitrate and perchlorate ions having very similar

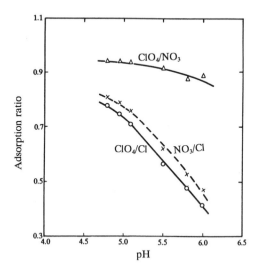

Fig. 4.13. Adsorption ratio of anions in Rhodic Ferralsol in relation to pH (Ji and Kong, 1992).

Table 4.10   Effect of Anions on Zeta Potential of Variable Charge Soils
(Ji and Kong, 1992)

Treatment	Zeta Potential (mV)		
	Ali–Haplic Acrisol	Rhodic Ferralsol	Hyper–Rhodic Ferralsol
HCl	+14.9	+26.8	+31.2
HNO$_3$	+17.8	+29.6	+34.4
HClO$_4$	+18.0	+31.5	+35.9

hydration radius, the zeta potential of the treated colloids should be equal to one another. However, as is seen in Table 4.10, the zeta potential of soil colloids adsorbed with chloride ions is the lowest while that adsorbed with perchlorate ions the highest. The difference in zeta potential between soils adsorbed with chloride ions and those adsorbed with the other two ion species is especially remarkable. Thus, at least for chloride ions, there is the possibility that they may enter the inner Helmoholtz layer, inducing the change in surface properties of soil colloids.

Blok and De Bruyn (1970), in studying the differential capacitance of the electric double layer of ZnO in 1:1-type electrolyte solutions, found that, as in the adsorption by Fe_2O_3 and TiO_2, the adsorption of monovalent anions was of the order $ClO_4^- \leq NO_3^- < I^- < Br^- < Cl^-$. For imogolite,

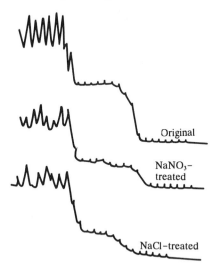

Original

NaNO$_3$–
treated

NaCl–treated

Fig. 4.14. Effect of anions on surface acidity of Fe_2O_3-coated Ferrali-Haplic Acrisol (Ph.D. thesis of G. S. Hu).

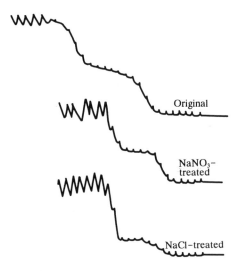

Fig. 4.15. Effect of anions on surface acidity of Al_2O_3-coated Ferrali-Haplic Acrisol (Ph.D. thesis of G. S. Hu).

Clark and McBride (1984) found that more Cl than ClO_4 was adsorbed. Su and Harsh (1993) observed that Cl reduced the positive charge to a greater extent than did ClO_4, and Su et al. (1992) proposed a specific adsorption of Cl. Hu in this laboratory, using X-ray photoelectron spectroscopy, observed that when goethite, kaolinite, and Fe_2O_3-coated Ferrali-Haplic Acrisol were treated with NaCl solution or $NaNO_3$ solution, chloride ions reacted with the surface of these adsorbents to some extent, whereas nitrate ions did not. The effects of adsorption of chloride ions and nitrate ions on surface acidity of two soil samples as studied with program gas chromatography are shown in Figs. 4.14 and 4.15, respectively. The figures show that adsorption of chloride ions induced the change in surface properties of the soil, while the effect of nitrate was not distinct. These results once again indicate that a specific reaction between chloride ions and the surface of soil colloids may occur.

Summarizing the above discussions it can be concluded that, while the adsorption of such anions as chloride, nitrate, and perchlorate ions by variable charge soils is controlled chiefly by electrostatic force, in the adsorption some specific force may also be involved, at least for chloride ions. At present, the mechanism of such specific adsorption is not exactly known. However, it is apparent that such adsorption is related to both the nature of the anions and the properties of the adsorbents.

One of the important reasons for regarding chloride and nitrate as purely electrostatically adsorbed ions in soil science is that most of the research were conducted with constant charge soils or clay minerals. In this case, it may be difficult to distinguish the minute difference in properties between these ion species. Even though there was a limited amount of research using oxides or variable charge soils as the object, no attempt was made to compare these two important anion species in detail under strictly identical conditions. Actually, as presented in this chapter, the nature of these ions differ materially when reacting with variable charge soils. As shall be seen in later chapters of this book, such difference may reflect itself in many other properties.

In 1949 Schofield suggested to use the adsorption of chloride as a measure of the quantity of positive surface charge of soils. This method has been adopted widely. Actually, the results obtained in this way represent the surface status of a soil that has already been modified, and therefore they are conditional to some extent.

BIBLIOGRAPHY

Arnold, P. W. (1977) *J. Soil Sci.*, 28:393-402.

Barber, R. G. and Rowell, D. L. (1972) *J. Soil Sci.*, 23:135-146.

Barrow, N. J. (1985) *Adv. Agron.*, 38:183-230.

Barrow, N. J. (1989) *J. Soil Sci.*, 40:415-425.

Barrow, N. J. (1992) *J. Soil Sci.*, 43:421-428.

Barrow, N. J., Bowden, J. W., Posner, A. M., and Quirk, J. P. (1980) *Aust. J. Soil Res.*, 18:37-47.

Bear, F. E. (1955) *Chemistry of the Soil*. Reinhold, N. Y.

Berg, W. A. and Thomas, G. W. (1959) *Soil Sci. Soc. Am. Proc.*, 23:348-350.

Black, A. S. and Waring, S. A. (1979) *Aust. J. Soil Res.*, 17:271-282.

Blok, L. and De Bruyn, P. L. (1970) *J. Colloid Interface Sci.*, 32:527-532.

Bolt, G. H. (1955) *Soil Sci.*, 79:267-276.

Borggaard, O. K. (1984) *J. Soil Sci.*, 35:71-78.

Bowden, J. W., Posner, A. M., and Quirk, J. P. (1977) *Aust. J. Soil Res.*, 15:121-136.

Bowden, J. W., Nagarajah, S., Barrow, N. J., Posner, A. M., and Quirk, J. P. (1980a) *Aust. J. Soil Res.*, 18:49-60.

Bowden, J. W., Posner, A. M. and Quirk, J. P. (1980b) in *Soils with Variable Charge* (B. K. G. Theng, ed.). New Zealand Society of Soil Science, pp. 147-166.

Clark, C. J. and McBride, M. B. (1984) *Clays Clay Miner.*, 32:291-299.

Conway, B. E. (1981) *Ionic Hydration in Chemistry and Biophysics*. Elsevier, Amsterdam.

Davis, J. A., James, R. O., and Leckie, J. O. (1978) *J. Colloid Interface. Sci.*, 63:480-499.

Heald, W. R., Frere, M. H., and De Wit, C. T. (1964) *Soil Sci. Soc. Am.*

Proc., 28:622–627.

Gillman, G. P. and Sinclair, D. F. (1989) *Aust. J. Soil Res.*, 25:275–285.

Hingston, F. J., Posner, A. M. and Quirk, J. P. (1972) *J. Soil Sci.*, 23: 177–192.

Hingston, F. J. (1981) in *Adsorption of Inorganics at Solid–Liquid Interfaces* (M. A. Anderson and A. J. Rubin, eds.). Ann Arbor Science, Ann Arbor, MI, pp. 51–90.

Hiroki, I. and Hideo; O. (1980) *J. Sci. Soil Manure Japan*, 51:102–106.

Ji, G. L. (1986) *Acta Pedol. Sinica*, 23:220–227.

Ji, G. L. and Kong, X. L. (1992) *Pedosphere*, 2:317–326.

Karmarkar, S. V., Dudley, L. M., Jurinak, J. J., and James, D. W. (1991) *Soil Sci. Soc. Am. J.*, 55:1268–1274.

Kinjo, T. and Pratt, P. F. (1971a) *Soil Sci. Soc. Am. Proc.*, 35:722–725.

Kinjo, T. and Pratt, P. F. (1971b) *Soil Sci. Soc. Am. Proc.*, 35:725–728.

McBride, M. B. (1989) in *Minerals in Soil Environments* (J. B. Dixon and S. B. Weed, eds.) Soil Science Society of America, Madison, WI, pp. 35–88.

Nightingale, E. R. (1959) *J. Phys. Chem.*, 63:1381–1387.

Parfitt, R. L. and Russell, J. D. (1977) *J. Soil Sci.*, 28:297–305.

Rao, T. S., Page, A. L., and Coleman, N. T. (1968) *Soil Sci. Soc. Am. Proc.*, 32:639–643.

Rhue, R. D. and Reve, W. H. (1990) *Soil Sci. Soc. Am. J.*, 54:705–708.

Schalscha, E. B., Pratt, P. F., and Domecq, T. C. (1974) *Soil Sci. Soc. Am. Proc.*, 38:44–45.

Schofield, R. K. (1949) *J. Soil Sci.*, 1:1–8.

Shainberg, I. and Kemper, W. D. (1967) *Soil Sci.*, 103:4–9.

Singh, B. R. and Kanehiro, Y. (1969) *Soil Sci. Soc. Am. Proc.*, 33:681–683.

Sparks, D. L. (ed.) (1986) *Soil Physical Chemistry*. CRC Press, Boca Raton, FL.

Sposito, G. (1983) *J. Colloid Interface Sci.*, 91:329–340.

Sposito, G. (1991) *Soil Sci. Soc. Am. J.*, 55:965–967.

Stumm, W. and Morgan, J. J. (1970) *Aquatic Chemistry*. Wiley–Interscience, New York.

Su, C. M. and Harsh, J. B. (1993) *Clays Clay Minerals*, 41:461–471.

Su, C. M., Harsh, J. B., and Bertsch, P. M. (1992) *Clays Clay Minerals*, 40:280–286.

Uehara, G. and Gillman, G. P. (1981) *The Mineralogy, Chemistry and Physics of Tropical Soils with Variable Charge Clays*. Westview Press, Boulder, CO.

Wang, P. G., Ji, G. L., and Yu, T. R. (1987) *Z. Pflanzenernähr. Bodenk.*, 150:17–23.

Westall, J. and Hohl, H. (1980) *Adv. Colloid Interface Sci.*, 12:265–294.

Wong, M. T. F., Hughes, R., and Rowell, D. L. (1990) *J. Soil Sci.*,

41:655–663.

Yu, T. R. (1976) *Electrochemical Properties of Soils and Their Research Methods*. Science Press, Beijing.

Yu, T. R. (1981) *Soils Cir.*, 5:40–45.

Zhang, X. N. and Jiang, N. H. (1964) *Acta Pedol. Sinica*, 12:120–130.

Zhang, X. N., Jiang, N. H., Shao, Z. C., Pan, S. Z., and Zhang, W. G. (1979) *Acta Pedol. Sinica*, 16:145–156.

5

SPECIFIC ADSORPTION OF CATIONS

T. R. Yu, H. Y. Sun and H. Zhang

In the adsorption of cations by soils, except for the electrostatic force discussed in Chapter 3, specific forces between the surface of soil particles and the cations may sometimes be involved. The adsorption caused by these specific forces is called *specific adsorption*. Apparently, this adsorption would be related to both the nature of the cations and the surface properties of the soil. Generally speaking, various oxides are the principal materials responsible for specific adsorption of cations in soils. Variable charge soils contain large amounts of iron and aluminum oxides. At the same time, their negative surface charge sites that can attract cations electrostatically are small in quantity. It can be expected that, compared to constant charge soils, their specific adsorption for cations would have more significance.

Most of the cation species that can be adsorbed specifically by soils, such as copper, zinc, cobalt, and cadmium, belong to heavy metals. A large part of these heavy metals are transition elements in the periodic table. Alkali metal and alkaline earth metal ions can also be adsorbed specifically to some extent by soils under certain circumstances. However, this kind of adsorption is of less importance when compared to electrostatic adsorption, and the mechanism involved may be different from that for heavy metals. Among heavy metals, zinc occupies a special position in soil science because it is one important nutrient element for plants.

In this chapter, after treatment on the principles of specific adsorption of heavy metal ions, detailed discussions will be presented for both the relative importance of specific adsorption and electrostatic adsorption of these ions and the consequences of specific adsorption, using the adsorption of zinc ions as an example.

5.1 PRINCIPLES OF SPECIFIC ADSORPTION OF CATIONS

5.1.1 Properties of Transition Metal Ions

The cause of the difference in properties between the transition metal ions and the alkali metal and alkaline earth metal ions with respect to adsorption lies primarily in the difference in their atomic structure. Alkali metal and alkaline earth metal ions are characterized by a small amount of electric charge in the atomic nucleus, large ionic size, and weak polarizability. Therefore, the atom is difficult to deform. Besides, because the electronic structure of the outer shell is of the ns^0 type, electrostatic force plays the dominant role when forming coordination compounds. By contrast, for the IB group, the IIB group, and many other transition metal ions in the periodic table, the amount of electric charge in the atomic nucleus is larger and the ionic size is smaller than that of alkali and alkaline earth metals. Therefore, the polarizability and the ease of deformation are comparatively great. The structure of the electronic shell is of the type $(n - 1)d^9ns^0$, $(n - 1)d^{10}s^0$ or $(n - 1)d^{10}ns^0$ and unsaturated d layer. Hence, most of these kinds of ions can form an inner–sphere complex with ligands, leading to the increase in their stability. In this case, electrostatic force may not be the dominant contributing factor.

Because of the above-mentioned features in the structure of the electronic shell, transition metal ions have more hydration heat. These ions exist in the form of hydrated ions in solution and easily undergo hydrolysis, forming hydroxyl cations such as MOH^+. This hydration leads to the decrease in the average amount of electric charge per ion and correspondingly the decrease in energy barrier that must be overcome when the ions approach the surface of oxides, thus facilitating interactions between these ions and the oxide surface. This is the principal reason why hydrolytic cations are generally adsorbed by iron and aluminum oxides in larger amount than nonhydrolyzable cations (James and Healy, 1972 a–c; Kinniburgh et al., 1975, 1976).

The first hydrolytic constant of these cations is quite small. For zinc ions, it is about 10^{-9}. Therefore, within the pH range where the adsorption of zinc ions occurs markedly, the ratio of hydroxyl zinc ions to total zinc ions as calculated from the theoretical first hydrolytic constant is very small. However, it is possible that the degree of hydrolysis of metal ions in the electric double layer near the surface of colloids is higher than that in the bulk solution, because it has been observed that colloidal surfaces could induce hydrolysis, and therefore the pH at which metal ions began to hydrolyze was much lower than the corresponding pH occurring in solution (Leckie and James, 1974; Quirk and Posner, 1975).

The order of affinity for the surface of oxides among various transition metal ions is generally in conformity with the order of their tendency to form hydroxyl metal ions in aqueous solution (Baran, 1971; Benjamin and Leckie, 1981; Davis et al., 1978; Forbes et al., 1976; Grimme, 1968; Hodgson et al., 1964; James and Healy, 1972a–c; Kinniburgh et al., 1976).

This can be explained through the following theoretical derivations.

Metal ions M^{2+} undergo hydrolysis in aqueous solution, forming hydroxyl metal ions:

$$M^{2+} + H_2O \rightleftharpoons MOH^+ + H^+ \tag{5-1}$$

Assuming that the total amount of various forms of M ions in solution is $M_{liq.}{}^{n+}$, the chemical equilibrium of the reaction between these ions and the surface of the oxide can be written as:

$$SOH_m + M_{liq.}{}^{n+} \rightleftharpoons SOM_{ads.} + mH^+ \tag{5-2}$$

where SOH_m is $-OH$ or $-OH_2$ group at the surface of the oxide, and $SOM_{ads.}$ is the total amount of adsorbed M ions. According to the mass action law we have

$$K = \frac{[SOM_{ads.}][H^+]^m}{[SOH_m][M_{liq.}^{n+}]} \tag{5-3}$$

When expressed in logarithmic form we obtain

$$\log \frac{[M_{liq.}^{n+}]}{[SOM_{ads.}]} = pK - mpH \tag{5-4}$$

where K is the equilibrium constant in equation (4-2) and is a measure of the affinity of metal ions for the surface of the oxide. When the proportion of adsorbed ions is equal to 50% of the total added ions, the left side of equation (5-4) is equal to zero. Then, we can get

$$pH_{50} = \frac{1}{m} pK \tag{5-5}$$

A small pK or a low pH_{50} indicates that the metal ions can be adsorbed strongly even at a low pH. Because it has been found that the relationship between pH_{50} and pK is essentially in conformity with the relationship between pH_{50} and the first hydrolytic constant (Grimme, 1968; Kinniburgh et al., 1976; Kinniburgh and Jackson, 1981; McKenzie, 1980; Sposito, 1984), it can be inferred that it is the hydroxyl metal ion that is preferentially adsorbed on the surface of the oxide. Actually it is not difficult to derive the relationship between pH_{50} and pK_{h1}, if the $M_{liq.}{}^{n+}$ term in equation (5-2) is replaced by MOH^+.

In most cases the above derivations can explain the experimental data quite well. However, there are exceptions. For example, sometimes different adsorbents can have different orders of adsorption for various cations (Kinniburgh and Jackson, 1981).

5.1.2 Soil Components Capable of Specific Adsorption

In soils, the oxides of iron, aluminum, and manganese and their hydrates are the principal materials capable of adsorbing cations specifically. Layer silicates can also have this type of adsorbing ability under certain circumstances.

The structural features of various oxides of metals, including the hydroxides and hydroxyl compounds of these metals, are that in the structure one or several metal ions combine with the oxygen or hydroxyl group, and at the surface there is the presence of dissociable water molecule or hydroxyl group, caused by hydration of unsaturated bonds of the cation. Because of this feature, the sign and quantity of electric charge at the surface can change with the adsorption or desorption of protons. Transition metal ions can react with the hydroxyl group of the surface, forming surface complexes. Since in the formation of this kind of complex both electrostatic force and chemical force are involved, the free energy of adsorption is high.

The apparent charge density of the surface of many metal oxides may be very high but at the same time may not be accompanied by a high electrokinetic potential (Davis and Leckie, 1978; Lyklema, 1968; Perram et al., 1974; Wright and Hunter, 1973). This phenomenon may be related to the high porosity of the surface.

The zero point of charge of most metal oxides and their hydrates lies well above pH 8. On the other hand, the adsorption of transition metal ions occurs frequently at pH 3-7. Thus, it may be assumed that this kind of adsorption may occur when the surface of the oxides carries net positive charge or zero charge. This phenomenon cannot be explained simply by the operation of electrostatic force. At present it is generally conceived that metal oxides can be regarded as amphoteric compounds and that the hydroxyl or water groups on their surface are the sites of adsorption. In this case, H^+ ions, OH^- ions and specifically adsorbed cations may function as the potential-determining ions. The specific adsorption of these cations can lead to the change in surface properties of the oxides, such as the shift of zero point of net charge, the irregular change in electrokinetic potential, and so on, as shall be discussed later.

The capacity of different oxides in specifically adsorbing cations differs markedly. For the adsorption of zinc ions, for example, the order is birnessite > amorphous aluminum oxide > amorphous iron oxide (Brümmer et al., 1983). For the same kind of oxide, the adsorption capacity for cations varies with the degree of crystallization of the oxide, because during the aging process some changes in the form of the crystal, surface area, and chemical properties of the surface may occur. Generally speaking, amorphous substances possess a large specific surface area and a large adsorption capacity. By contrast, the reaction activity of well–crystallized substances is comparatively low. Freshly prepared iron hydroxide and aluminum hydroxide can adsorb zinc ions to an amount of 10 mg per gram, whereas the adsorbed amount decreases to about 10% of the original value after aging (Shuman, 1977).

Layer silicates can also have the ability of specific adsorption for heavy metal ions. Some authors supposed that the exposed Al-OH and Si-OH groups at the edges of the layer silicates are similar in properties to hydroxyl groups on the surface of oxides, which also possess the ability of specific adsorption to some degree. This type of specific adsorption is related to the retention of these cations in the form of MOH^+ (Koppelman and Dillard, 1977). Another possible mechanism is that these cations might form hydroxide precipitates in the presence of layer silicates (Farrah and Pickering, 1979). It has been frequently observed that during the adsorption of transition metal ions by layer silicates, two steps may occur, one initial rapid reaction and a subsequent slow reaction that can last several months. Some cations that have been adsorbed for a long time cannot be desorbed completely even under acid conditions. Therefore, some authors supposed that some cations with a size similar to that of the structural cations or the space length of the layer silicates can enter the crystal lattice or the interlayer of the mineral (Elgabaly, 1950; McBride and Mortland, 1974).

Variable charge soils contain both oxides and layer silicates. It is difficult to directly distinguish the relative contributions of these two kinds of adsorbent to specific adsorption of metal cations (Sposito, 1984). In order to solve this problem, Barrow (1986, 1987) proposed a model for describing the adsorption of zinc ions by variable charge soils. In this model, instead of considering the minute detail in the adsorption of ions by various components of the soil, a concept referring to the normal distribution of potential in the adsorption layer at the surface is introduced.

5.1.3 Mechanisms of Specific Adsorption

In order to interpret the specific adsorption of transition metal ions by the surface of metal oxides, a variety of mechanisms have been suggested (Kinniburgh and Jackson, 1981; McBride, 1989).

Some authors suspected that the surface groups of these oxides have the ability of coordination adsorption or complexation with these cations (Bleam and McBride, 1986; Clark and McBride, 1984; Forbes et al., 1976; Grimme, 1968; Kalbasi et al., 1978; Kinniburgh, 1983; Loganathan and

Burns, 1973; McKenzie, 1979; Morgan and Stumm, 1964; Murray, 1975a,b; Quirk and Posner, 1975; Schlindler, 1981). The proton of the hydroxyl group or the water molecule on the surface of these oxides may undergo ligand exchange with these metal cations, forming monodentate, bidentate, or even tridentate surface complexes in which the metal ions combine with the oxygen atoms through chemical bonds (McBride, 1978, 1982). It was supposed that the interactions between M ions and surface hydroxyl groups proceed in a fashion of constant charge, that is, the surface charge of the colloid does not change after the reaction. Actually, specific adsorption of cations can induce the shift in zero point of charge and the change in electrokinetic potential of the oxides, and during the reaction the exchange ratio H^+/M^{2+} is not 2, but instead a noninteger between 1 and 2. According to the theory of surface complexation, the hydroxyl group on the surface of oxides may be regarded as an amphoteric functional group. Therefore, when cations react with these groups it is possible to form either a monodentate complex or a bidentate chelate, resulting in the release of one or two protons. This is the reason why the exchange ratio H^+/M^{2+} varies within the range of 1-2.

Other authors regarded the adsorption of metal ions M^{2+} by the surface of oxides as an ion-exchange reaction, in which M^{2+} ions can be adsorbed on a site closely adjacent to the surface layer or in the Stern layer, resulting in the change of electric charge of the diffuse layer at the surface of oxides to a positive sign (Huang and Stumm, 1973; Kinniburgh et al., 1975).

At present, one widely accepted theory is that these oxides can preferentially adsorb hydroxyl metal ions (Benjamin and Leckie, 1981; Bowden et al., 1973; Davis et al., 1978; Kinniburgh, 1983; Kinniburgh and Jackson, 1981; Leckie and James, 1974; McKenzie, 1979; Sposito, 1984; Tiller et al., 1984). James and Healy (1972a-c) suggested that when ions are adsorbed specifically, the change in free energy consists essentially of three parts, namely the coulombic reaction term ΔG^0_{coul}, the solvation term $\Delta G^0_{solv.}$ and the chemical reaction term $\Delta G^0_{chem.}$. The higher the free energy of solvation, the more difficult for the ions to approach the surface of the adsorbent. The free energy of solvation is proportional to the square of electric charge carried by the ion. Since the amount of charge of transition metals is generally quite large, their secondary solvation energy is high. As a result, they can hardly enter the inner Helmholtz layer. On the other hand, when the pH is sufficiently high, due to the formation of hydroxyl metal ions as a result of the hydrolysis of the ions, the average amount of electric charge carried by the ions would decrease, inducing a sharp decrease in free energy of secondary solvation. This would cause the decrease in energy barrier that must be overcome when ions approach the surface of the oxides. Therefore, for each oxide-aqueous solution interface there is a characteristic critical pH range, generally less than one pH unit, within which the adsorption percentage can increase sharply from a very small value at low pH to 100% at high pH. Obviously, this critical pH range is related to the hydrolysis constant of the metal ions. This theory can explain the dependence of adsorption on pH and can also explain experimental results that the exchange ratio H^+/M^{2+} is not exactly 2, but instead a value varying between 1 and 2. Furthermore, the sequence of selectivity in the adsorption of various cations can also be interpreted more reasonably by this theory.

5.2 FACTORS AFFECTING SPECIFIC ADSORPTION BY SOILS

Variable charge soils contain both clay minerals and the oxides of iron, aluminum, and manganese, and therefore in the adsorption of cations both electrostatic attraction and specific adsorption may occur. In order to study the specific adsorption of cations by soils or oxides, many authors conducted experiments in which an electrolyte solution was present, so that the contribution of electrostatic adsorption could be reduced to a minimum. In practice, this is difficult to do. Since the selectivity of most adsorbents in soils to heavy metals may be larger than that to alkali and alkaline earth metals by an order of 3–4 in magnitude (Abd–Elfattah and Wada, 1981), and since the difference may be even larger when the coverage (occupancy) of heavy metal ions on the adsorbent is very low, which is generally the case, it would be very difficult to completely exclude the effect of electrostatic force, except when the concentration of the electrolyte is very high. On the other hand, if the electrolyte concentration is sufficiently high, the tendency of forming ion–pairs or complexes with the anions of the electrolyte may affect the existing form of the metal ions, which in turn would affect the adsorption. Therefore, one alternative and more meaningful way may be to study the adsorption of cations in the presence of a relatively low concentration (such as 0.01 M) of electrolyte, so that the relative contributions of the two mechanisms under different conditions can be examined.

Among the factors that can affect the specific adsorption of cations by variable charge soils, the most important ones are the pH of the medium, the nature of the cation species, the surface properties of the soil and accompanying ions. These will be discussed in the following sections.

5.2.1 pH

As in the case for the adsorption of heavy metal ions by oxides (Ankomah, 1992; Barrow, 1986; Bolland et al., 1977; Brümmer et al., 1988; Forbes et al., 1976; Gadde and Laitinen, 1974; Grimme, 1968; Harter, 1983; Jeffery and Uren, 1983; Kalbasi et al., 1978; Kinniburgh et al., 1975, 1976; Kinniburgh and Jackson, 1981; McKenzie, 1979, 1980), the adsorption of these cations by variable charge soils is strongly pH-dependent. Actually, the relationship is more complicated than that for pure oxides. This is due to several reasons. For specific adsorption, as seen in equation (5-1), a high pH is favorable to the hydrolysis of the ions, resulting in the increase in the proportion of hydroxyl metal ions MOH^+. For example, the quantity of $ZnOH^+$ ions at pH 6 should be 100 times larger than that at pH 4. It has been mentioned in Section 5.1.3 that, because of the decrease in electric charge carried by these hydroxyl metal ions, the energy barrier that must be overcome when these ions approach the surface of soil particles is decreased. This would be favorable to the adsorption of these ions on the surface through short-range forces. Although at pH 6 $ZnOH^+$ ions account for only about 0.1% of the total zinc ions, the amount of adsorption may increase markedly with the rise in pH at this time, because the hydrolysis reaction can proceed continuously and the surface of soil colloid may have an accelerating (catalyst) effect on hydrolysis of metal ions. Another reason is that during the adsorption of metal ions by soils, hydrogen ions may be released, as can be seen from equation (5-2). According to the mass action law, the increase in pH of the medium would favor the adsorption process. Besides, as has been discussed in Chapter 2, for variable charge soils the negative surface charge density increases with the rise in pH. This would cause the increase in adsorption of metal ions through electrostatic attraction force.

The adsorption of zinc ions by two variable charge soils as a function of pH is graphically shown in Fig. 5.1. Because in a $1M$ KCl solution electrostatic adsorption can be regarded as having been suppressed nearly completely, the adsorption curve should reflect the feature of specific adsorption. In this case, the quantity of adsorption is small when the pH is lower than about 6, whereas it may amount to 90% of the total adsorption at pH 7. In a 0.01 M KCl solution, because electrostatic adsorption also occurs, Zn adsorption still accounts for 25% of the total adsorption at pH 4, and the adsorption-pH curve lies well above that in $1M$ KCl solution. A similar Cd adsorption-pH curve was obtained for an Australian Oxisol (R. Naidu and M. E. Sumner, personal communication).

The adsorption curve in Fig. 5.1 can be distinguished into three regions. Within the lower pH region I with an upper pH limit at which the amount of adsorbed zinc corresponds to 50% of total zinc ions (i.e., pH_{50}), the adsorbed amount is strongly dependent on pH. Within region II from pH_{50} to pH_{80-90} the dependence becomes weaker than that in region I. Within region III from pH_{80-90} to pH_{100}, superficially the amount of adsorbed zinc ions changes only by 10-20%. Actually the dependence of adsorbed amount on pH is the strongest among the three regions, if calculated on the basis of the quantity of zinc ions remaining in solution.

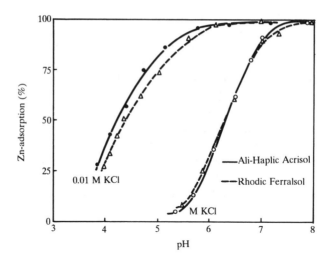

Fig. 5.1. Effect of pH on adsorption of zinc ions by two variable charge soils.

Utilizing the log D–pH relationship proposed by Kurbatov et al. (1951) based on the mass action law, it is possible to transform the S-shaped curve in Fig. 5.1 into a simple two-parameter linear equation. Here D is equal to $\%_{ads.}/(100 - \%_{ads.})$. Because specific adsorption is accompanied by the release of hydrogen ions, under certain conditions the slope of the straight line may approximately correspond to the ion-exchange ratio H^+/Zn^{n+} (Grimme, 1968; Kolarik, 1962; Posselt et al., 1968).

For variable charge soils capable of adsorbing cations specifically as well as nonspecifically, a parameter n may be used to express the fraction of specific adsorption in the total adsorption of zinc ions at a given pH. Thus, if the adsorption occurs in the presence of an electrolyte (such as KCl) with a certain concentration (such as 0.01 M), the reaction can be written as

$$[\text{soil}]_K^K + Zn^{2+} \rightleftharpoons [\text{Soil}]Zn + nxH^+ + 2(1 - n)K^+ \qquad (5\text{-}6)$$

where [soil] represents all the adsorption sites of the soil for zinc ions, including both specific adsorption and nonspecific adsorption, and x represents the number of released H^+ ions caused by specific adsorption of one zinc ion. In the equation the term $2K^+$ means that during nonspecific adsorption of one zinc ion there are two K^+ ions replaced.

According to the mass action law, the following relationship can be written:

$$K = \frac{[Zn]_{ads.}[H^+]^{nx}[K^+]^{2(1-n)}}{[Zn]_{soln.}[Soil]} \qquad (5\text{-}7)$$

where K is the equilibrium constant. When the quantity of adsorbed zinc ions is low, [soil] is a constant and [K^+] is also essentially a constant. Then, equation (5-7) may be modified as follows:

$$\log\frac{[Zn]_{ads.}}{[Zn]_{soln.}} = \log K' + nx\text{pH} \qquad (5\text{-}8)$$

This equation describes the relationship between the adsorption of zinc ions and pH in variable charge soils. In the absence of the effect of other factors, the relationship is only determined by the quantity of H^+ ions released in the adsorption process. When $\log\{[Zn]_{ads.}/[Zn]_{soln.}\}$ is plotted against pH, the slope of the straight line should correspond to the nx value.

The relevant curves for a Rhodic Ferralsol and a Ferrali–Haplic Acrisol are shown in Figs. 5.2 and 5.3, respectively. It can be seen that these curves can be distinguished into three parts. The slope is comparatively steep within both the low pH range and the high pH range, and there is a relatively flat region within the middle pH range. These three pH ranges correspond to regions I, II, and III shown in Fig. 5.1, respectively. Based on experimental data, it has been calculated that the adsorption of one zinc ion corresponds to the release of 1 and 1.3–1.5 H^+ ions in regions I and III, respectively, whereas the ratio is only 0.5–0.7 in region II.

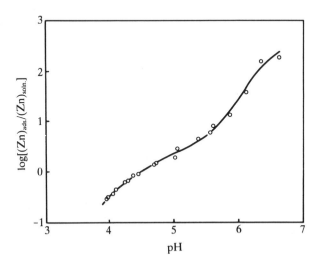

Fig. 5.2. $\log[(Zn)_{adsorption}/(Zn)_{solution}]$–pH curve for Rhodic Ferralsol.

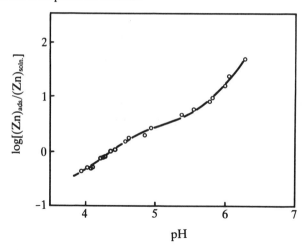

Fig. 5.3. $\log[(Zn)_{adsorption}/(Zn)_{solution}]$–pH curve for Ferrali–Haplic Acrisol.

If zinc ions are specifically adsorbed in the form of Zn^{2+} in region I, according to the reaction equation

$$[Soil]OH + Zn^{2+} \rightleftharpoons [Soil]OZn^+ + H^+ \qquad (5-9)$$

the theoretical value of x should be 1. In region I, because the numeral value of n is approximately 1, specific adsorption should account for 100% of the total adsorption. In region II, n is about 0.5-0.7, which means that specific adsorption accounts for 50-70% of the total adsorption. In region III, because the pH is high, zinc ions are specifically adsorbed in the form of $ZnOH^+$. In this latter case, according to equations

$$Zn^{2+} + H_2O \rightleftharpoons ZnOH^+ + H^+ \qquad (5-1a)$$

$$[Soil]OH + ZnOH^+ \rightleftharpoons [Soil]OZnOH + H^+ \qquad (5-10)$$

the theoretical value of n should be 2, and about 70% of zinc ions are specifically adsorbed in the form of $ZnOH^+$. Thus, under the specified condition that there are 0.01 M K^+ ions present in the solution, the relative contributions of specific adsorption and electrostatic adsorption would be different at different pH. In view of the mechanisms of the adsorption of heavy metal ions by variable charge soils mentioned previously, this difference is understandable. In region I, specific adsorption accounts for most of the adsorption. In region II, because of the increase in negative surface charge, electrostatic adsorption contributes to a certain extent. In region III where the degree of hydrolysis of zinc ions increases markedly, specific adsorption of zinc ions in the form of $ZnOH^+$ becomes the dominant mechanism.

From the above example it can be inferred that many factors are involved in the effect of pH on the adsorption of heavy metal ions by variable charge soils. Therefore, the relative importance of electrostatic adsorption and specific adsorption varies with environmental conditions.

5.2.2 Nature of Cation Species

The adsorption of Pb, Zn, and Cd by two Rhodic Ferralsols as affected by pH is shown in Fig. 5.4. At a given pH, the adsorption is of the order Pb > Zn > Cd. This order is the same as that found for the oxides of iron, aluminum, and manganese (Abd–Elfattah and Wada, 1981; Forbes et al., 1976; Gadde and Laitinen, 1974; Kinniburgh et al., 1976; Kinniburgh and Jackson, 1981; McKenzie, 1989; Schwertmann and Taylor, 1989).

Apparently, the difference in affinity of the three kinds of metal ions for soils is related to their tendency of hydrolysis. If the first hydrolytic constant K_1 is large, there would be more hydroxyl metal ions present in the solution at a given pH, which in turn would be favorable to adsorption by the soil. For the three kinds of metal ions shown in Fig. 5.4, the pK_1 values are 7.7 for Pb, 9.0 for Zn, and 10.1 for Cd (Kinniburgh et al., 1976). Thus, the first hydrolytic constant of Pb^{2+} ions is about 250 times larger than that of Cd^{2+}

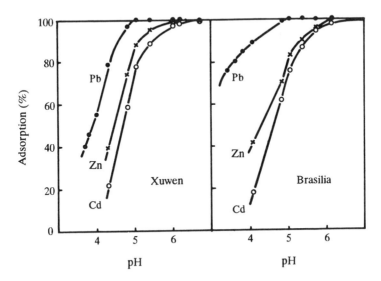

Fig. 5.4. Adsorption percentage of Pb, Zn, and Cd by Rhodic Ferralsols of Xuwen and Brasilia at different pH (electrolyte solution: 0.01 M NaCl) (data of G. Y. Zhang).

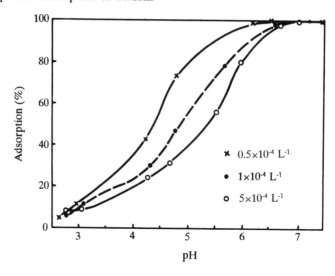

Fig. 5.5. Adsorption percentage of Pb by Ali–Haplic Acrisol at different pH as affected by initial concentration of Pb ions (electrolyte solution: 0.1 M NaNO$_3$) (data of G. S. Hu).

ions.

For a given cation species, the adsorption–pH curve varies with the initial concentration of the cations in the system. The general trend is that the lower the concentration, the larger the percentage of adsorption at a given pH. One example for the adsorption of Pb by an Ali–Haplic Acrisol is shown in Fig. 5.5. In the figure, the pH$_{50}$ values for the three initial concentrations varying from dilute to concentrated are 4.40, 4.87, and 5.37, respectively. This may be caused by differences in extent of hydrolysis of Pb^{2+} ions and in affinity of adsorption sites for Pb.

5.2.3 Surface Properties of Soil

Because both specific adsorption and electrostatic adsorption may occur in the adsorption of cations by variable charge soils, the quantity of adsorbed ions by different soils should be related to the surface charge and other properties of the soil, with the relationship depending on pH. Such a relationship for three representative variable charge soils in the adsorption of zinc ions is shown graphically in Fig. 5.6. Qualitatively speaking, both the net negative surface charge density and the quantity of adsorbed zinc ions are of the order Ali–Haplic Acrisol > Ferrali–Haplic Acrisol > Rhodic Ferralsol. If the curves are distinguished into three regions, it can be seen that at the low pH region both the difference in surface charge density and the difference in the quantity of adsorbed zinc ions among the three soils

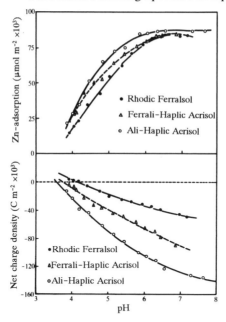

Fig. 5.6. Relationship between adsorption of zinc ions and net surface charge in variable charge soils (electrolyte solution: 0.01 M NaCl) (Sun, 1993b).

are comparatively small. Following the rise in pH, both the differences increase. When the pH is higher than about 6, the quantity of adsorbed zinc ions does not increase markedly, and the difference among the three soils decreases gradually until vanishing, although the difference in surface charge density increases continuously. This implies that at this time most of the adsorption of zinc ions are not caused by electrostatic attraction, but are controlled by specific force.

After the removal of iron oxides from the soil, because at low pH the quantity of positive surface charge is smaller than that of the original soil while at high pH some negative charge sites originally masked by iron oxides are released, as a whole the quantity of net negative surface charge increases. Conversely, after the addition of aluminum oxide coatings, the quantity of net negative surface charge would decrease, due to the increase in positive surface charge at low pH and the masking of some original negative charge sites at high pH. These changes in surface charge properties of the soil should affect the adsorption of cation ions through electrostatic attraction. As can be seen from Fig 5.7, the quantity of zinc ions adsorbed by Rhodic Ferralsol increased after the removal of iron oxides, while the quantity decreased after the addition of aluminum oxide coatings. This reflects the role of surface charge of the soil in electrostatic

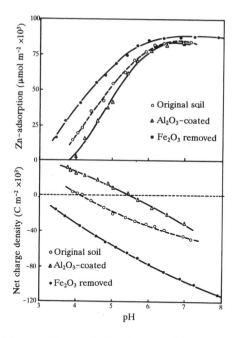

Fig. 5.7. Effects of iron oxides and aluminum oxides on net surface charge density and zinc adsorption of Rhodic Ferralsol (electrolyte solution: 0.01 M NaCl) (Sun, 1993b).

adsorption. Note that after the removal of iron oxides both the zero point of net charge and the zero point of adsorption lie at about pH 3. On the contrary, for the untreated soil the zero point of adsorption is lower than the zero point of net charge, which corresponds to about pH 4.1. This implies that the adsorption of zinc ions by iron oxides at low pH may exist. The zero point of net charge increased to 5.6 after the addition of aluminum oxide coatings. In this case the point of zero adsorption became much lower than the zero point of net charge of the soils. The above phenomena illustrate that the oxides of iron and aluminum can contribute to the adsorption of zinc ions at low pH, presumably through specific force.

At a pH higher than about 6, the difference is quite small in quantity of adsorbed zinc ions between soils free from iron oxides and those added with aluminum oxide coatings, although the surface charge density differs markedly. Apparently, at this time the adsorption is caused chiefly by specific adsorption, and therefore it is less dependent on the surface charge of the soil.

The effect of surface properties of the adsorbent on cation adsorption may be more clearly illustrated by a comparison of the adsorption of yttrium by a Rhodic Ferralsol with the adsorption by some minerals shown

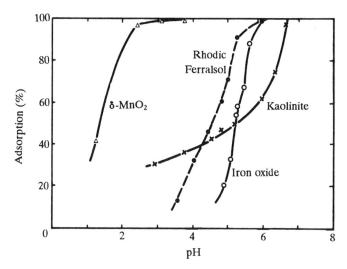

Fig. 5.8. Adsorption of Y ions by Rhodic Ferralsol and minerals at different pH (initial concentration of Y^{3+} is 1.64×10^{-4} M; electrolyte solution is 0.058 M NaCl) (Ran and Liu, 1992).

in Fig. 5.8. For this kind of trivalent rare earth metal ion, the tendency of hydrolysis is even higher than that of the majority of divalent heavy metal ions. As seen in the figure, for the synthetic iron oxide the slope of the adsorption–pH curve is very sharp. By contrast, the adsorption by kaolinite at low pH, which is presumably caused by cation–exchange with adsorbed Na^+ ions, is rather insensitive to pH. Only when the pH is higher than about 6, at which a large portion of Y ions is present in forms of $Y(OH)^{2+}$ and $Y(OH)_2^+$, the adsorption increases rapidly. It is interesting that the pH_{50} for the Rhodic Ferralsol is 4.39, lower than the value 5.04 for the iron oxide. This means that the affinity of the surface of this soil for Y ions is higher than that of the iron oxide. Because the affinity of aluminum oxides for heavy metal ions is generally lower than that of iron oxides (Kinniburgh and Jackson, 1981), it would be unlikely that aluminum oxides and hydroxides contained in the soil can induce a lower pH_{50} for this soil. The soil contains 1.79 g kg^{-1} of total Mn and 0.87 g kg^{-1} of amorphous Mn. Since it has been well-established that manganese oxides possess a high affinity and a large adsorption capacity for heavy metal ions (Kinniburgh and Jackson, 1981; McKenzie, 1989), as can also be clearly seen for δ-MnO$_2$ in Fig. 5.8, a logical suggestion may be that the manganese oxides in this variable charge soil play an important role in inducing specific adsorption of Y.

For variable–valency metals, such as cerium, the oxidation state of the ions can be changed after adsorption. A comparison between Fig. 5.9 and

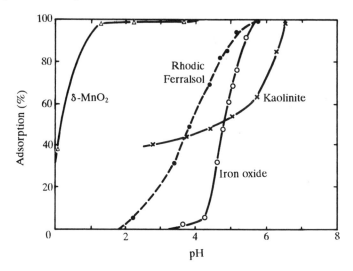

Fig. 5.9. Adsorption of Ce ions by Rhodic Ferralsol and minerals at different pH (initial concentration of Ce^{3+} is 1.64×10^{-4} M; electrolyte solution is 0.058 M NaCl) (Ran and Liu, 1992).

Fig. 5.8 shows that the pH_{50} values of Ce-adsorption for δ-MnO_2, Rhodic Ferralsol, and iron oxide are 0.08, 3.72, and 4.92, respectively, lower than the corresponding values 1.34, 4.39, and 5.04 for Y-adsorption, especially for the δ-MnO_2. This phenomenon might be caused by the oxidation of Ce^{3+} to Ce^{4+} by δ-MnO_2 directly or by air in the presence of iron oxide, which in turn enhances the adsorption of Ce (Ran and Liu, 1992).

From the above discussions it can be concluded that the adsorption of heavy metal ions by variable charge soils is affected by surface properties of the soil in a complicated manner. The oxides of iron, aluminum, and manganese are the principal causes for inducing specific adsorption, and they also affect the adsorption by affecting the surface charge of the soil. For a given adsorbent, the relative importance of specific adsorption and electrostatic adsorption is determined chiefly by the pH of the medium.

5.2.4 Accompanying Ions

In the adsorption of heavy metal ions by variable charge soils, both cations and anions coexisting with these ions may affect the reaction, and the situation may be more complex than the effect on alkali metal or alkaline earth metal ions.

Various accompanying ions can affect the adsorption of cations in a variety of ways. The effect of ionic strength induces the change in the activity coefficient of the metal ions, H^+ ions, and OH^- ions in solution, with the activity coefficient of heavy metal ions, especially of trivalent ions, more sensitive to changes in ionic strength than that of alkali metal ions such as sodium and potassium. Some anions can form ion pairs or complexes with the metal ions. The presence of electrolytes induces the contraction of the electric double layer at the surface of soil colloid, resulting in the lowering of electrokinetic potential at the slip surface. This would affect the adsorption of metal ions. Cations can compete with the adsorbed ions for exchange sites on the surface of the soil. If the adsorption of other ion species can alter the surface charge of the soil, this alteration may also affect the adsorption of metal ions indirectly.

We have already seen (Fig. 5.1) the difference in position of the adsorption–pH curve of Zn ions in two concentrations of KCl solution. The

Table 5.1 Effect of $NaNO_3$ Concentration on Adsorption of Pb and Cu by Soils and Minerals[a] (Hu, 1994)

Adsorbent	$NaNO_3$ (mol L^{-1})	Adsorption (cmol kg^{-1})	
		Pb	Cu
Ali–Haplic Acrisol	0.01	0.25	0.25
	0.1	0.13	0.25
	0.5	0.09	0.24
	1	0.02	0.24
Rhodic Ferralsol	0.01	0.25	0.25
	0.1	0.22	0.25
	0.5	0.20	0.22
	1	0.17	0.21
Mn–coated kaolinite	0.01	7.78	7.74
	0.1	7.73	5.71
	0.5	7.53	4.66
	1	7.19	4.23
Goethite	0.01	6.32	5.15
	0.1	7.07	6.38
	0.5	7.72	7.24
	1	7.78	7.56

[a]Initial concentration of Pb or Cu is 0.25 cmol kg^{-1} for soils and 8.00 cmol kg^{-1} for minerals; pH of the suspensions is 5–6.

effect of $NaNO_3$ concentration on adsorption of Cu ions and Pb ions by two variable charge soils and two minerals is shown in Table 5.1. For the two soils and the Mn-coated kaolinite, within the pH range (5-6) higher than their zero point of charge at which the adsorbent carries negative net surface charge, the adsorption decreases with the increase in $NaNO_3$ concentration. This phenomenon can be interpreted as caused by the combined effects mentioned in the previous paragraph. Similar effect has been observed for oxides (Benjamin and Leckie, 1982) and Italian soils (Petrezelli et al., 1985).

Conversely, for goethite, the adsorption increases with the increase in $NaNO_3$ concentration. This should be related to the positive net surface charge carried by this mineral. In this case, the contraction of the electric double layer in the presence of high concentrations of electrolyte may be favorable to the approach of the metal ions to the surface of the mineral and thus favorable to their adsorption, and this effect may exceed the ionic strength effect occurring for negatively charged adsorbents.

If the accompanying anions can also be adsorbed by the soil specifically, this specific adsorption can induce the change in surface charge properties of the soil, as has been seen in Chapter 2 and shall be seen in Chapter 6. If these anions can form ion pairs or complexes with the metal ions, they may affect the quantity of free metal ions in solution. Besides, if the anions can form complexes with other metal species that can compete with the adsorbate, the situation may be more complicated. These factors may affect the adsorption of the metal ions in a complex manner.

The effects of phosphate, sulfate, and fluoride ions on the adsorption of

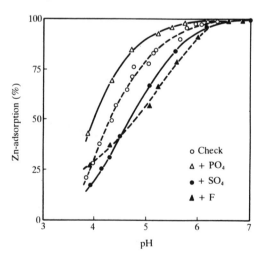

Fig. 5.10. Effect of phosphate, sulfate, and fluoride ions on zinc adsorption by Ali-Haplic Acrisol (electrolyte solution is 0.01 M NaCl).

zinc ions by three variable charge soils are shown in Figs. 5.10 to 5.12, respectively. A common phenomenon occurred for the three soils: The three anion species affected the adsorption of zinc ions differently. The adsorption increased markedly in the presence of phosphate ions. A similar effect has been observed for minerals and other soils (Barrow, 1987b; Diaz–Barrientos et al., 1990; Ghanem and Mikkelsen, 1988; Madrid et al., 1991; Naidu et al., 1994). The exact mechanism is not known, although some suggestions have been made in the literature. For the three variable charge soils shown in Figs. 5.10 to 5.12, since the quantity of negative surface charge of the soils increased after the adsorption of phosphate ions, the increase in zinc adsorption should at least be partly caused by increased contribution of electrostatic adsorption.

On the other hand, the adsorption of sulfate ions and fluoride ions, both of which induced an increase in negative surface charge of the soil, affected the adsorption of zinc ions in a complicated manner. At pH lower than about 4 the effect was small or positive in direction. This may be the effect of the increase in negative surface charge of the soil, because the lower the pH the higher the adsorption of the two anion species, as shall be shown in Chapter 6. Within the pH range of 4 to 6 the presence of the two kinds of anions induced the decrease in the adsorption of zinc ions, particularly for fluoride ions. Apparently, in this case one factor that retards the adsorption of Zn ions by the solid phase, or more exactly one factor that tends to keep Zn ions in solution, plays the dominant role in controlling the distribution of these ions between the two phases. Probably, this factor is

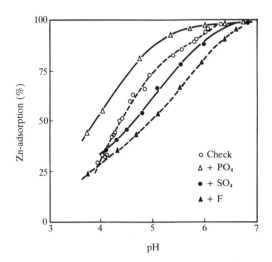

Fig. 5.11. Effect of phosphate, sulfate, and fluoride ions on zinc adsorption by Ferrali–Haplic Acrisol (electrolyte solution is 0.01 M NaCl)

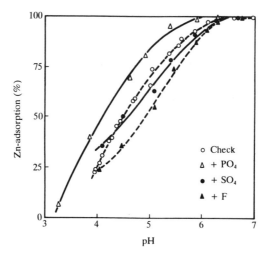

Fig. 5.12. Effect of phosphate, sulfate, and fluoride ions on zinc adsorption by Rhodic Ferralsol (electrolyte solution is 0.01 M NaCl).

the formation of ion pairs or complexes of zinc ions with fluoride or sulfate ions. The principles of complex formation have been discussed in the literature (Kinniburgh and Jackson, 1981; Sposito, 1984; Stumm and Morgan, 1971). Some authors (Barrow and Cox, 1992a,b; Boekhold et al., 1993) supposed that even chloride ions could affect the adsorption of heavy metals by oxides and soils.

From the above discussions it can be concluded that in the adsorption of heavy metal ions by variable charge soils, accompanying ions can affect the adsorption directly through competing for negative surface charge sites, or indirectly through affecting the surface properties of the soil or the ionic composition of the solution. By contrast, in the adsorption of these ions by constant charge soils, this kind of indirect effect is secondary in importance.

5.3 CONSEQUENCES OF SPECIFIC ADSORPTION

5.3.1 Release of Hydrogen Ions

In the presence of a large amount of alkali metal ions, and especially when the pH is higher than about 5.5, the essence of electrostatic adsorption of heavy metal ions by variable charge soils is probably the exchange with adsorbed alkali metal ions without the release of hydrogen ions. On the other hand, during specific adsorption there should be the release of hydrogen ions, with the quantity of release depending on the reaction mechanism.

If the metal ion is adsorbed specifically in the form of M^{2+}, it can form a monodentate surface complex:

$$\begin{array}{l} \text{OH} \mid -1 \\ \quad \mid \quad + M^{2+} \rightarrow \\ \text{OH} \mid \end{array} \quad \begin{array}{l} \text{O-M} \mid 0 \\ \quad \mid \quad + H^+ \\ \text{OH} \mid \end{array} \qquad (5\text{-}11)$$

In this reaction, one H^+ ion is released and the surface charge of the soil changes by one unit.

If the metal ion is adsorbed specifically in the form of MOH^+, two H^+ ions would be released and accompanied by the change in surface charge of the soil:

$$M^{2+} + H_2O \rightleftharpoons MOH^+ + H^+ \qquad (5\text{-}1a)$$

$$\begin{array}{l} \text{OH} \mid -1 \\ \quad \mid \quad + MOH^+ \rightarrow \\ \text{OH} \mid \end{array} \quad \begin{array}{l} \text{OH} \quad\quad \mid -1 \\ \quad \mid \quad + H^+ \\ \text{O-MOH} \mid \end{array} \qquad (5\text{-}12)$$

Thus, the quantity of H^+ ions released during the adsorption of metal ions by variable charge soils is determined by the relative magnitudes of electrostatic adsorption and specific adsorption as well as the form of metal ions in the latter type of adsorption. Apparently, this quantity would be closely related to the pH of the medium. In Section 5.2.1, the release of H^+ ions within different pH ranges has been inferred to through graphic method, based on the dependence of the quantity of adsorption on pH. It is also possible to neutralize the released H^+ ions by an alkali with a back-titration method, so that the release of H^+ ions during adsorption of metal ions by soils under constant pH conditions can be directly examined.

Kinetic studies on the release of H^+ ions during adsorption of zinc ions by variable charge soils showed that the adsorption attained a nearly steady state after about 4 min. On the other hand, the release of H^+ ions consisted of a rapid process and a slow process. Thus, the exchange ratio H/Zn varied with the time of reaction. The ratio was small at first. It then increased gradually, and it finally attained a nearly constant value after about 1 hr (Fig. 5.13). This seems to indicate that the initial reaction between zinc ions and the surface of the soil is chiefly related to a rapid ion-exchange reaction. At the same time, part of the adsorbed zinc ions begin to react with those hydroxyl groups having a low energy level on the surface of the soil, releasing H^+ ions. Afterward, part of these zinc ions react with high-energy-level hydroxyl groups to release additional H^+ ions. Therefore, this slow reaction is the rate-determining step in affecting the H/Zn exchange kinetics.

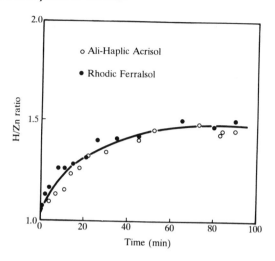

Fig. 5.13. Change of exchange ratio H/Zn with time during adsorption of zinc ions by variable charge soils (pH of Ali-Haplic Acrisol, 6.83; pH of Rhodic Ferralsol, 6.98).

Table 5.2 Effect of Quantity of Zn^{2+} Ions on Exchange Ratio H/Zn
(Electrolyte solution: 0.01 M NaCl)

Treatment	Zn^{2+} Added (cmol kg^{-1})	Ali-Haplic Acrisol pH	H/Zn	Rhodic Ferralsol pH	H/Zn
Original soil	2.62	6.07	1.42	5.86	1.49
	5.24	6.08	1.30	5.79	1.41
	7.34	6.12	1.23	5.83	1.39
	9.43	6.13	1.23	5.86	1.29
Al_2O_3-coated	2.62	6.52	1.41	7.00	1.39
	5.24	6.56	1.40	6.97	1.45
	7.34	6.67	1.36	6.99	1.44
	9.43	6.66	1.36	6.89	1.36
	10.5	6.69	1.35	–	–
Fe_2O_3 removed	2.62	6.02	1.76	6.15	1.56
	5.24	6.01	1.68	6.13	1.55
	7.34	6.03	1.54	6.12	1.43
	9.43	6.02	1.48	6.08	1.40
	10.5	6.00	1.43	–	–

The data given in Table 5.2 show that within the pH range of 5.5–7.0 most of the exchange ratios H/Zn lie in the range of 1.3–1.5, consistent with the extrapolated value based on the dependence of the quantity of adsorbed zinc ions on pH. If examined in more detail, it can be found that the ratio showed a slightly decreasing tendency with the increase in amount of added zinc. This seems to imply that if the quantity of zinc ions is large, the proportion of electrostatic adsorption or the proportion of the specific adsorption of $ZnOH^-$ may increase.

For trivalent ions La, Ce, and Y adsorbed by variable charge soils and iron oxide, the ratios were around 2, implying that the ions were adsorbed chiefly in the form of $M(OH)^{2+}$, whereas when adsorbed by Mn oxide the ratios were about 1.2 (Ran and Liu, 1992).

In the literature, reports differed greatly on the exchange ratio H/M during the adsorption of zinc ions and other specifically adsorbed cations by variable charge soils and various oxides (Gaddle and Laitinen, 1974; Grimme, 1968; Kinniburgh, 1983; Kinniburgh and Jackson, 1981; Kolarik, 1962; Kozawa and Takai, 1978; McKenzie, 1980; Murray, 1975a,b; Quirk and Posner, 1975). The value obtained by different authors varied even for the same kind of adsorbent. This discrepancy could be caused by differences in surface conditions of the adsorbent and experimental conditions.

5.3.2 Effect on Adsorption of Anions

In addition to ionic strength effect, metal adsorption may show a cooperative adsorption effect on adsorption of some anions. For example, zinc ions

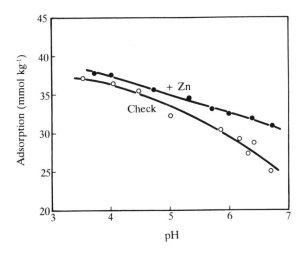

Fig. 5.14. Effect of zinc ions on phosphate adsorption by Ferrali–Haplic Acrisol.

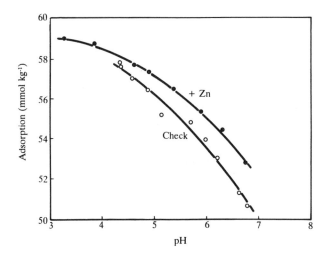

Fig. 5.15. Effect of zinc ions on phosphate adsorption by Rhodic Ferralsol.

may exert a remarkable enhancing effect on the adsorption of phosphate by variable charge soils, similar to the effect for goethite (Bolland et al., 1977). The extent of effect increases with the rise in pH (Figs. 5.14 and 5.15). Since the quantity of adsorbed phosphate decreased with the increase in pH, the effect of zinc ions at high pH would be more pronounced when

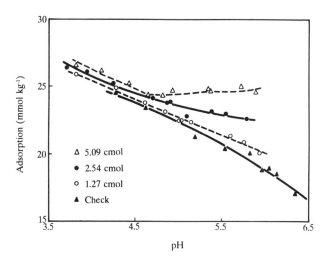

Fig. 5.16. Effect of quantity of zinc ions on phosphate adsorption by Ferrali-Haplic Acrisol (numerals in the figure are quantity of zinc per kilogram of soil).

compared on the basis of relative magnitude. The effect is also related to the quantity of zinc ions. As can be seen in Fig. 5.16, while the adsorption of phosphate increased with the increase in zinc ions, when the quantity of the latter ions was sufficiently high the regularity of the decrease in phosphate adsorption with the increase in pH was disturbed; that is, when the pH exceeded a certain level the adsorption of phosphate increased with the increase in pH. This reflects the strong dependence of the adsorption of phosphate on zinc ions.

Based on the phenomena shown in Figs. 5.14 to 5.16, it can be suggested that, for variable charge soils with high contents of iron and aluminum oxides, at low pH it is chiefly the specific adsorption of phosphate by these oxides that determines the quantity of phosphate adsorbed, whereas at high pH there is a coordinative adsorption between zinc ions and phosphate ions.

The presence of zinc ions can also lead to the increase in adsorption of fluoride ions by variable charge soils. Unlike the case for the adsorption of phosphate ions, the effect becomes more pronounced when the pH is decreased (Figs. 5.17 and 5.18). The pH below which the effect of zinc adsorption becomes remarkable coincides approximately with the pH at which the adsorption of fluoride ions is at its maximum on the typical fluoride adsorption–pH curve for oxide surfaces. Since the adsorption of zinc ions itself decreases with the decrease in pH, this effect of zinc ions on the adsorption of fluoride ions deserves notice. Because in addition to the chemical equilibrium $HF \rightleftharpoons H^+ + F^-$ there is the occurrence of another chemical equilibrium $Zn^{2+} + F^- \rightleftharpoons ZnF^+$ in the solution phase, it may be suggested that it is this latter reaction that promotes the adsorption of

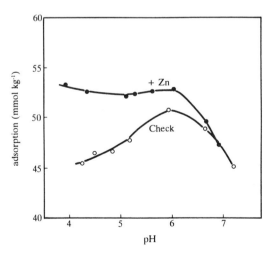

Fig. 5.17. Effect of zinc ions on fluoride adsorption by Ali–Haplic Acrisol.

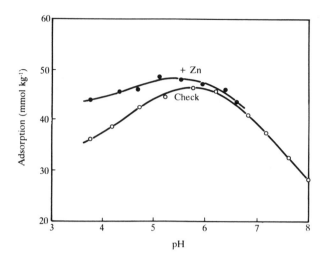

Fig. 5.18. Effect of zinc ions on fluoride adsorption by Ferrali–Haplic Acrisol.

fluoride ions by the soil. Thus, the specific adsorption of zinc ions and other heavy metal ions may affect the adsorption of anions in a complicated manner, with the extent of effect more pronounced either at low pH or at high pH, depending on the nature of the anion species.

5.3.3 Change in Surface Properties of Soil

According to the discussions in Section 5.1.3, specific adsorption of metal ions can affect the surface charge of the adsorbent. For oxides, it has been observed that such adsorption resulted in the shift of zero point of net charge and the change in isoelectric point (Breenwsma and Lyklema, 1973; Kinniburgh and Jackson, 1981; Lyklema, 1984; Stumm and Morgan, 1970). For variable charge soils, such phenomena may also occur.

The changes in zeta potential of an Ali–Haplic Acrisol with the change in pH in $CuCl_2$ solution and in $ZnCl_2$ solutions are shown in Fig. 5.19 and 5.20, respectively. It can be seen that the zeta potentials in the two solutions are more positive than those in NaCl solution. This difference increases with the increase in pH, and in the $ZnCl_2$ solution the zeta potential tends to increase in a more marked manner when the pH is higher than 7. This latter point is of important significance.

It was shown in Chapter 2 that the variability in surface charge carried by the above Ali–Haplic Acrisol derived from Quaternary red clay is not very great. For Ferralsols with a greater proportion of variable charge component, the characteristics of hydrolytic metal cations in adsorption may

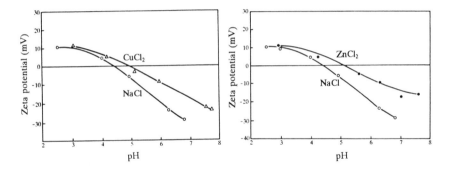

Fig. 5.19. (Left) Zeta potentials of Ali–Haplic Acrisol in CuCl$_2$ solution and in NaCl solution (Zhang and Zhang, 1991).

Fig. 5.20. (Right) Zeta potentials of Ali–Haplic Acrisol in ZnCl$_2$ solution and in NaCl solution.

be reflected in electrokinetic properties more remarkably. This point can be seen from the effect of CoCl$_2$ on electrokinetic properties of the Rhodic Ferralsol from Brasilia shown in Fig. 5.21, where the zeta potentials are positive even in the high pH range. Thus, no isoelectric point occurs. Another point deserving notice is that on the zeta potential–pH curve there appears a valley at about pH 6.8. For another Rhodic Ferralsol (Xuwen) in CuCl$_2$ solution, a valley also appears when the CuCl$_2$ concentration is

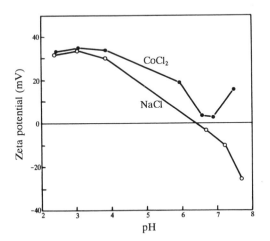

Fig. 5.21. Effect of CoCl$_2$ on electrokinetic properties of Rhodic Ferralsol of Brasilia.

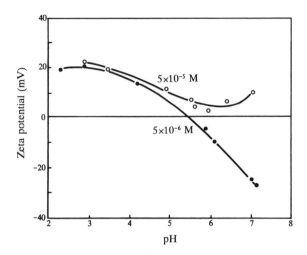

Fig. 5.22. Electrokinetic properties of Rhodic Ferralsol in $CuCl_2$ solutions of different concentrations (Zhang and Zhang, 1991).

high (Fig. 5.22). The pattern of the zeta potential–pH curve of this soil in $ZnCl_2$ solution is even more complex. It can be seen in Fig. 5.23 that the valley of the curve in 5×10^{-5} M $ZnCl_2$ solution occurs in the negative potential region, and consequently two isoelectric points occur, one at pH 5.8 and another at pH 7.3.

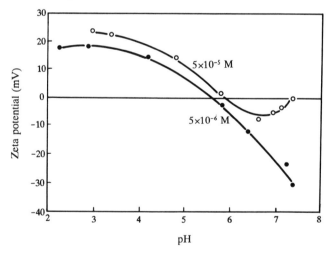

Fig. 5.23. Electrokinetic properties of Rhodic Ferralsol in $ZnCl_2$ solutions of different concentrations.

In summary, owing to the adsorption of hydrolytic heavy metal cations, the zeta potential–pH curve of Ferralsols becomes distinctly different from that in ordinary electrolyte solutions. Even the sign of the potential may change. This clearly indicates that these metal ions have been strongly adsorbed specifically by these Ferralsols.

Many reports have appeared in the literature about the occurrence of more than one turnover in sign of zeta potential of oxides and clay minerals after adsorption of hydrolytic metal ions. These include adsorption of Ca^{2+} by γ-Al_2O_3 (Huang and Stumm, 1973), adsorption of Co^{2+}, La^{3+} and Th^{4+} by SiO_2 (James and Healy, 1972b), adsorption of Co^{2+} by TiO_2 (James and Healy, 1972b), adsorption of Fe^{3+}, Co^{2+} and Pb^{2+} by quartz (Leckie and James, 1974), adsorption of Hg(II) by SiO_2 (MacNaughton and James, 1974), adsorption of Co^{2+} by TiO_2 and ZrO_2 (Tewari and Lee, 1975), adsorption of Ba^{2+} and Ca^{2+} by γ-Al_2O_3 (Escudey and Gil–Llambias, 1985), adsorption of Mn^{2+} by TiO_2 (Bleam and McBride, 1986), adsorption of Zn^{2+} by aluminosilicate (Huang and Rhoads, 1989), and adsorption of Co^{2+}, Cd^{2+}, and Cu^{2+} by kaolinite (Hunter and James, 1992). In order to explain these phenomena, James and Healy (1972b) proposed the surface precipitation concept. They considered that the IEP corresponding to the first change in sign of charge from "+" to "-" with the change in pH ought to be governed, in principle, by the equilibrium between poten-tial–determining ions of the solid phase, H^+ ions and OH^- ions. After adsorption of specific metal ions, however, this IEP may shift toward higher pH. The IEP corresponding to the second change in sign of charge, which occurs at a rather high pH, is caused by the coating of the hydrolytic product of metal ions on the surface of the solid phase. If the coating covers the surface completely, the apparent IEP is the characteristic IEP of the hydrolyzed product of the metal ions, and is not related to the metal oxides per se. If the covering is not complete, the IEP is then generally lower than that of the hydrolytic product and would be a reflection of the overall characteristics in electrokinetic properties of both the oxides and the hydrolytic product.

With regard to the change in sign of zeta potential for variable charge soils, no report is available in the literature. From the material shown in Figures 5.21 to 5.23 it can be seen that under certain conditions variable charge soils may also exhibit a feature similar to that of oxides–that is, the occurrence of two changes in sign of charge and two corresponding IEPs. Of course, the mechanisms of interactions of metal ions with variable charge soils are much more complex than those with pure oxides, because soils are mixtures with both variable and constant charge surfaces. Nevertheless, the materials cited above demonstrate that the electrokinetic properties of variable charge soils present certain characteristics similar to those of pure oxides. It can thus also be postulated that iron and aluminum oxides in variable charge soils play a very important role in determining the behavior of these soils.

5.3.4 Practical Implications

The specific adsorption of cations, particularly of trace elements, is of practical importance in soils and environment.

This adsorption leads to the accumulation of heavy metals in certain fractions of the soil. For example, it has been found that the contents of Mn, Cu, Co, Ni, Zn, Pb, Cr, and V in iron-manganese concretions, which are common occurrences in variable charge soils, may be many times higher than those of the whole soil (Childs, 1975; Sidhu et al., 1977, Taylor and McKenzie, 1966). The content of Co may be up to 1.4%, and that of Cr may be up to 0.24%. For Pb, the content in Mn concretions may be as high a. 2% (Norrish, 1975).

Specifically adsorbed cations cannot participate in ordinary cation-exchange reaction. They can only be replaced by another kind of cation with a higher affinity for the adsorbent or extracted by acids. Generally the &concentration of many heavy metals in soil solution is much lower than that predicated by the solubility product principle. By contrast, specific adsorption plays an important role in controlling the concentration and thus the availability and the toxicity of some metal ions to plants in soils. Actually, the addition of manganese oxide to a lead-polluted soil could retard the uptake of Pb by plant (McKenzie, 1978).

The significance of specific adsorption in environment lies in the fact that soil is a sink of toxic metal elements. Once entered into the soil, generally it is very difficult for them to leach out. On the other hand, some soil components can function as the immobilizer of these metals through specific adsorption. Since the toxicity of these elements is to a large extent determined by factors that affect electrostatic adsorption and specific adsorption, in the national environmental quality standard for soils of many countries, such as in China, CEC and pH are taken as the chief criteria for evaluating the environmental quality with respect to heavy metals Cd, Hg, Cu, Pb, Cr, Zn, and Ni.

BIBLIOGRAPHY

Abd-Elfattah, A. and Wada, K. (1981) *J. Soil Sci.*, 32:271-283.
Ankomah, A. B. (1992) *Soil Sci.*, 154:206-213.
Baran, V. (1971) *Coord. Chem. Rev.*, 6:65-93.
Barrow, N. J. (1986) *J. Soil Sci.*, 37:295-302.
Barrow, N. J. (1987a) *Fert. Res.*, 14:16-29.
Barrow, N. J. (1987b) *J. Soil Sci.*, 38:453-459.

Barrow, N. J. and Cox, V. C. (1992a) *J. Soil Sci.*, 43:295-304.

Barrow, N. J. and Cox, V. C. (1992b) *J. Soil Sci.*, 43:305-312.

Benjamin, M. M. and Leckie, J. O. (1981) *J. Colloid Interface Sci.*, 79:209-242.

Benjamin, M. M. and Leckie, J. O. (1982) *Environ. Sci. Technol.*, 16:162-170.

Bleam, W. F. and McBride, M. B. (1986) *J. Colloid Interface Sci.*, 110:335-346.

Boekhold, A. E., Temminghoff, E. C. M., and van der Zee, S. E. A. T. M. (1993) *J. Soil Sci.*, 44:85-96.

Bohn, H. L., McNeal, B. and O'Connor, G. A. (1979) *Soil Chemistry*. John Wiley & Sons, New York.

Bolland, M. D. A., Posner, A. M., and Quirk, J. P. (1977) *Aust. J. Soil Res.*, 15:279-286.

Bowden, J. W., Bolland, M. D. A., Posner, A. M., and Quirk, J. P. (1973) *Nature*, 245:81-83.

Bowden, J. W., Posner, A. M., and Quirk, J. P. (1980) in *Soils with Variable Charge* (B. K. G. Theng, ed.). New Zealand Society of Soil Science, Lower Hutt, pp. 147-160.

Breenwsma, A. and Lyklema, J. (1973) *J. Colloid Interface Sci.*, 43:437-448.

Brümmer, G., Tiller, K. G., Herms, U., and Clayton, P. M. (1983) *Geoderma*, 31:337-354.

Brümmer, G., Gerth, J. and Tiller, K. G. (1988) *J. Soil Sci.*, 39:37-52.

Childs, C. W. (1975) *Geoderma*, 13:141-152.

Clark, C. J. and McBride, M. B. (1984) *Clays Clay Miner.*, 32:300-310.

Davis, J. A., James, R. O., and Leckie, J. O. (1978) *J. Colloid Interface Sci.*, 63:480-499.

Davis, J. A. and Leckie, J. O. (1978) *Environ. Sci. Technol.*, 12:1309-1315.

Diaz-Barrientos, E., Madrid, L., Contreras, M. C., and Morillo, E. (1990) *Aust. J. Soil Res.*, 28:549-557.

Egozy, Y. (1980) *Clays Clay Miner.*, 28:311-318.

Elgabaly, M. M. (1950) *Soil Sci.*, 69:167-173.

Escudey, M. and Gil-Llambias, G. (1985) *J. Colloid Interface Sci.*, 107:272-275.

Farrah, H. and Pickering, W. F. (1979) *Chem. Geol.*, 25:317-326.

Farrah, H. and Pickering, W. F. (1980) *Chem. Geol.*, 28:55-68.

Forbes, E. A., Posner, A. M., and Quirk, J. P. (1974) *J. Colloid Interface Sci.*, 49:403-409.

Forbes, E. A., Posner, A. M. and Quirk, J. P. (1976) *J. Soil Sci.*, 27:154-188.

Gaddle, R. R. and Laitinen, H. A. (1974) *Anal. Chem.*, 46:2022-2026.

Ghanem, S. A. and Mikkelsen, D. S. (1988) *Soil Sci.*, 146:15-21.

Grimme, H. H. (1968) *Z. Pflanzenernähr. Bodenk.*, 121:58-65.

Harter, R. D. (1983) *Soil Sci. Soc. Am. J.*, 47:47-51.

Hodgson, J. F., Tiller, K. G., and Fellows, M. (1964) *Soil Sci. Soc. Am. Proc.*, 28:42-46.

Hu, G. S. (1994) *Pedosphere*, 4:153-164.

Huang, C. P. and Rhoads, E. A. (1989) *J. Colloid Interface Sci.*, 131:289-306.

Huang, C. P. and Stumm, W. (1973) *J. Colloid Interface Sci.*, 43:409-420.

Hunter, R. J. (1981) *Zeta Potential in Colloid Science.* Academic Press, London.

Hunter, R. J. and James, M. (1992) *Clays Clay Miner.*, 40:644-649.

James, R. O. and Healy, T. W. (1972a) *J. Colloid Interface Sci.*, 40:42-52.

James, R. O. and Healy, T. W. (1972b) *J. Colloid Interface Sci.*, 40:53-64.

James, R. O. and Healy, T. W. (1972c) *J. Colloid Interface Sci.*, 40:65-81.

Jeffery, J. J. and Uren, N. C. (1983) *Aust. J. Soil Res.*, 21:479-488.

Kalbasi, M., Racz, G. J., and Loewen-Rudgers, L. H. (1978) *Soil Sci.*, 125:146-150.

Kinniburgh, D. G. (1983) *J. Soil Sci.*, 34:759-768.

Kinniburgh, D. G. and Jackson, M. L. (1981) in *Adsorption of Inorganics at Solid-Liquid Interfaces* (M. A. Anderson and A. J. Rubin, eds.). Ann Arbor Science, Ann Arbor, MI, pp. 91-160.

Kinniburgh, D. G., Syers, J. K. and Jackson, M. L. (1975) *Soil Sci. Soc. Am. Proc.*, 39:464-470.

Kinniburgh, D. G., Jackson, M. L., and Syers, J. K. (1976) *Soil Sci. Soc. Am. Proc.*, 40:796-799.

Kishk, F. M. and Hassan, M. N. (1973) *Plant and Soil*, 39:497-505.

Koppelman, M. H. and Dillard, J. G. (1977) *Clays Clay Miner.*, 25:457-462.

Kolarik, Z. (1962) *Coll. Czech. Chem. Commun.*, 27:951-958.

Kozawa, A. and Takai, T. (1978) in *Surface Electrochemistry, Advanced Methods and Concepts* (T. Tokamura and A. Kozawa, eds.). Scientific Society Press, Japan, pp. 85-113.

Kuo, S. and Baker, A. S. (1980) *Soil Sci. Soc. Am. J.*, 44:969-974.

Kurbatov, M. H., Wood, G. B., and Kurbatov, J. D. (1951) *J. Phys. Chem.*, 55:1170-1182.

Leckie, J. O. and James, R. O. (1974) in *Aqueous-Environmental Chemistry of Metals* (A. J. Rubin, ed.). Ann Arbor Science, Ann Arbor, MI, pp. 1-76.

Loganathan, P. and Burns, R. G. (1973) *Geochim. Cosmochim. Acta*, 37:1277-1293.

Lyklema, J. (1968) *J. Electroanal. Chem.*, 18:341-348.

Lyklema, J. (1984) *J. Colloid Interface Sci.*, 99:109-117.

MacNaughton, M. G. and James, R. O. (1974) *J. Colloid Interface Sci.*, 47:431-440.

Madrid, L., Diaz-Banientos, E., and Contreras, M. C. (1991) *Aust. J. Soil*

Res., 29:239–247.

Marsh, K. B., Tillman, R. W., and Syers, J. K. (1987) *Soil Sci. Soc. Am. J.,* 51:318–323.

Marshall, C. E. (1964) *The Physical Chemistry and Mineralogy of Soils.* vol. I. John Wiley & Sons, New York.

Mattson, S. (1932) *Soil Sci.,* 34:209–240.

McBride, M. B. and Blasiak, J. J. (1979) *Soil Sci. Soc. Am. J.,* 43:866–870.

McBride, M. B. (1978) *Soil Sci. Soc. Am. J.,* 42:27–31.

McBride, M. B. (1982) *Clays Clay Miner.,* 30:21–28.

McBride, M. B. (1989) *Adv. Soil Sci.,* 10:1–56.

McBride, M. B. and Mortland, M. M. (1974) *Soil Sci. Soc. Am. Proc.,* 38:408–415.

McKenzie, R. M. (1978) *Aust. J. Soil Res.,* 16:209–214.

McKenzie, R. M. (1979) *Geochim. Cosmochim. Acta,* 43:1855–1857.

McKenzie, R. M. (1980) *Aust. J. Soil Res.,* 18:61–73.

McKenzie, R. M. (1989) in *Minerals in Soil Environments* (J. B. Dixon and S. B. Weed, eds.). Soil Science Society of America, Madison, WI, pp. 439–465.

McLaren, R. G. and Crawford, D. V. (1973) *J. Soil Sci.,* 24:443–452.

Morgan, J. J. and Stumm, W. (1964) *J. Colloid Sci.,* 19:347–359.

Murray, J. W. (1975a) *Geochim. Cosmochim. Acta,* 39:505–519.

Murray, J. W. (1975b) *Geochim. Cosmochim. Acta,* 39:635–647.

Naidu, R., deLacy, N. J., Bolan, N. S., Kookana, R. S., and Tiller, K. G. (1994) *Trans. 15th World Congr. Soil Sci.* 3b:190–191.

Norrish, K. (1975) in *Trace Elements in Soil–Plant–Animal Systems* (D. J. D. Nicholas and A. R. Egan, eds.). Academic Press, New York, pp. 55–81.

Padmanabham, M. (1983) *Aust. J. Soil Res.,* 21:309–320.

Perram, J. W., Hunter, R. J., and Wright, H. J. L. (1974) *Aust. J. Chem.,* 27:461–475.

Petrezelli, G., Guide, G. and Lubran, L. (1985) *Commun. Soil Sci. Plant Anal.,* 16:971–986.

Posselt, H. A., Anderson, F. J., and Weber, W. J. (1968) *Environ. Sci. Technol.,* 2:1087–1093.

Quirk, J. P. and Posner, A. M. (1975) in *Trace Element Soil–Plant–Animal Systems* (D. J. D. Nicholas and A. R. Egan, eds.). Academic Press, New York, pp. 97–107.

Ran, Y. and Liu, Z. (1992) *Pedosphere,* 2:13–22.

Schlindler, P. W. (1981) in *Adsorption of Inorganics at Solid–Liquid Interfaces* (M. A. Anderson and A. J. Rubin, eds.). Ann Arbor Science, Ann Arbor, MI, pp. 1–49.

Schulthess, C. P. and Sparks, D. L. (1988) *Soil Sci. Soc. Am. J.,* 52:92–97.

Schwertmann, U. and Taylor, R. M. (1989) in *Minerals in Soil Environments*

(J. B. Dixon and S. B. Weed, eds.). Soil Science Society of America, Madison, WI, pp. 379–438.

Shuman, L. M. (1977) *Soil Sci. Soc. Am. J.*, 41:703–706.

Sidhu, P. S., Sehgal, J. S., Sinha, M. K., and Randhawa, N. S. (1977) *Geoderma*, 18:241–249.

Sposito, G. (1984) *The Surface Chemistry of Soils*. Oxford University Press, New York.

Strahl, R. S. and James, B. R. (1991) *Soil Sci. Soc. Am. J.*, 55:1287–1290.

Stumm, W., Hohl, He, and Dalang, F. (1976) *Croatica Chem. Acta*, 48:491–504.

Stumm, W. and Morgan, J. (1970) *Aquatic Chemistry*. John Wiley & Sons, New York.

Sun, H. Y. (1993a) *Pedosphere*, 3:23–34.

Sun, H. Y. (1993b) *Pedosphere*, 3:239–246.

Taylor, R. M. and McKenzie, R. M. (1966) *Aust. J. Soil Res.*, 4:29–39.

Tewari, P. H. and Lee, W. (1975) *J. Colloid Interface Sci.*, 52:77–88.

Tiller, K. G., Gerth, J., and Brümmer, G. (1984) *Geoderma*, 34:17–35.

Wada, K. (1989) in *Minerals in Soil Environments* (J. B. Dixon and S. B. Weed, eds.). Soil Science Society of America, Madison, WI, pp. 1051–1087.

Wakatsuki, T., Furukawa, H., and Kawaguchi, K. (1974) *Soil Sci. Plant Nutr.*, 20:353–362.

Wright, H. J. L. and Hunter, R. J. (1973) *Aust. J. Chem.*, 26:1191–1206.

Zhang, H. and Zhang, X. N. (1991) *Pedosphere*, 1:41–60.

6

COORDINATION ADSORPTION OF ANIONS

G. Y. Zhang and T. R. Yu

In Chapter 4, when the electrostatic adsorption of anions by variable charge soils is discussed, another type of adsorption, specific adsorption, has already been mentioned, although it is not very remarkable for chloride ions and nitrate ions. For some other anions, specific adsorption can be very important. Specific adsorption is determined by the nature of the anions and is also related to the kind of functional groups on the surface of soils. In general, this type of adsorption is more pronounced in soils containing large amounts of iron and aluminum oxides. Therefore, specific adsorption of anions is one of the important characteristics of variable charge soils.

Specific adsorption is a common term. For anions, the mechanism of specific adsorption is ligand exchange between these ions and some groups that have already been coordinately linked on the surface of soil particles. Therefore, the term coordination adsorption may be more appropriate than the term specific adsorption.

For variable charge soils, phosphate is the strongest specifically adsorbed anion species. Phosphate adsorption is also the most intensively studied anion adsorption in soil science. However, the valence status of phosphate ions is apt to change with the change in environmental conditions. In the adsorption of phosphate by soils, in addition to ligand exchange, other mechanisms, such as chemical precipitation, may also be involved. Therefore, the phenomenon of phosphate adsorption is rather complex, and it is often difficult to make definitive interpretations of experimental results.

In the present chapter, the coordination adsorption of anions will be discussed, mainly taking sulfate as the example, because sulfate is only secondary to phosphate in importance for agricultural production among anions capable of undergoing coordination adsorption. For the purpose of comparison, the adsorption of fluoride ions will also be mentioned.

6.1 PRINCIPLES OF COORDINATION ADSORPTION OF ANIONS

6.1.1 Causes of Coordination Adsorption of Anions in Soils

On the surface of soil particles there are functional groups such as hydroxyl groups (M-OH) and water molecules (M-OH$_2$) that can participate in ligand exchange with anions. Al-OH, Fe-OH, Al-OH$_2$, and Fe-OH$_2$ groups on the surface of soil particles are the important sites for coordination adsorption of anions. Therefore, when a soil contains large amounts of iron and aluminum oxides, the phenomenon of coordination adsorption of anions will be more pronounced.

Besides surface hydroxyl groups and surface water molecules, -ol [-M(OH)$_2$M-] groups and oxy-[-M(O)$_2$M-] groups can also function as ligands in ligand exchange reactions. However, in this case it is necessary for the surface of the oxides to have ruptures so that anions can permeate, or one bond of the bridging -ol group or oxy- group is broken to produce hydroxyl group or coordinated water molecule, and then they participate in ligand exchange with anions.

6.1.2 Difference Between Coordination Adsorption and Electrostatic Adsorption

Anion adsorption in soils can be classified into two types, namely, electrostatic adsorption and coordination adsorption. Anions adsorbed electrostatically are subject to attractive force of positive charges and repulsive force of negative charges carried by the soil simultaneously, and they stay in the outer layer of the electric double layer through dynamic balance between electrostatic attraction and thermal motion. Electrostatic adsorption is reversible, is stoichiometric in exchange, and obeys the mass action law. During this type of adsorption there is neither electron transfer nor the sharing of electron pair between anions and adsorption sites on the surface of solid phase.

Anions in coordination adsorption are not merely controlled by electrostatic forces. They can be adsorbed not only on positive charge surfaces, but also on negative charge surfaces or zero charge surfaces. They can enter the coordination shell of the metal atoms on the surface of the oxides, recoordinating with hydroxy groups or water molecules of the shell, and are combined on the surface directly through covalent bond or coordination bond (Hingston et al., 1967; Ryden et al., 1977a,b). Coordination adsorption occurs in the inner layer of the electric double layer or the Stern layer, depending on the model being used. Anions in coordination adsorption are nonexchangeable. They cannot be replaced by those ions capable of electrostatic adsorption under conditions of identical ionic strength and pH (Hingston et al., 1968). They can only be replaced or partially replaced by anions possessing a stronger adsorptive property.

Coordination adsorption of anions can lead to a decrease in positive surface charge and an increase in negative surface charge of the soil, as well as a rise in pH of the medium. Under certain circumstances the anions coordinately adsorbed can become potential-determining ions on the surface of the oxides (Ryden et al., 1977a,b).

However, the distinction between electrostatic adsorption and coordination adsorption is not absolute. It is not possible to distinguish the adsorption of a given anion species sharply based on these two mechanisms in ordinary experiments. Most often, in actual cases, the two mechanisms occur simultaneously during an anion adsorption, with the relative importance between them depending on concrete conditions.

6.1.3 Mechanism of Coordination Adsorption on the Surface of Metal Oxides

The mechanisms of coordination adsorption of anions on metal oxides can be treated from two aspects: the electric double layer model (Curtin and Syers, 1990a–c) and the change in surface structure of the oxides. The electric double layer model describes the distribution of ions in the region adjacent to the surface in a surface–ion–solution system after making some reasonable assumptions. When treating the change in surface structure of the oxides, concerns are mainly on the adsorption sites, the exchange between two ligands, and the influence on surface properties.

Barrow (1985) proposed a four-layer model for ion adsorption. This model is derived from the three-layer model by making two modifications. First, in the new model a new layer is introduced. However, the position of this layer is not definitive. Because of differences in geometric factors and affinity, some anion species, such as fluoride, can approach the surface more closely than do other anion species such as sulfate. If the mean position after adsorption is also different among various anion species, these ions can be further distinguished into different sublayers. For example, SO_4^{2-}, F^- and PO_4^{3-} ions may be considered as staying in different sublayers. Second, in the derivation of the adsorption equation the existing form of adsorption sites are purely imaginary.

In the derivation of the model, the neutral sites can be considered as $M(OH)(OH_2)$. Therefore, the possible reaction may be written as

$$\left[M \genfrac{}{}{0pt}{}{\diagup OH_2}{\diagdown OH_2} \right]^+ \rightleftharpoons \left[M \genfrac{}{}{0pt}{}{\diagup OH}{\diagdown OH_2} \right]^0 + H^+ \rightleftharpoons \left[M \genfrac{}{}{0pt}{}{\diagup OH}{\diagdown OH} \right]^- + 2H^+ \qquad (6\text{-}1)$$

This assumption is important when examining the ligand exchange reaction between surface groups and ions.

When considering the change in surface structure of the oxides, for monovalent anions such as F^- ions, the ligand exchange reactions at positive charge sites and zero charge sites on the surface of oxides may be classified as follows:

$$\left[\genfrac{}{}{0pt}{}{M-OH_2}{M-OH} \right]^+ + F^- \rightarrow \left[\genfrac{}{}{0pt}{}{M-OH_2}{M-F} \right]^+ + OH^- \qquad (6\text{-}2)$$

$$\left[\genfrac{}{}{0pt}{}{M-OH_2}{M-OH} \right]^+ + F^- \rightarrow \left[\genfrac{}{}{0pt}{}{M-F}{M-OH} \right]^0 + H_2O \qquad (6\text{-}3)$$

$$\left[\genfrac{}{}{0pt}{}{M-OH_2}{M-OH} \right]^+ + 2F^- \rightarrow \left[\genfrac{}{}{0pt}{}{M-F}{M-F} \right]^0 + OH^- + H_2O \qquad (6\text{-}4)$$

$$\left[M \genfrac{}{}{0pt}{}{\diagup OH_2}{\diagdown OH} \right]^0 + F^- \rightarrow \left[M \genfrac{}{}{0pt}{}{\diagup OH_2}{\diagdown F} \right]^0 + OH^- \qquad (6\text{-}5)$$

$$\left[M \genfrac{}{}{0pt}{}{\diagup OH_2}{\diagdown OH} \right]^0 + F^- \rightarrow \left[M \genfrac{}{}{0pt}{}{\diagup F}{\diagdown OH} \right]^- + H_2O \qquad (6\text{-}6)$$

For divalent anions such as SO_4^{2-} ions, the ligand exchange reactions may be classified as follows (Parfitt and Smart, 1978; Rajan, 1978; Ryden et al., 1977b):

$$\left. \begin{array}{l} M-OH_2 \\ \\ M-OH \end{array} \right]^+ + SO_4^{2-} \rightarrow \left. \begin{array}{l} M-OH_2 \\ \\ M-SO_4 \end{array} \right]^0 + OH^- \qquad (6\text{-}7)$$

$$\left. \begin{array}{l} M-OH_2 \\ \\ M-OH \end{array} \right]^+ + SO_4^{2-} \rightarrow \left. \begin{array}{l} M-SO_4 \\ \\ M-OH \end{array} \right]^- + H_2O \qquad (6\text{-}8)$$

$$\left. \begin{array}{l} M-OH_2 \\ \\ M-OH \end{array} \right]^+ + SO_4^{2-} \rightarrow \left. \begin{array}{c} M-O \quad O \\ \diagdown S \diagup \\ M-O \quad O \end{array} \right]^0 + OH^- + H_2O \qquad (6\text{-}9)$$

$$\left. M \begin{array}{l} {}^{\diagup OH_2} \\ {}_{\diagdown OH} \end{array} \right]^0 + SO_4^{2-} \rightarrow \left. M \begin{array}{l} {}^{\diagup OH_2} \\ {}_{\diagdown SO_4} \end{array} \right]^- + OH^- \qquad (6\text{-}10)$$

$$\left. M \begin{array}{l} {}^{\diagup OH_2} \\ {}_{\diagdown OH} \end{array} \right]^0 + SO_4^{2-} \rightarrow \left. M \begin{array}{l} {}^{\diagup SO_4} \\ {}_{\diagdown OH} \end{array} \right]^{2-} + H_2O \qquad (6\text{-}11)$$

$$\left. \begin{array}{l} M-OH \\ \\ M-OH \end{array} \right]^0 + SO_4^{2-} \rightarrow \left. \begin{array}{c} M-O \quad O \\ \diagdown S \diagup \\ M-O \quad O \end{array} \right]^0 + 2OH^- \qquad (6\text{-}12)$$

In addition to the characteristics of the surface of the oxides and the nature of the anions, coordination adsorption is also affected by the concentration of the anions, the kind of supporting electrolytes, the pH of the system, and so on. Taking the adsorption of phosphate on goethite as an example, it has been suggested that the bridging complexes formed by phosphate on the surface are ionized at high pH, which renders the surface to carry negative charge, and are protonated at low pH, so that the surface carries positive charges (Parfitt and Atkinson, 1976). When the pH is high, phosphate ions will replace surface $-OH$ groups and transfer more negative charges to the surface. When the pH is low, phosphate ions will replace $-OH_2$ groups as well as $-OH$ groups, neutralizing a part of positive surface charges. Thus, it is obvious that pH has a significant effect on the pathway of ligand exchange of anions and the surface properties of the solid.

6.2 COORDINATION ADSORPTION BY SOIL COMPONENTS

Because the composition of soils is very complex, in order to elucidate the mechanism of coordination adsorption of anions by soils, most of the studies in soil science were conducted with some soil components, including iron and aluminum oxides. This way, interference by other components of the soil can be avoided. The information and conclusions obtained from this kind of research have provided a basis for the study of coordination adsorption of anions by soils. In the following sections, the coordination adsorption of anions by these soil components shall be examined.

6.2.1 Goethite

Synthesized goethite is crystalline with well–defined crystal surfaces. On the (001) plane of goethite there are three types of –OH groups, denoted as A, B, and C, which can coordinate with one, three, and two iron ions (shown as o in the structure below), respectively (Parfitt, 1978):

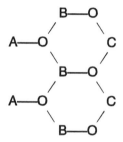

It is generally believed that only the A–type –OH group capable of coordinating with one iron ion can participate in ligand exchange, while the B– and C–type –OH groups can form hydrogen bonds with the adsorbing ligands. When anions phosphate, sulfate, and fluoride ions are adsorbed on goethite through coordination adsorption, they can exchange with all of the A– type –OH groups, while oxalate and halide ions can only exchange with these groups partially (Parfitt et al., 1976; Parfitt and Russell, 1977; Parfitt and Smart, 1978).

Among anion adsorption by goethite, the adsorption of phosphate is the most extensively studied. The adsorption proceeds with a fast reaction and is followed by a slow reaction (Torrent et al., 1992). The exchange rate of ^{32}P on goethite adsorbed with phosphate ions is rather low. Therefore, under ordinary experimental conditions, adsorbed phosphate ions can be distinguished as two types: isotopically exchangeable and isotopically nonexchangeable. The amount of the latter type of phosphate is not affected by the concentration of phosphate ions in the equilibrium solution.

When the coverage of phosphate ions on the surface of goethite is low, most of the adsorbed phosphate ions are isotopically nonexchangeable (Ji, 1986). When $H_2PO_4^-$ ions exchange with A–type –OH groups, a binuclear bridging complex [FeOP(O)$_2$OFe], but not monodentate complex [FeOP(O-)$_3$], is formed through oxolinkage (Atkinson et al., 1972, 1974; Parfitt et al., 1976). This mechanism is consistent with the experimental results that phosphate ions are strongly adsorbed (Yates and Healy, 1975) and cannot be leached out from goethite (Hingston et al., 1974).

The adsorption of sulfate ions proceeds rapidly (Zhang and Sparks, 1990) and is highly irreversible (Turner and Kramer, 1992). Sulfate ions can replace all the A–type –OH groups on the surface of goethite and can form mononuclear complex or binuclear bridging complex [FeOS(O)$_2$OFe] through the replacement of two A–type –OH groups by the oxygen atom of one ligand (Parfitt and Russell, 1977; Turner and Kramer, 1990), with binuclear bridging decreasing in importance as pH is lowered. This latter kind of complex can also be formed on other iron oxides such as hematite and lepidocrocite (Parfitt and Smart, 1978; Turner and Kramer, 1990).

Fluoride ions can also be adsorbed on goethite through the ligand exchange reaction (Bowden et al., 1974; Hingston et al., 1974). While fluoride ions are capable of replacing A–type –OH groups, they are inert to B–type and C–type groups (Parfitt et al., 1976). Unlike phosphate ions and sulfate ions, fluoride ions cannot form a bridging complex but can only form a monodentate complex. This may be the reason why the adsorption of fluoride ions is reversible (Hingston et al., 1974).

If there is a competitive adsorption between two anion species capable of participating in coordination adsorption, the relative quantities of adsorption of each anion species is determined by the nature of the ions and their relative concentrations as well as experimental conditions including sequence of adsorption. Taking the competitive adsorption between phosphate ions and fluoride ions as an example, if phosphate ions are adsorbed prior to fluoride adsorption or the two kinds of ions are adsorbed simultaneously, the presence of fluoride ions will not affect the adsorption of phosphate ions, even though the amount of fluoride ions exceeds the maximum adsorption capacity of goethite calculated from the ligand exchange theory. On the other hand, if fluoride ions are adsorbed prior to phosphate adsorption, the amount of adsorbed phosphate will be slightly lower than that in the absence of fluoride ions. There is also the tendency that the more fluoride ions present in the system, the less the adsorption of phosphate ions (Fig. 6.1). This indicates that the adsorption energy of phosphate ions on goethite is much stronger than that of fluoride ions. It also implies that on the surface of goethite there are certain adsorption sites that can be occupied by fluoride ions, interfering with the adsorption of phosphate ions.

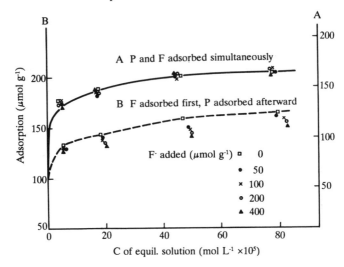

Fig. 6.1. Effect of sequence of fluoride adsorption and phosphate adsorption on adsorption isotherm of phosphate ions (Ji, 1986).

Conversely, adsorption sequence of the two ion species may have remarkable influence on the adsorption of fluoride ions. Figure 6.2 shows this point clearly. It can be seen from the figure that, at a given fluoride to

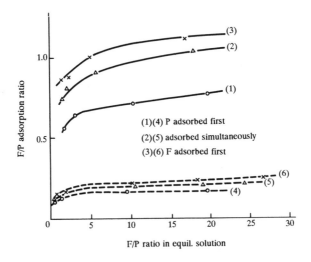

Fig. 6.2. Effect of sequence of adsorption on adsorption ratio F/P (Ji, 1986). (Solid line: pH=3.2; amount of fluoride added, 400 μmol g^{-1}, dash line: pH=4.2; amount of fluoride added, 50 μmol g^{-1}.)

phosphate ratio in equilibrium solution, the adsorbed amount of fluoride ions is the highest when they are adsorbed prior to phosphate adsorption, intermediate when the two kinds of ions are adsorbed simultaneously, and the lowest when phosphate ions are adsorbed prior to fluoride adsorption.

When phosphate ions and fluoride ions are adsorbed competitively, the sum of adsorbed quantities can exceed the quantity of A–type –OH groups calculated according to ligand exchange theory (Ji, 1986). It is still not known as to whether coordinated water molecules and even bridging hydroxyl groups can participate in ligand exchange reactions.

Other anions such as silicate, molybdate, arsenate, borate and certain organic anions such as benzoate can also be adsorbed on goethite through ligand exchange.

6.2.2 Hematite

The surface of an ideal crystalline hematite consists of hydroxyl groups coordinated with one Fe atom and those coordinated with two Fe atoms (Parfitt et al., 1975). The actual degree of order of the surface of hematite is determined by the method of preparation. Unlike the case for the surface of goethite, after the disappearance of hydroxyl groups this kind of group can be regenerated if hematite is heated to 500⁰C and exposed to water vapor (Jurinak, 1966).

Phosphate ions can undergo ligand exchange with surface hydroxyl groups of hematite, forming a mondentate complex:

$$\text{Fe-OH}]^0 + \left[\begin{array}{c} \text{O OH} \\ \| / \\ \text{O-P} \\ \backslash \\ \text{OH} \end{array} \right]^{-} \rightarrow \left[\begin{array}{c} \text{O OH} \\ \| / \\ \text{Fe-O-P} \\ \backslash \\ \text{OH} \end{array} \right]^{0} + \text{OH}^{-} \quad (6\text{-}13)$$

There are also reports that phosphate ion may form binuclear bridging complexes on hematite (Parfitt et al., 1975). Hematite is slightly permeable to anions. The manner of adsorption of phosphate ions on hematite may be different at different pH. Some authors thought that $H_2PO_4^-$ ions replace OH_2 at low pH and replace OH^- groups at high pH (Breeumsma and Lyklema, 1973).

Sulfate ions can also be adsorbed on hematite through ligand exchange (Aylmore et al., 1967; Breeumsma and Lyklema, 1973). It is suggested (Turner and Kramer, 1990) that in the adsorption one to two sulfate ions exchange with hydroxyl groups to form mononuclear and binuclear bridging complexes, with binuclear exchange increasing in importance as pH is raised.

6.2.3 Aluminum Hydroxides

The situation in the adsorption of anions by aluminum hydroxides is more complex than that by iron oxides. The surface of synthesized gibbsite is constructed of regular hexagonal crystals. In an ideal structure, on the edge face every Al^{3+} ion coordinates with one H_2O molecule and one $-OH$ group, while on the (001) face every $-OH$ group coordinates with two Al^{3+} ions below the surface layer.

The adsorption of phosphate ions from dilute phosphate solutions by gibbsite occurs on the edge face. Earlier, it was thought that phosphate ions are adsorbed in the form of monodentate complex. It was found later that they can also be adsorbed in the form of bridging complexes (Kyle et al., 1975). Parfitt et al. (1977) observed that $Al(OH)(H_2O)$ on the edge face of gibbsite can adsorb phosphate ions and oxalate ions strongly, forming binuclear or bidentate complexes.

The mechanism of the adsorption of phosphate by aluminum hydroxide is related to the concentration of phosphate ions. At low concentrations, phosphate ions replace water molecules on the surface of aluminum hydroxide, leading to a decrease in positive surface charge. With the increase in phosphate concentration, Al-OH groups become the principal adsorbing sites gradually, and the proportion of replaced hydroxyl groups becomes larger. When the concentration is further increased, bonds of bridging hydroxyl groups (Al-OH-Al) can break, producing new adsorbing sites. Thus, phosphate ions can undergo ligand exchange with the bridging hydroxyl groups. These three mechanisms can be expressed as follows (Rajan et al., 1974; Rajan 1975, 1976):

$$Al-OH_2]^+ + H_2PO_4^- \rightarrow Al-OPO(OH)_2]^0 + H_2O \qquad (6-14)$$

$$Al-OH]^0 + H_2PO_4^- \rightarrow Al-OPO(OH)_2]^0 + OH^- \qquad (6-15)$$

$$
\begin{array}{ll}
Al\cdots H_2PO_4^- & Al-OPO(OH)_2]^{\frac{1}{2}-} \\
\quad\backslash & \\
\quad OH]^0 & \rightarrow \\
\quad/ & \\
Al & Al-OH]^{\frac{1}{2}-} + H^+ \rightarrow \\
Al-OH_2]^{\frac{1}{2}+} + H_2PO_4^- \rightarrow Al-OPO(OH)_2^{\frac{1}{2}-} + H_2O
\end{array}
\qquad (6-16)
$$

In the above reactions, the relationship between the release of OH^- and the adsorption of phosphate varies with the quantity of adsorbed phosphate ions. According to the experimental results obtained by Rajan (1976), the dOH^-/dP value was 0.11 initially, attained 1.0 when the amount of adsorbed phosphate reached 900 μmol g^{-1}, but decreased sharply to 0.5 and then tended to become steady when the amount of adsorbed phosphate was further increased. These results indicate the possibility of the functioning of the three mechanisms mentioned above. However, it is difficult to interpret the adsorption of phosphate ions precisely, because both phosphate ions and aluminum hydroxides can accept or release protons. Under a given condition, when phosphate ions form binuclear bridging complexes on the surface of aluminum oxides, the dOH^-/dP value is 2 (Parfitt, 1978).

The mechanisms for the adsorption of phosphate ions presented above can also be applied to the interpretation of the adsorption of sulfate ions (Rajan, 1978). SO_4^{2-} ions can replace water molecules from positive charge sites and replace OH^- ions from zero charge sites, leading to an increase in negative surface charge:

$$\left[\begin{array}{c} OH_2 \\ / \\ Al \\ \backslash \\ OH_2 \end{array}\right]^+ + SO_4^{2-} \rightleftharpoons \left[\begin{array}{c} OH_2 \\ / \\ Al \\ \backslash \\ SO_4 \end{array}\right]^- + H_2O \qquad (6\text{-}17)$$

$$\left[\begin{array}{c} OH_2 \\ / \\ Al \\ \backslash \\ OH \end{array}\right]^0 + SO_4^{2-} \rightleftharpoons \left[\begin{array}{c} OH_2 \\ / \\ Al \\ \backslash \\ SO_4 \end{array}\right]^- + OH^- \qquad (6\text{-}18)$$

Adsorbed SO_4^{2-} ions can further replace the OH group of an adjacent neutral aluminum–oxygen group or the $-OH_2$ group of a positively charged aluminum–oxygen group, forming a six–member ring. In the two cases, the quantity of negative surface charge may either decrease or remain unchange.

Thus, in the formation of a six–member ring structure, one SO_4^{2-} ion replaces either two water molecules or two hydroxyl groups, or one water molecule and one hydroxyl group.

Rajan (1978) studied the adsorption mechanism of sulfate ions on poorly crystallized pseudoboehmite [AlO(OH)] and bayerite [Al(OH)$_3$]. The adsorption of sulfate ions proceeded rapidly, attaining 98% of reaction within 10 minutes. The adsorption led to a release of OH^- from the surface of the aluminum oxides. The $d(OH)/dSO_4^{2-}$ value started at 0.3 when the amount of adsorbed sulfate was 25 μmol g^{-1}, and it increased to a maximum 0.7 when the adsorbed amount was 400 μmol g^{-1}. This can be interpreted as follows: At low concentrations, SO_4^{2-} ions were preferentially adsorbed on positive charge sites, replacing OH_2. Following the increase in surface coverage, the proportion of the replacement of OH^- groups on zero charge sites increased, and as a result the value $d(OH)/dSO_4^{2-}$ increased.

6.2.4 Amorphous Hydroxides

The surface of amorphous hydroxides of iron and aluminum is not as definitive and clear as the surface of crystallized hydrated oxides. When phosphate ions and sulfate ions are adsorbed on amorphous $Fe(OH)_3$, a binuclear bridging complex can be formed (Parfitt et al., 1975; Parfitt and Smart, 1978).

Rajan (1979) has studied the adsorption and desorption of sulfate ions on hydrated aluminum oxide and synthetic and natural allophanes. Hydrated aluminum oxide and synthetic allophane still carried positive surface charges even though they were nearly saturated by sulfate ions. By contrast, the positive surface charges of natural allophane could be neutralized by the negative charge of adsorbed sulfate ions entirely. On hydrated aluminum oxides, almost all adsorbed sulfate ions formed a six-member ring. On synthetic and natural allophanes, 60% of the adsorbed sulfate ions formed such rings, leaving other ions forming single-bond complexes and rendering the surface negatively charged.

6.2.5 Clay Minerals

Phosphate ions can exchange with Al-OH groups on the edge of clay minerals (Parfitt, 1978). Because the distance between two adjacent Al-OH groups is 0.296 nm, some phosphate ions can be fixed on kaolinite in the form of a binuclear complex (Kafkafi et al., 1967). The exchange with Al-H$_2$O groups on kaolinite is also possible (Kuo and Lotse, 1972).

When there are Al-OH and Fe-OH groups on the surface of mica, the adsorption of phosphate ions and sulfate ions is enhanced (Perrott et al., 1974a,b).

6.3 COORDINATION ADSORPTION BY SOILS

The composition of soils is much more complex than that of natural or synthesized soil components. Also, when these components are present in soils their composition, structure, and properties may be different from those when they exist alone. Therefore, the actual mechanisms for the adsorption of anions by soils may be similar to those by certain soil components in some respects and may be different in other respects. This is the main point that is of concern in this chapter. In the following sections, based on experimental results, this question shall be examined.

6.3.1 Anion Adsorption in Relation to Soil Type

The adsorption of sulfate ions by variable charge soils may be due to two reasons. One is electrostatic attraction by positive surface charge of the soil, and the other is ligand exchange between SO_4^{2-} ions and hydroxyl groups or water molecules on soil surface. The latter mechanism is of interest in the present chapter. In order to avoid complexities in the interpretation of experimental results caused by electrostatic attraction, research on the coordination adsorption of sulfate ions by soils is frequently carried out with solutions containing a certain concentration of salt, such as 0.2 M NaCl. It can be calculated that under such conditions the adsorption of SO_4^{2-} ions caused by electrostatic attraction should account for less than 1% of the total adsorption in most cases and thus can be neglected. Variable charge soils may adsorb chloride ions specifically to a certain extent (cf. Chapter 4). This may affect the adsorption of sulfate ions to a certain degree. However, if the concentration of chloride ions is kept identical in all the experiments, this effect will not be large enough to invalidate the comparisons among various treatments. Thus, when experimental conditions, such as pH of the solution, electrolyte concentration, accompanying ions and temperature, are kept identical, it would be possible to compare the difference in adsorption of sulfate ions by various types of soils.

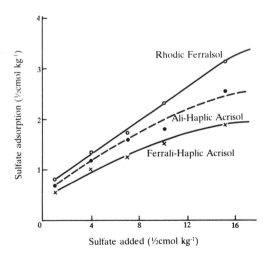

Fig. 6.3. Adsorption isotherm of sulfate ions by three representative variable charge soils of China (pH 5.0) (Zhang et al., 1987).

If the adsorption of sulfate ions by three representative variable charge soils of China (Fig. 6.3) is examined, it can be seen that for all three soils the amount of adsorbed SO_4^{2-} ions increases with the increase in amount of added SO_4^{2-} ions. The Rhodic Ferralsol possesses the largest adsorption capacity. This is related to its mineralogical composition, characteristics in surface charge and clay content, and in particular the high contents of free iron oxides and gibbsite, because SO_4^{2-} ions can be adsorbed through ligand exchange with $M-OH_2$ and $M-OH$ groups on the surface of iron and aluminum oxides. Besides, the high content of clay (77%) is also one important factor in causing the high adsorption capacity.

Figure 6.4 shows the adsorption of sulfate ions by four soils with a wider difference in mineralogical composition. It can be seen from the figure that the Rhodic Ferralsol of Brasilia possesses the largest adsorption capacity. The adsorption isotherm of the Hyper–Rhodic Ferralsol (Kunming) is similar to that of the Brazilian soil when the adsorbed amount is small, but becomes slightly lower than the latter curve when the adsorbed amount is further increased. The adsorption capacity of this soil, derived from paleo–crust, is about twice as large as that of the Rhodic Ferralsol of Xuwen derived from Basalt shown in Fig. 6.3. This soil is the one that possesses the largest adsorption capacity for sulfate ions known in China. The adsorption isotherm of the Rhodic Ferralsol of Hawaii is similar to that of the Ferrali–Haplic Acrisol of Xishuangbanna derived from sand-stone–shale. Both of the soils possess a relatively smaller adsorption capacity.

According to the X–ray diffraction analysis of these four soils, the Hyper–Rhodic Ferralsol contains, besides kaolinite, large amounts of

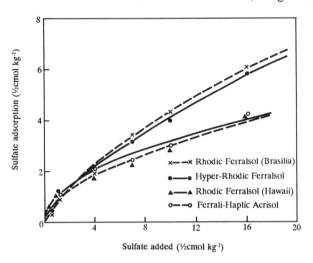

Fig. 6.4. Adsorption of sulfate ions by four soils (pH 5.0).

gibbsite and iron oxides. The diffraction intensity of gibbsite is quite high. This soil contains free iron oxides in amounts as high as 22.3%. It is suggested that the high contents of iron oxides and aluminum oxides of this soil are the principal causes of the large adsorption capacity for sulfate ions. However, iron oxide is only one of the important factors in inducing the adsorption of sulfate ions. For instance, the Rhodic Ferralsol of Brazilia with kaolinite as the principal clay mineral has a free iron oxide content (7.7%) of only one third of that of the Hyper–Rhodic Ferralsol, whereas it possesses a rather large adsorption capacity. It is assumed that this is related to the surface characteristics of this soil. The pH of this soil in a $1M$ KCl solution is 5.75, higher by 0.3 than the pH in water. This is an indication of a large amount of easily replaceable hydroxyl groups. Likewise, the Rhodic Ferralsol of Hawaii has a free iron oxide content (14.8%) about twice as high as that of the Ferrali–Haplic Acrisol (6.6%), yet the two soils have a similar adsorption capacity. Thus, it follows that the adsorption capacity of a soil for sulfate ions cannot be judged merely based on the free iron oxide content.

If the surface area is also taken into account and the adsorption density of sulfate is calculated based on the surface area of the soil, one can examine the relationship between sulfate adsorption and soil type from another viewpoint (Fig. 6.5). The difference in adsorption density among the four soils is larger than that when the comparison is made on a weight basis. The adsorption density is not always proportional to the amount of sulfate added. Instead, the increment in adsorption density decreases with the increase in coverage of sulfate ions on the surface of soils. This implies that the mechanism involved at low surface coverage may be different from

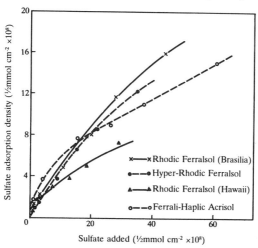

Fig. 6.5. Adsorption density of sulfate ions by four soils (pH 5.0).

that at high surface coverage. Turner and Kramer (1990) made a similar suggestion for goethite and hematite.

6.3.2 Role of Iron Oxides in Coordination Adsorption

Iron oxides is one important component in variable charge soils. They can affect the surface area, surface groups, and surface charges and thus play an important role in the adsorption of anions (Fuller et al., 1985; Guadalix and Pardo, 1991; Johnson and Todd, 1983; Parfitt, 1978).

Figure 6.6 shows that the adsorption of sulfate ions by three soils decreased remarkably after the removal of free iron oxides. The decrease in sulfate adsorption was quite pronounced for the Rhodic Ferralsol of Brasilia, although the amount of iron oxides is not high. This means that iron oxides played a very important role in the adsorption of sulfate ions by this soil. It may be assumed that, in this soil derived from diluvium, the age of the iron oxides is young and their crystallinity is poor, and therefore the form of these oxides is favorable for the adsorption of sulfate ions. For the Rhodic Ferralsol of Hawaii, the decrement in sulfate adsorption was not large after the removal of iron oxides, owing presumably to the good crystalinity of these oxides, although their content is high. X-ray diffraction patterns show that the crystallinity of kaolinite in Rhodic Ferralsol of Hawaii and in Hyper-Rhodic Ferralsol is poorer than that in Rhodic Ferralsol of Brasilia. These poorly crystallized kaolinites may have abundant broken bonds on their edge faces, and should be favorable to ligand exchange reactions. The iron-free Rhodic Ferralsol of Brasilia with well-crystallized kaolinite adsorbed the least amount of sulfate ions.

The specific area of soils decreased to different extent when iron oxides were removed. Consequently, the adsorption density of sulfate ions on the surface of soils changed (Fig. 6.7). Compared to original samples, the adsorption density decreased remarkably for the Rhodic Ferralsol of Brasilia, did not change significantly for the Rhodic Ferralsol of Hawaii, and increased for the Hyper-Rhodic Ferralsol. Such change in adsorption density is caused by the combined effects of the reduction in adsorption and the decrease in surface area after the removal of iron oxides. Since free iron oxides have a large specific surface area (cf. Chapter 2), the removal of these oxides would lead to a decrease in specific surface area of the soil. However, the decrement in specific surface area is not simply proportional to the content of free iron oxide in the soil, but is also related to how iron oxides and clay minerals are combined. These complicated factors are the reasons for the difference in change of adsorption density in three soils from one another when iron oxides were removed.

When iron oxides were removed from the soil, following the decrease in the amount of SO_4^{2-} ions adsorbed, the amount of released OH^- ions also

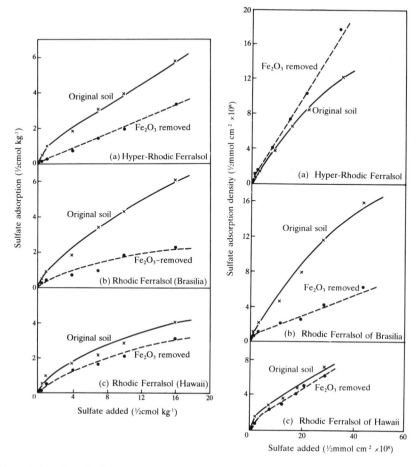

Fig. 6.6. (Left) Adsorption of sulfate ions by soils before and after the removal of iron oxides (pH 5.0) (Zhang et al., 1991).

Fig. 6.7. (Right) Adsorption density of sulfate ions by soils before and after the removal of iron oxides (pH 5.0).

decreased (Fig. 6.8). Such a tendency was most pronounced for the Hyper–Rhodic Ferralsol. This indicates that OH^- ions released from this soil were chiefly provided by Fe–OH of iron oxides.

When comparing the relationship between the release of OH^- ions and the adsorption of SO_4^{2-} ions among the three iron–free soils (Fig. 6.9), it can be seen that the increment in release with the increase in sulfate adsorption for the Rhodic Ferralsol of Brasilia was the largest, even though this soil released a similar amount of OH^- ions as the Hyper–Rhodic

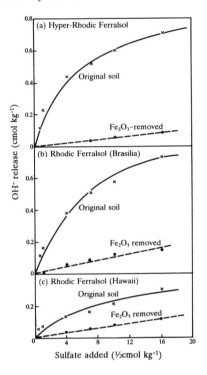

Fig. 6.8. Release of OH⁻ ions during sulfate adsorption by soils before and after the removal of iron oxides (pH 5.0).

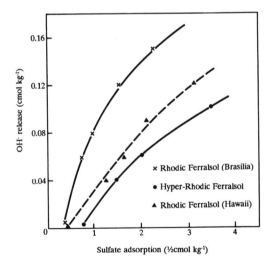

Fig. 6.9. Release of OH⁻ ions during sulfate adsorption by three iron-free soils (pH 5.0).

Ferralsol before the removal of iron oxides. This indicates that aluminum hydroxides and the broken bonds on the edge face of kaolinite in the former soil can provide more hydroxyl groups to undergo ligand exchange with sulfate ions than those in the latter soil. For the Rhodic Ferralsol of Hawaii, the contribution of aluminum hydroxides and clay minerals in providing hydroxyl groups was almost equivalent to that of iron oxides. Possibly, the iron oxides in this soil are mostly in the form of separate particles and cannot provide a large amount of Fe-OH groups for the adsorption of sulfate ions.

6.3.3 Anion Adsorption in Relation to Ion Species

The manner of coordination adsorption of anions is related to both the characteristics of the soil and the nature of the anions. There have been numerous studies on the adsorption of phosphate (Parfitt, 1978; Zhang and Zhang, 1989), and the adsorption is stronger than that of sulfate (Gelbhardt and Coleman, 1974; Marsh et al., 1987; Yates and Healy, 1975). Generally speaking, fluoride ions also have a stronger power for ligand exchange than sulfate ions. Table 6.1 shows the adsorption of fluoride ions and the corresponding release of hydroxyl groups in three variable charge soils. It can be seen from the table that, under conditions when the amounts of ions added were the same, the soils adsorbed more fluoride ions and released more hydroxyl groups than when sulfate ions were adsorbed. This is also true when adsorption occurred at different pH (Table 6.2). Besides, viewed from the percentage of adsorbed amount in total amount of ions, it was found that when the amount of fluoride ions added was less than 8 cmol kg^{-1}, almost all the fluoride ions were adsorbed by the Hyper-Rhodic Ferralsol. When the amount of fluoride ions added reached 32 cmol kg^{-1}, the Rhodic Ferralsol of Brasilia also adsorbed almost all fluoride ions. These results suggest that fluoride ions possess a much stronger ligand exchange power than do sulfate ions. The reason for this is that, except for the advantageous geometric factor in which the diameter of fluoride ions (0.136 nm) is smaller than that of sulfate ions (0.36 nm) but is similar to that of OH$^-$ ions (0.176 nm), there are also other factors involved. This will be discussed latter.

6.3.4 Effect of pH

Soil pH may affect the valence status of ions in solution. For instance, for phosphate ions, the relative proportions among $H_2PO_4^-$, HPO_4^{2-}, and PO_4^{3-} are different at different pH. One more important point is that, for variable charge soils, the change in pH may directly affect the protonation and deprotonation of hydroxyl groups on the surface of oxides, leading to

Table 6.1 Adsorption of Fluoride Ions and Release of Hydroxyl Ions in Variable Charge Soils (pH 5.0) (Zhang et al., 1987)

Soil	F⁻ Added (cmol kg⁻¹)	F⁻ Adsorbed (cmol kg⁻¹)	OH⁻ Released (cmol kg⁻¹)	OH/F
Ali–Haplic	1	0.91	0.95	1.04
Acrisol	4	2.77	2.47	0.89
	7	5.79	5.27	0.91
	10	8.43	6.34	0.75
	20	15.7	8.79	0.56
Ferrali–Haplic	1	0.93	0.80	0.86
Acrisol	2	1.78	1.44	0.81
	4	3.37	2.84	0.84
	7	5.89	5.76	0.98
	14	12.3	7.34	0.60
	20	17.9	7.94	0.42
Rhodic Ferralsol	1	0.99	0.64	0.65
	4	3.92	2.09	0.53
	7	6.74	3.77	0.56
	10	9.31	5.22	0.56
	15	13.5	8.16	0.60
	20	18.8	10.8	0.61

the changes in the ratio between hydroxyl groups and water molecules on the surface of soil inorganic colloids and in surface charge of soils. As a result, the adsorption and desorption of anions and the release of hydroxyl groups will be affected. Therefore, many studies have shown that this type of anion adsorption is strongly pH–dependent (Guadalix and Pardo, 1991; Marsh et al., 1987; Nodwin et al., 1986; Parfitt, 1978; Shanley, 1992; Zhang and Sparks, 1990).

Figures 6.10 and 6.11 show the adsorption of sulfate by three representative variable charge soils of China and another four soils at different pH. It can be seen from the figures that an increase in pH invariably led to a decrease in the adsorbed amount of sulfate. The decrease was more remarkable within the pH range below about 5.5. This indicates that the adsorption mechanisms may be different for different pH ranges. Within the first steeper part of the curve, because the pH is relatively low,

Table 6.2 Effect of pH on Adsorption of Fluoride Ions and Release of
Hydroxyl Ions (Zhang et al., 1987)

F⁻ Added (cmol kg⁻¹)	Soil	pH	F⁻ Adsorbed (cmol kg⁻¹)	OH⁻ Released (cmol kg⁻¹)	OH/F
3.5	Ali–Haplic Acrisol	4.87	2.65	2.18	0.85
		5.77	2.67	2.07	0.77
		6.92	2.94	1.64	0.56
		7.22	3.20	1.38	0.43
		7.62	2.90	0.99	0.34
	Ferrali–Haplic Acrisol	4.45	2.77	2.07	0.75
		5.09	2.84	1.96	0.69
		6.00	3.11	1.85	0.59
		6.77	3.35	1.56	0.46
		7.81	2.87	1.30	0.45
	Rhodic Ferralsol	4.81	3.41	1.78	0.52
		5.30	3.50	1.60	0.46
		5.93	3.50	1.49	0.43
		7.03	3.50	1.27	0.36
		7.83	3.50	1.22	0.35
20	Ali–Haplic Acrisol	4.62	18.7	9.90	0.53
		6.40	19.4	8.30	0.43
		6.88	19.2	7.54	0.39
		7.25	18.7	7.41	0.40
		7.50	18.3	7.44	0.41
	Ferrali–Haplic Acrisol	4.64	18.6	8.40	0.45
		6.56	19.6	8.00	0.41
		7.34	18.0	7.73	0.43
		8.02	17.4	7.14	0.44
	Rhodic Ferralsol	5.74	19.5	7.59	0.39
		6.49	19.5	7.00	0.36
		6.93	18.7	6.32	0.34
		7.26	17.0	6.22	0.37
		7.43	16.8	5.07	0.30

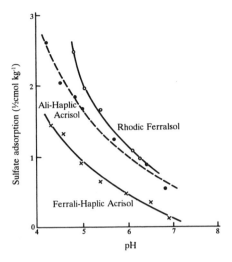

Fig. 6.10. Effect of pH on sulfate adsorption by three representative variable charge soils of China (amount of SO_4^{2-} added is 3.5 cmol kg^{-1}) (Zhang et al., 1987).

the surface of the soil can accept protons and becomes positively charged, leading to the increase in $-OH_2$ groups. Under such circumstances, SO_4^{2-} ions are preferentially adsorbed on positive charge sites to replace $-OH_2$ molecules. Besides, at low pH the surface groups of iron and aluminum

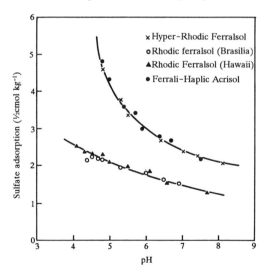

Fig. 6.11. Effect of pH on amount of sulfate adsorbed (amount of SO_4^{2-} added, 50 cmol kg^{-1}).

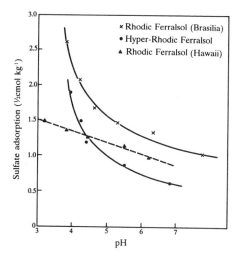

Fig. 6.12. Effect of pH on sulfate adsorption by iron-free soils (amount of SO_4^{2-} added is 50 cmol kg^{-1}).

oxides are more active and are easier to undergo ligand exchange. Following the increase in pH, the proportion of positively charged surface of the oxides would decrease while the proportions of zero charge surface and negatively charged surface would increase. Even though at high pH some water molecules on the surface of oxides may change to surface hydroxyl groups through the release of protons, the energy required for the replacement of surface OH$^-$ groups also increases. Therefore, the overall result is that the amount of sulfate adsorbed at high pH is lower than that at low pH.

For the three iron-free soils, the effect of pH on the adsorption of sulfate ions was similar to that of the original soils (Fig. 6.12). For the Rhodic Ferralsol of Brasilia and the Hyper-Rhodic Ferralsol, the adsorbed amount decreased rather sharply with the rise in pH within the pH range of 3.5–4.5. This indicates that, for these two soils, the number of positively charged sites capable of adsorbing sulfate ions through ligand exchange decreased with the rise in pH in the lower pH range. Thereafter, following the increase in pH, because the energy required for the adsorption of sulfate ions increased, the adsorbed amount decreased. The effect of pH on the adsorption of sulfate ions for the iron-free Rhodic Ferralsol of Hawaii was relatively small, a situation similar to that of the original soil.

The release of OH$^-$ ions decreased with the rise in pH. The relationship between the amount of OH$^-$ released and pH for five soils was approximately linear (Figs. 6.13 and 6.14). When the pH was higher than 7.5–8.0, all the soils did not release OH$^-$ ions, although in this case the soils still

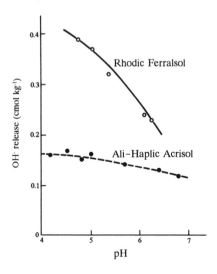

Fig. 6.13. Effect of pH on release of OH⁻ ions (Zhang et al., 1987).

adsorbed large amounts of sulfate ions (cf. Fig. 6.11). It can be calculated from Fig. 6.12 that, for the iron–free Hyper-Rhodic Ferralsol, Rhodic Ferralsol of Brasilia, and Rhodic Ferralsol of Hawaii, the changes in adsorbed amount of SO_4^{2-} ions caused by the change in pH by 1 pH unit were 0.061, 0.043, and 0.014 cmol kg⁻¹, respectively. This suggests a remarkable difference in properties among the three iron–free soils.

When iron oxides were removed from the soils, the release of OH⁻ ions was reduced markedly and also became less affected by pH. This suggests that surface Fe–OH groups of iron oxides are the major groups undergoing ligand exchange with SO_4^{2-} ions because, according to the principle of chemical equilibrium, the lower the pH of the medium, the more favorable the conditions for the release of hydroxyl groups. On the other hand, the release of hydroxyl ions during adsorption of sulfate by iron–free soils was still affected by pH, especially for the Hyper-Rhodic Ferralsol. This indicates that the surface hydroxyl groups of other components of the soil, such as gibbsite, may also participate in ligand exchange. It has already been shown that the Hyper-Rhodic Ferralsol contains a large amount of gibbsite.

6.4 CONSEQUENCES OF COORDINATION ADSORPTION

When anions are adsorbed on the surface of soil colloids, they can undergo ligand exchange reactions with surface ligands including hydroxyl groups and water molecules, leading to the release of these ligands into solution.

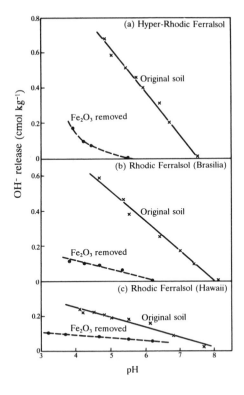

Fig. 6.14. Effect of pH on release of OH⁻ ions during sulfate adsorption from original soil and iron–free soil.

Meanwhile, the surface charge properties of the soil may also change. This change in surface charge will in turn affect the next step of ligand exchange. Therefore, this kind of ion adsorption proceeds under conditions in which the amount of ions and the surface status are undergoing constant change. Thus, the amounts of ions adsorbed and ligands released as well as the change in surface charge as determined in ordinary experiments are average values for a given condition. In the following sections, these consequences of coordination adsorption will be examined.

6.4.1 Release of Hydroxyl Groups

The release of hydroxyl groups is the most important direct consequence of coordination adsorption of anions in variable charge soils. Because the release of hydroxyl groups will lead to the increase in pH of the solution, it is easy to monitor the amount of hydroxyl groups released during the course of reaction, which can be calculated from the amount of hydrogen

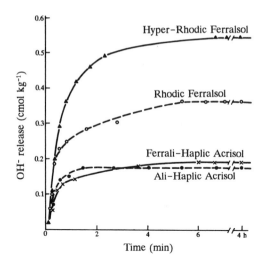

Fig. 6.15. Release of OH⁻ ions with time during sulfate adsorption from four soils (amount of SO_4^{2-} added is 7 cmol kg⁻¹; pH = 5.0).

ions required to neutralize the released OH⁻ ions. Taking the adsorption of sulfate ions as an example, it has been found that the rate of release of OH⁻ ions accompanying the adsorption is very rapid. It can be seen from Fig. 6.15 that within the first minute of reaction the amount of OH⁻ ions released was 70% of the total release, reaching more than 80% within 4 minutes, and the release practically finished at the seventh minute. Rajan (1978), when studying the release of OH⁻ ions caused by the adsorption of SO_4^{2-} ions in hydrated aluminum oxide, found that the releases were 85% in 10 minutes and 95% in 1 hour. Thus, it appears that the release of OH⁻ ions in soils is faster than that of aluminum oxides.

Since the release of OH⁻ ions is caused by the replacement by SO_4^{2-} ions, the ligand exchange reaction between OH⁻ ions and SO_4^{2-} ions would also be very rapid. Hence, the amount of OH⁻ ions released in the first half hour may be taken as the released amount at the steady state of reaction. The data presented in the following sections were obtained under such an experimental condition.

Figures 6.16 and 6.17 show the amount of released OH⁻ ions as a function of the amount of SO_4^{2-} ions adsorbed for three representative variable charge soils of China and four other soils, respectively. It can be seen from the figures that the amount of released hydroxyl ions increased with the increase in the amount of adsorbed SO_4^{2-} ions. When comparing with Figs. 6.3 and 6.4 it can be found that soils that can adsorb more sulfate ions can also release more hydroxyl ions. However, the release of OH⁻ ions was generally not proportional to the adsorption of SO_4^{2-} ions within a wide

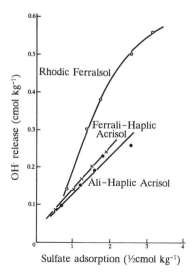

Fig. 6.16. Relationship between release of OH⁻ ions and adsorption of SO₄²⁻ ions for three representative variable charge soils of China (pH 5.0) (Zhang et al., 1987).

range of sulfate adsorbed. This again suggests that different mechanisms may operate for different soils and under different conditions.

The adsorption of fluoride ions caused the release of more OH⁻ ions

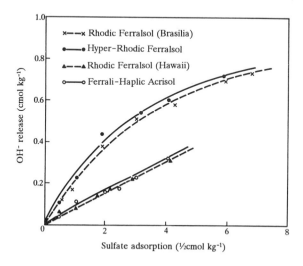

Fig. 6.17. Relationship between release of OH⁻ ions and adsorption of SO₄²⁻ ions for four soils (pH 5.0).

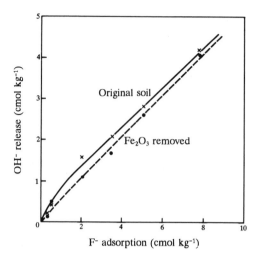

Fig. 6.18. Effect of removal of iron oxides on release of OH⁻ ions during fluoride adsorption for a Hyper–Rhodic Ferralsol (pH 5.0).

than did the adsorption of sulfate ions (Tables 6.1 and 6.2; Figs. 6.18 and 6.19). This suggests that the mechanisms through which the two ion species undergo ligand exchange with variable charge soils may be different.

Another phenomenon deserving notice is that the release of hydroxyl ions was reduced considerably after the removal of iron oxides from the soil,

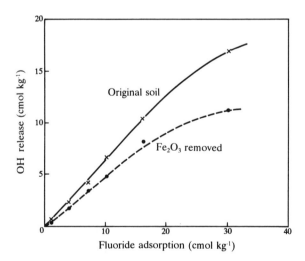

Fig. 6.19. Effect of removal of iron oxides on release of OH⁻ ions during fluoride adsorption for Rhodic Ferralsol of Brasilia (pH 5.0).

although the adsorption of fluoride ions decreased only slightly. This is particularly noticeable for the Rhodic Ferralsol of Brasilia when adsorbing large amounts of fluoride ions (Fig. 6.19). This indicates that when reacting with iron–free soils, fluoride ions may be adsorbed through mechanisms other than ligand exchange with hydroxyl groups.

6.4.2 OH/A Ratio

In this section, the relationship between the amount of released hydroxyl ions (OH) and the amount of adsorbed anions (A) in variable charge soils will be examined in a quantitative way. Such relationship can be used to deduce the mechanisms of adsorption of anions.

When studying the ratio between the amount of hydroxyl ions released and that of SO_4^{2-} ions adsorbed in three representative variable charge soils of China, it was found that the equivalent ratio OH/SO_4 was rather small. As shown in Table 6.3, the average ratios for the Ali–Haplic Acrisol and the Ferrali–Haplic Acrisol were 0.12 and 0.14, respectively. For the Rhodic Ferralsol with a higher content of iron and aluminum oxides, the average ratio was 0.20. For the other four variable charge soils, namely, the Hyper–Rhodic Ferralsol, Rhodic Ferralsol of Brasilia, Rhodic Ferralsol of Hawaii, and Ferrali–Haplic Acrisol of Xishuangbanna, the ratios of OH/SO_4 were also small (Table 6.4), being 0.19, 0.22, 0.085, and 0.087, respectively, which approximately correspond the ratios of 0.12–0.20 given in Table 6.3. However, it is rather rough if the characteristic feature in the adsorption of sulfate ions is examined merely based on the average value of OH/SO_4, because this value is not always a constant. For instance, when the adsorption density of sulfate increased, the ratio may either change very

Table 6.3 OH/SO_4 Equivalent Ratio During Adsorption of Sulfate Ions by Three Soils at pH 5

SO_4^{2-} Added ($\frac{1}{2}$cmol kg^{-1})	OH/SO$_4$		
	Ali–Haplic Acrisol	Ferrali–Haplic Acrisol	Rhodic Ferralsol
1	0.17	0.16	0.18
4	0.10	0.13	0.22
7	0.12	0.14	0.22
10	0.13	0.13	0.19
15	0.10	0.13	0.18
Mean	0.12	0.14	0.20

Table 6.4   Comparison of OH/SO$_4$ Equivalent Ratio Between Fe$_2$O$_3$-
-Removed Soil and Original Soil

| | OH/SO$_4$ | | | | | | | |
| | Soil 1[b] | | Soil 2 | | Soil 3 | | Soil 4 | |
SO$_4^{2-a}$	OS[c]	FR[c]	OS	FR	OS	FR	OS	FR
0.5	0.24				0.42		0.120	
1	0.23		0.18		0.26		0.080	
4	0.24		0.22	0.051	0.22	0.082	0.081	0.033
7	0.17	0.027	0.22	0.049	0.17	0.086	0.078	0.039
10	0.15	0.030	0.19	0.048	0.14	0.064	0.077	0.044
16	0.12	0.026	0.18	0.042	0.12	0.066	0.076	0.039
Mean	0.19	0.028	0.20	0.048	0.22	0.075	0.085	0.039
Rel.	1	0.15	1	0.24	1	0.34	1	0.46

[a]Amount added in unit of ½cmol kg^{-1}.
[b]Soil 1, Hyper–Rhodic Ferralsol; Soil 2, Rhodic Ferralsol of Xuwen;
Soil 3, Rhodic Ferralsol of Brasilia; Soil 4, Rhodic Ferralsol of
Hawaii.
[c]OS, original soil; FR, Fe$_2$O$_3$ removed.

little, as in the cases of Rhodic Ferralsol of Xuwen, Rhodic Ferralsol of
Hawaii, and Ferrali–Haplic Acrisol of Xishuangbanna, or decrease
gradually, as in the cases of Hyper–Rhodic Ferralsol and Rhodic Ferralsol
of Brasilia. The phenomenon that the ratio of released OH$^-$ ions to
adsorbed anions varies with the change in the latter parameter also
occurred when phosphate ions were adsorbed by aluminum hydroxide, in
which the ratio increased from an original value of 0.11 to 1.0, and then
declined to a steady value 0.5 (Rajan et al., 1974; Rajan 1975, 1976), or was
2 under certain conditions (Parfitt, 1978). These results suggest that an
anion species may have different adsorption mechanisms at different
adsorption levels.

From Table 6.4 it can also be seen that the ratio OH/SO$_4$ became
smaller when iron oxides were removed from the soil. For Hyper–Rhodic
Ferralsol, Rhodic Ferralsol of Xuwen, Rhodic Ferralsol of Brasilia, and
Rhodic Ferralsol of Hawaii, the average values of this ratio were 0.028,
0.048, 0.075, and 0.035, respectively, accounting for 15%, 24%, 34%, and
46% of the respective ratios before the removal of iron oxides, respectively.
The extent of decrease in the ratio just correlated with the content of iron

oxides of the soils. These results indicate that, on the one hand, the broken bonds on the edge face of clay minerals such as kaolinite and the surface Al-OH groups of gibbsite can also provide some hydroxyl groups for ligand exchange with sulfate ions, and, on the other hand, the more important point is that iron oxides are the major carriers of hydroxyl groups capable of undergoing ligand exchange with sulfate ions in variable charge soils.

In the past, most of the experimental results on the relationship between the release of OH^- ions and the adsorption of SO_4^{2-} ions were semiquantitative (Chao et al., 1965; Hasan et al., 1970; Mekaru and Uehara, 1972). Rajan (1978, 1979) reported that the ratio $d(OH)_{rel}/d(SO_4)_{ads.}$ for synthetic hydrated aluminum oxide increased from 0.3 when adsorbed SO_4^{2-} was 25 μmol g^{-1}, and it rose to 0.7 when it was 400 μmol g^{-1}. For three variable charge soils of Spain in the presence of a dilute KCl solution, a ratio of about 0.5 was reported (Guadalix and Pardo, 1991). Obviously, from the data presented in this chapter, the ratio is much smaller, and it varies with the type of the soil, particularly with the iron oxides content.

In order to examine the behavior of sulfate ions with respect to adsorption, one can determine and compare the release of OH^- ions during the adsorption of fluoride ions by soils. The data presented in Table 6.5 show that the ratio OH/F for the Rhodic Ferralsol of Brasilia was not affected by the amount of added F^- ions. The average value was 0.61. For the Hyper-Rhodic Ferralsol, the ratio only decreased slightly with the increase in adsorption of F^- ions. The average value was 0.71. These ratios were much larger than the corresponding OH/SO_4 ratios when sulfate ions were adsorbed by the two soils. This indicates that the behavior of SO_4^{2-} ions is quite different from that of F^- ions when undergoing ligand exchange with variable charge soils. Another point deserving attention is that, after the removal of iron oxides from the soils, the extent of decrease in release of OH^- ions during F^- adsorption was much smaller than that when SO_4^{2-} ions were adsorbed. The average values of the ratio OH/F for the two soils were 0.47 and 0.59, respectively. This again indicates that the behaviors of the two anion species are different.

pH can influence the OH/SO_4 ratio. As shown in Fig. 6.20, the ratio decreased with the rise in pH for all of the four soils. This trend was more remarkable for the Hyper-Rhodic Ferralsol and the Rhodic Ferralsol of Brasilia with a larger ratio, especially when the pH was higher than about 6. This can be interpreted as follows. The higher the pH, the higher the concentration of OH^- ions in solution, and thus the less favorable for the release of surface hydroxyl groups. However, other mechanisms involving coordination adsorption of sulfate ions, such as the release of water molecules, may not be affected or less affected by pH. Therefore, the overall result is that the ratio OH/SO_4 decreases with the rise in pH.

Table 6.5 OH/F Ratio During Adsorption of Fluoride Ions by Soils (pH 5.0)

Soil	F⁻ Added (cmol kg⁻¹)	OH/F Original Soil	OH/F Fe₂O₃ Removed
Rhodic Ferralsol	1	0.57	0.42
(Brasilia)	4	0.60	0.45
	7	0.59	0.50
	10	0.67	0.51
	16	0.67	0.54
	32	0.57	0.37
	Mean	0.61	0.47
Hyper–Rhodic	0.25	0.96	0.75
Ferralsol	0.5	0.82	0.67
	2	0.79	0.56
	3.5	0.60	0.49
	5	0.56	0.53
	8	0.53	0.53
	Mean	0.71	0.59

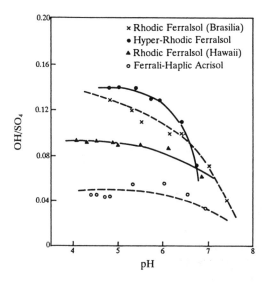

Fig. 6.20. Effect of pH on OH⁻-release/SO₄²⁻-adsorption.

Table 6.6   Change in Surface Charge Properties of a Rhodic Ferralsol (Xuwen) After Adsorption of Sulfate Ions (pH 5.0) (Zhang et al., 1987)

SO$_4^{2-}$ Adsorbed (½cmol kg^{-1})	OH$^-$ Released (cmol kg^{-1})	Decrease in Pos. Charge (cmol kg^{-1})	Increase in Neg. Charge (cmol kg^{-1})	Percentage of SO$_4^{2-}$ Adsorbed		
				OH$^-$	Charge	Sum
0.78	0.14	0.32	0.19	18	65	83
1.43	0.19	0.58	0.44	13	71	84
2.44	0.30		0.68	12		
2.84	0.37		1.11	13		
2.94	0.50		1.43	17		

6.4.3   Change in Surface Charge of Soil

In addition to the release of hydroxyl ions, the change in surface charge of the soil is another important consequence of coordination adsorption of anions. Since surface charge is the basic factor in affecting other surface chemical properties of soils, such change in surface charge of variable charge soils deserves close consideration.

It can be seen from Tables 6.6 and 6.7 that, following the increase in adsorption of sulfate ions, besides the increase in release of hydroxyl groups, the quantity of positive surface charge decreased while that of negative surface charge increased for all three variable charge soils. At pH 5, the contribution of changes in surface charge to the total consequence of sulfate adsorption was much larger than the contribution of the release of hydroxyl ions. This is an interesting phenomenon.

The situation in the adsorption of fluoride ions is different from that of sulfate adsorption. For the same soil, when fluoride ions were adsorbed the contribution of changes in surface charge to the total consequence of adsorption was much smaller than that when sulfate ions were adsorbed. As shown in Table 6.8, for the Rhodic Ferralsol of Brasilia, the contribution of the release of OH$^-$ ions was 60-70%, the increase in negative surface charge was 25-30%, and the decrease in positive surface charge to the total consequence of the adsorption of fluoride ions was 5-25%. The corresponding figures were about 40-50%, 40-60% and 5-15%, respectively, after the removal of free iron oxides. These results suggest that, when fluoride ions were adsorbed, the decrease in positive surface charge only accounted for a small proportion of the total consequence of adsorption. By contrast, this proportion may be as high as 50-60% when sulfate ions were adsorbed by the same soil.

Table 6.7 Change in Surface Charge Properties of Ferralsols After
Adsorption of Sulfate Ions (pH 5.0)

Soil[a]	Adsorption[b]	Change (cmol kg^{-1})			Percentage of Adsorption			
		A[c]	B	C	A	B	C	Sum
Rhodic	1.0	0.08	0.35	0.34	8	35	34	77
Ferralsol	1.72	0.14	0.55	0.85	8	32	49	89
(Hawaii)	2.18	0.17	0.60	1.14	8	27	52	88
	2.85	0.22	0.64	1.16	8	22	41	72
Same soil,	0.30	0	0.07	0.24	0	23	80	103
Fe$_2$O$_3$	0.47	0	0.19	0.29	0	41	62	103
removed	1.22	0.04	0.20	0.59	3	17	48	85
	1.52	0.06	0.26	0.83	4	17	55	76
	2.05	0.09	0.37	1.00	4	18	49	71
Rhodic	0.50	0.12	0.28	0.03	24	56	6	86
Ferralsol	0.84	0.17	0.46	0.19	20	55	23	88
(Brasilia)	2.93	0.39	1.81	0.52	13	62	18	102
	3.40	0.51	1.93	0.82	15	57	24	96
	4.26	0.58	1.98	1.00	14	46	23	83

[a]Positive surface charge carried by three samples, 0.52, 0.51 and 2.58 cmol kg^{-1}, respectively; negative surface charge carried by three samples, 4.53, 5.76 and 1.82 cmol kg^{-1}, respectively.
[b]In unit of ½cmol kg^{-1}.
[c] A, OH$^-$ released; B, decrease in positive surface charge; C, increase in negative surface charge.

Figure 6.21 also shows that for the Rhodic Ferralsol of Brasilia the relationship between positive surface charge density and adsorption density of fluoride ion is not close and that the effect of iron oxides is small. By contrast, the slopes of the curves for negative surface charge density versus adsorption density of fluoride ions are about 0.26 and 0.36 for the original soil and the iron-free soil, respectively. This indicates that the adsorption of fluoride ions has a remarkable influence on the negative surface charge of the soil.

pH not only directly affects surface charges of soils and thus the adsorption of sulfate ions, but can also affect the relative magnitudes of

Table 6.8 Effect of Fluoride Adsorption on Surface Charge Properties
of a Rhodic Ferralsol of Brasilia

Treatment	Adsorption[b]	Change (cmol kg^{-1})			Percentage of Adsorption			
		A[c]	B	C	A	B	C	Sum
Original	0.99	0.56	0.21	0.24	57	21	24	102
Soil[a]	3.98	2.37	1.00	1.05	60	25	26	111
	6.97	4.09	1.61	1.87	59	23	27	109
	9.90	6.65	1.65	2.69	67	17	27	111
	15.6	10.4	1.66	4.91	67	11	32	110
Fe_2O_3	1.00	0.42	0.17	0.57	42	17	57	116
Removed	3.98	1.79	0.26	1.77	45	7	44	96
	6.89	3.47	0.31	2.68	50	5	39	94
	9.72	4.97	0.37	4.30	51	4	44	99
	15.3	8.23	0.39	6.77	54	3	44	101

[a]Positive surface charge carried by original soil and Fe_2O_3-removed soil, 2.20 and 0.60 cmol kg^{-1}, respectively; negative surface charge carried by the soils, 1.75 and 2.31 cmol kg^{-1}, respectively.
[b]In unit cmol kg^{-1}.
[c]A, OH$^-$ release; B, decrease in positive surface charge; C, increase in negative surface charge.

changes in three surface properties of the soil after the adsorption of sulfate ions. As shown in Table 6.9, the proportion of the release of OH$^-$ ions in total change tended to decrease with the rise in pH. Based on the principle of chemical equilibrium, this is easy to understand. The proportions of the change in surface charges varied with the soil type.

6.4.4 Release of Water Molecules

It has been mentioned several times in the previous sections that when anions are adsorbed by variable charge soils through ligand exchange, they can replace surface hydroxyl groups as well as coordinately adsorbed water molecules. In general, these water molecules are easier to replace by anions than are hydroxyl groups. The significance of this point in coordination adsorption of anions may be greater than generally thought previously, at least for sulfate adsorption, because this may be the primary cause of the

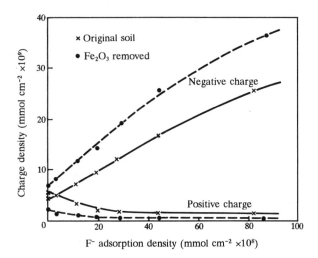

Fig. 6.21. Effect of fluoride adsorption on surface charge density of Rhodic Ferralsol of Brasilia.

Table 6.9   Change in Surface Charge Properties After Adsorption of Sulfate Ions by Two Rhodic Ferralsols at Different pH

Location[a]	pH	Adsorption[b]	Change (cmol kg⁻¹) A[c]	B	C	Percentage of Adsorption A	B	C	Sum
Brasilia	5.0	4.26	0.58	1.98	1.00	14	47	23	84
	6.8	2.41	0.18	1.35	0.84	7	56	35	98
Hawaii	4.2	2.55	0.24	1.09	1.36	9	43	53	105
	6.4	1.53	0.09	0.09	1.21	6	6	79	91

[a]The Rhodic Ferralsol from Brasilia carries positive surface charge 2.58 and 1.94 cmol kg⁻¹ and negative surface charge 1.82 and 1.95 cmol kg⁻¹ at pH 5.0 and 6,8, respectively. The Rhodic Ferralsol from Hawaii carries positive surface charge 1.55 and 0.92 cmol kg⁻¹ and negative surface charge 3.76 and 4.73 cmol kg⁻¹ at pH 4.2 and 6.4, respectively.
[b]($\frac{1}{2}$cmol kg⁻¹)
[c]A, OH⁻ release; B, decrease in positive surface charge; C, increase in negative surface charge.

easy desorption of adsorbed sulfate. It has been found that the desorption rate of adsorbed sulfate ions for three variable charge soils in water was very fast. The desorption reached a steady state within half of an hour. The desorbed amount accounted for 60-80% of the total adsorbed sulfate, varying with the type of the soil. The desorbed amount was even higher in the presence of NaCl. Similar results have been obtained by other authors.

However, up to the present, it is still not possible to directly measure the quantity of water released during the coordination adsorption of anions experimentally. If required, one can make calculations in this respect (Parfitt and Smart, 1978; Rajan, 1978).

6.5 MECHANISMS OF COORDINATION ADSORPTION OF ANIONS IN SOILS

In principle, models for describing ligand exchange on oxides may also be applicable to variable charge soils, because oxides are the principal carriers on which ligand exchange occurs in variable charge soils. However, the forms and aggregation status of these oxides in soils may be different from those when they exist separately. In this section, based on the reaction equations and the experimental results regarding changes in surface properties after anion adsorption presented in the last section, the mechanisms for the adsorption of anions by soils will be explored.

6.5.1 Adsorption of Sulfate Ions

To our knowledge, when the pH is below the zero point of charge of the oxides, there may be several kinds of reactions in the ligand exchange of sulfate ions on positive charge sites and zero charge sites. According to equations (6-7) to (6-12), SO_4^{2-} ions can undergo ligand exchange in different ways, replacing OH^- ions or H_2O molecules and causing the change in surface charge of soils. These relationships can be summarized in Table 6.10.

Parfitt and Russell (1977) suggested that one SO_4^{2-} ion can replace two $-OH$ or $-H_2O$ groups on the surface of iron oxides, forming binuclear bridging complex $FeOS(O)_2OFe$. This suggestion means that the reaction occurs according to equation (6-9) or equation (6-12). If the reaction takes place in the later way, there should only be the release of OH^- ions but not the change in surface charge. Apparently, according to the materials presented in this chapter, this type of adsorption can hardly occur in soils, because this is at odds with the experimental results presented in Tables 6.6 and 6.7 in which the amount of released OH^- ions is much smaller than that of changes in surface charge. It can be imagined that even though this type of adsorption can occur on pure iron oxides or aluminum oxides, the

Table 6.10 Relationship Between Sulfate Adsorption and Changes in
Surface Properties of Soil Expressed in Different Equations

Equation	Adsorption[a]	Change (mole)			Percentage of Adsorption	
		A^b	B	C	OH^-	Charge
(6-7),(6-9)	1	1	1	0	50	50
(6-8)	1	0	1	1	0	100
(6-10)	1	1	0	1	50	50
(6-11)	1	0	0	2	0	100
(6-12)	1	2	0	0	100	0

[a](mole).
[b]A, OH^- release; B, decrease in positive charge, C, increase in
negative charge.

possibility of the occurrence of such reaction in soils is small. This is
because iron and aluminum oxides are scatteringly distributed in soils, and
therefore the probability of the existence of favorable geometric positions,
at which one SO_4^{2-} ion can replace two ligands of the same kind simultaneously, should be much smaller than that in pure minerals. Besides, the
activity of the replaceable groups on the surface of these minerals may be
different from that in natural soils. Similarly, it can be inferred that the
reaction according to equation (6-9) should be small in extent in soils.

If equations (6-7), (6-8), (6-10), and (6-11) are applicable to coordination adsorption of SO_4^{2-} ions in soils, the adsorption of one mole of SO_4^{2-}
ions would cause the change in surface charge by two moles if water
molecules are replaced, and it would cause the change in surface charge by
one mole and the release of OH^- ions by one mole if surface hydroxyl
groups are replaced. Therefore, these equations can be classified into two
groups, in which there are both the release of OH^- ions and the change in
surface charge according to equations (6-7) and (6-10), or there is only the
change in surface charge but not the release of OH^- ions according to
equations (6-8) and (6-11). Only those parts of change in net surface
charge equivalent to the release of OH^- ions are caused by the adsorption
of SO_4^{2-} ions according to equation (6-7) or equation (6-10). Most of the
changes in net surface charge occur during the equivalent adsorption of
SO_4^{2-} ions according to equation (6-8) or (6-11) and is accompanied by
the simultaneous release of water molecules. Therefore, most of the
adsorption of SO_4^{2-} ions proceeds through the replacement of coordination
water molecules in soils. This is the basic reason why the equivalent ratio

OH/SO_4 for soils is small.

It has been seen that among the consequences of adsorption of sulfate ions the proportion of the change in surface charge is much larger than that of the release of hydroxyl groups (Tables 6.6 and 6.7). If analyzed in detail, these proportions vary with the type of the soil. For the Rhodic Ferralsol of Brasilia, the proportion of the decrease in positive surface charge is about 50–60%. This can be interpreted as the predominance of ligand exchange of SO_4^{2-} ions with water molecules on positive surface charge sites. For the Rhodic Ferralsol of Hawaii, the proportion of the increase in negative surface charge is relatively large, particularly for the iron–free soil in which it may be as high as 50%. This indicates that in this case a large part of SO_4^{2-} ions participates in ligand exchange with water molecules of the zero charge sites.

If the relationship between the adsorption density of sulfate ions on unit surface area and the change in surface charge density is examined, some interesting phenomena can be found. As seen in Figs. 6.22 and 6.23, the relationship is nearly linear. For the Rhodic Ferralsol of Brasilia, the slope of the straight line for positive surface charge density is –0.78, and that for negative charge is 0.30. In Table 6.4, the average value of OH/SO_4 for several soils is about 0.16. Thus, at pH 5, the contribution of individual reactions mentioned above to the overall reaction is of the order (6-8) > (6-11) > (6-7). For the Rhodic Ferralsol of Hawaii, the slope of the curve for negative charge is larger than that for positive charge, whereas the ratio of OH/SO_4 in Table 6.4 is smaller than 0.1. Thus, coordination adsorption

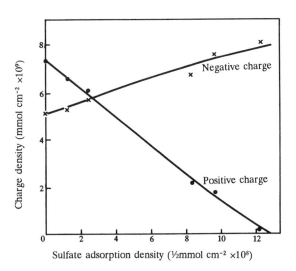

Fig. 6.22. Effect of sulfate adsorption on surface charge density of Rhodic Ferralsol of Brasilia.

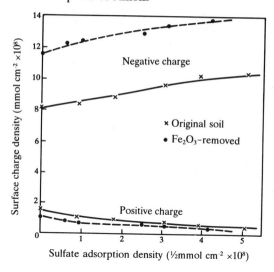

Fig. 6.23. Effect of sulfate adsorption on surface charge density of Rhodic Ferralsol of Hawaii.

of sulfate ions by this soil should proceed mainly according to equations (6-11) and (6-8), with the former contribution larger than the latter contribution.

6.5.2 Adsorption of Fluoride Ions

Similar to the adsorption of sulfate ions, at pH lower than the zero point of charge of the oxides, the ligand exchange of fluoride ion on positive charge sites and zero charge sites can also be classified into several types [cf. equations (6-2) to (6-6)]. These pathways of ligand exchange can also be applied to the adsorption of fluoride ions by soils. Because fluoride ions are monovalent, it is not possible for them to form binuclear bridging complex or exist in the form of six-member ring on the surface of oxides as in the case for phosphate ions and sulfate ions. They can only form monodentate complex. Thus, the reactions can be classified into three types. In one type there is the release of OH^- ions but not the change in surface charge, as expressed by equations (6-2) and (6-5). In another type there is the change in surface charge but not the release of OH^- ions, as represented by equations (6-3) and (6-6). In the third type of reaction there are both the release of OH^- ions and the change in surface charge, as shown in equation (6-4). For the adsorption of sulfate ions by soils, because the change in surface charge greatly exceeds the release of OH^- ions, it can be imagined that it is almost impossible for the first type of reaction to occur. But, for the adsorption of fluoride ions, the occurrence of such reaction may be possible, because for the Rhodic Feralsol the release of OH^- ions exceeded the change in surface charge (Table 6.8). The second type of reaction is the ligand exchange with water molecules. For the iron–free soils the amount of released OH^- ions was smaller than that for the original sample, but the change in surface charge remained the same. This may be related to the occurrence of the second type of adsorption of fluoride ions. However, considered comprehensively, because the difference in magnitude between the release of OH^- ions and the change in surface charge was not great, it can be inferred that for such soils as the Rhodic Ferralsol of Brasilia, the adsorption of fluoride ions should proceed chiefly in a fashion of the third type of reaction, that is, it is caused by ligand exchange with both surface OH^- groups and water molecules.

6.5.3 Concluding Remarks

The mechanisms for coordination adsorption of anions by variable charge soils vary with the nature of the anion species. The adsorption of sulfate ions is chiefly caused by ligand exchange with coordinately bonded water molecules on the surface of soil particles. The principal consequence of adsorption is the change in surface charge. The adsorption of fluoride and phosphate ions is stronger than that of sulfate ions, because flluoride and phosphate ions can approach the surface of soils more closely. Thus, they can replace more hydroxyl groups from the surface of soil particles as compared with sulfate ions. However, for variable charge soils with a complex composition because the amounts, distribution, and activity of various reactive surface groups differ greatly not only from those of model materials but also with the type of the soils themselves, at present there is no generalizable model applicable to the coordination adsorption of sulfate and that of other specifically adsorbed anions, especially when electrostatic forces are involved at the same time. This is particularly the case when a quantitative approach is attempted.

The characteristis in the coordination adsorption of sulfate ions by variable charge soils have some practical significance. Because the adsorption process is mainly the exchange with coordinated water molecules but not a reaction on the surface of soils itself as in the exchange with hydroxyl groups, the bonding energy is low. As a consequence, these adsorbed sulfate ions would be easy to desorb. Thus, under field conditions, particularly in regions where the temperature is high and the rainfall is abundant, although the coordination adsorption of sulfate ions may play an important role in weakening the leaching of these ions, it cannot prevent their eventual loss. This differs from the case of the adsorption of phosphate ions.

The quantity of negative surface charge of soils increases with coordination adsorption of sulfate ions. For variable charge soils carrying a low quantity of negative surface charge, this increase would be favorable for the retention of nutrient cations such as potassium, ammonium, and calcium, although this advantageous effect will disappear gradually with the leaching loss of sulfate ions themselves.

BIBLIOGRAPHY

Aylmore, L. A. G., Karim, M., and Quirk, J. P. (1967) *Soil Sci.*, 103:10–15.

Atkinson, R. J., Posner, A. M., and Quirk, J. P. (1972) *J. Inorg. Nucl. Chem.*, 34:2201–2211.

Atkinson, R. J., Parfitt, R. L., and Smart, R. St. C. (1974) *J. Chem. Soc. Faraday Trans.* I, 70:1472–1479.

Barrow, N. J. (1970) *Soil Sci.*, 109:282–288.

Barrow, N. J. (1984) *J. Soil Sci.*, 35:283–297.

Barrow, N. J. (1985) *Adv. Agron.*, 38:183–230.

Bolan, N. S. and Barrow, N. J. (1984) *J. Soil Sci.*, 35:273–281.

Bolan, N. S., Syers, J. K. and Tillman, R. W. (1986) *J. Soil Sci.*, 37:379–388.

Bolan, N. S., Scotter, D. R., Syers, J. K., and Tillman, R. W. (1987) *Soil Sci. Soc. Am. J.*, 50:1419–1424.

Bolan, N. S., Syers, J. K., and Sumner, M. E. (1993) *Soil Sci. Soc. Am. J.*, 57:691–696.

Bolland, M. D. A., Posner, A. M., and Quirk, J. P. (1976) *Aust. J. Soil Res.*, 14:197–216.

Bolt, G. H. (1979) *Soil Chemistry*. Elsevier, New York.

Bowden, J. W., Posner, A. M., and Quirk, J. P. (1974) *Trans. 10th Intern. Congr. Soil Sci.*, II:29–36.

Bowden, J. W., Posner, A, M., and Quirk, J. P. (1980) in *Soils with Variable Charge* (B. K. G. Theng, ed.). New Zealand Society of Soil Science, pp. 147–166.

Breeumsma, A. and Lyklema, J. (1973) *J. Colloid Interface Sci.*, 43:437–448.

Chao, T. T., Harward, M. E., and Fang, S. C. (1965) *Soil Sci.*, 99:104–108.

Curtin, D. and Syers, J. K. (1990a) *J. Soil Sci.*, 41:295–304.

Curtin, D. and Syers, J. K. (1990b) *J. Soil Sci.*, 41:305–312.

Curtin, D. and Syers, J. K. (1990c) *J. Soil Sci.*, 41:433–442.

Fuller, R. D., David, M. B., and Driscoll, C. T. (1985) *Soil Sci. Soc. Am. J.*, 49:1034–1040.

Gelbhardt, H. and Coleman, N. T. (1974) *Soil Sci. Soc. Am. Proc.*, 38:259–262.

Goldberg, S. and Sposito, G. (1985) *Commun. Soil Sci. Plant Anal.*, 16:801–821.

Guadalix, M. E. and Pardo, M. T. (1991) *J. Soil Sci.*, 42:607–614.

Hasan, S. M., For, R. L., and Boyd, C. C. (1970) *Soil Sci. Soc. Am. Proc.*,

34:897-901.

Hingston, F. J. (1981) in *Adsorption of Inorganics at Solid-liquid Interfaces* (M. A. Anderson and J. R. Rubins, eds.). Ann Arbor Science, Ann Arbor, MI, pp. 51-90.

Hingston, F. J., Posner, A. M., and Quirk, J. P. (1967) *Nature*, 215:1459-1461.

Hingston, F. J., Atkinson, A. M., Posner, A. M., and Quirk, J. P. (1968) *Trans. 9th Intern. Congr. Soil Sci.*, I:669-678.

Hingston, F. J., Posner, A. M., and Quirk, J. P. (1974) *J. Soil Sci.*, 25:16-26.

Hodges, S. and Johnson, G. C. (1987) *Soil Sci. Soc. Am. J.*, 51:323-331.

Inskeep, W. P. (1989) *J. Environ. Qual.*, 18:379-385.

Ji, G. L. (1986) *Acta Pedol. Sinica*, 23:220-227.

Johnson, D. W. and Todd, D. E. (1983) *Soil Sci. Soc. Am. J.*, 47:792-800.

Jurinak, J. J. (1966) *Soil Sci. Soc. Am. Proc.*, 30:559-562.

Kafkafi, U., Posner, A. M., and Quirk, J. P. (1967) *Soil Sci. Soc. Am. Proc.*, 31:348-352.

Kuo, S. and Lotse, E. G. (1972) *Soil Sci. Soc. Am. Proc.*, 36:725-729.

Kyle, J. H., Posner, A. M., and Quirk, J. P. (1975) *J. Soil Sci.*, 26:32-43.

Marcano-Martinez, E. and McBride, M. B. (1989) *Soil Sci. Soc. Am. J.*, 53:63-69.

Marsh, K. B., Tillman, R. W. and Syers, J. K. (1987) *Soil Sci. Soc. Am. J.*, 51:318-323.

Mekaru, T. and Uehara, G. (1972) *Soil Sci. Soc. Am. Proc.*, 36:296-300.

Mott, C. J. B. (1981) in *The Chemistry of Soil Processes* (D. J. Greenland and M. H. B. Hayes, eds.). John Wiley & Sons, Chichester, pp. 179-219.

Muljadi, D., Posner, A. M., and Quirk, J. P. (1966) *J. Soil Sci.*, 17:212-229.

Neal, R. H. and Sposito, G. (1981) *Soil Sci. Soc. Am. J.*, 53:70-74.

Nodwin, S. C., Driscoll, C. T., and Likens, G. E. (1986) *Soil Sci.*, 142:69-75.

Parfitt, R. L. (1978) *Adv. Agron.*, 30:1-50

Parfitt, R. L. and Atkinson, R. J. (1976) *Nature*, 264:740-742.

Parfitt, R. L. and Russell, J. D. (1977) *J. Soil Sci.*, 28:297-305.

Parfitt, R. L. and Smart, R. St. C. (1977) *J. Chem. Soc. Faraday Trans.* I, 73:796-802.

Parfitt, R. L. and Smart, R. St. C. (1978) *Soil Sci. Soc. Am. J.*, 42:48-50.

Parfitt, R. L., Atkinson, R. J., and Smart, R. St. C. (1975) *Soil Sci. Soc. Am. Proc.*, 39:837-841.

Parfitt, R. L., Russell, J. D., and Farmer, V. C. (1976) *J. Chem. Soc. Faraday Trans.* I, 72:1082-1087.

Parfitt, R. L., Fraser, A. R., Russell, J. D., and Farmer, V. C. (1977) *J. Soil Sci.*, 28:40-47.

Perrott, K. W., Langdon, A. G., and Wilson, A. T. (1974a) *Geoderma*, 12:223-231.

Perrott, K. W., Langdon, A. G., and Wilson, A. T. (1974b) *J. Colloid*

Interface Sci., 48:10-19.

Rajan, S. S. S. (1975) *J. Soil Sci.*, 26:250-256.

Rajan, S. S. S. (1976) *Nature*, 262:45-46.

Rajan, S. S. S. (1978) *Soil Sci. Soc. Am. J.*, 42:39-44.

Rajan, S. S. S. (1979) *Soil Sci. Soc. Am. J.*, 43:65-69.

Rajan, S. S. S., Perrot, K. W., and Saunders, W. M. H. (1974) *J. Soil Sci.*, 25:438-447.

Ryden, J. C., Syers, J. K., and McLaughlin, J. R. (1977a) *J. Soil Sci.*, 28:62-71.

Ryden, J. C., McLaughlin, J. R., and Syers, J. K. (1977b) *J. Soil Sci.*, 28:72-92.

Ryden, J. C., Syers, J. K. and Tillman, R. W. (1987) *J. Soil Sci.*, 38:211-218.

Schnabel, R. R. and Potter, R. M. (1991) *Soil Sci. Soc. Am. J.*, 55:693-698.

Schofield, R. K. (1949) *J. Soil Sci.*, 1:1-8.

Shanley, J. B. (1992) *Soil Sci.*, 153:499-508.

Sharpley, A. N. (1990) *Soil Sci. Soc. Am. J.*, 54:1571-1575.

Sposito, G. (1984) *The Surface Chemistry of Soils.* Oxford University Press, New York.

Sudhakar, M. R. and Sridharen, A. (1984) *Clays Clay Miner.*, 32:414-428.

Torrent, J., Schwertmann, U., and Barron, V. (1992) *Clays Clay Miner.*, 40:14-21.

Turner, L. L. and Kramer, J. R. (1990) *Soil Sci.*, 152:226-230.

Turner, L. L. and Kramer, J. R. (1992) *Water, Air and Soil Pollut.*, 63:23-32.

Wada, K. and Warward, M. E. (1974) *Adv. Agron.*, 26:211-254.

Wang, P. G., Ji, G. L., and Yu, T. R. (1986) *Z. Pflanzenernähr. Bodenk.*, 150:17-19.

Yates, D. E. and Healy, T. W. (1975) *J. Colloid Interface Sci.*, 52:222-228.

Yu, T. R. and Zhang, X. N. (1986) in *Proceedings of the International Symposium on Red Soils.* Science Press/Elsevier, Beijing/Amsterdam, pp. 409-441.

Zhang, G. Y. and Zhang, X. N. (1989) *Soils*, 21:239-242.

Zhang, G. Y., Zhang, X. N., and Yu, T. R. (1986) in *Current Progress in Soil Research in People's Republic of China.* Jiangsu Science and Technology Publishers, Nanjing, pp. 100-106.

Zhang, G. Y., Zhang, X. N., and Yu, T. R. (1987) *J. Soil Sci.*, 38:29-38.

Zhang, G. Y., Zhang, X. N., and Yu, T. R. (1991) *Pedosphere*, 1:17-28.

Zhang, P. C. and Sparks, D. L. (1990) *Soil Sci. Soc. Am. J.*, 54:1266-1273.

Zhang, X. N. and Zhao, A. Z. (1988) *Acta Pedol. Sinica*, 25:164-174.

Zhang, X. N., Zhang, G. Y., Zhao, A. Z., and Yu, T. R. (1989) *Geoderma*, 44:275-286.

7

ELECTROKINETIC PROPERTIES

H. Zhang

Cations and anions adsorbed by soil particles carrying surface charges are not present totally on the surface of the particles. Actually, in a soil-water system, a portion of adsorbed ions is distributed near the surface, forming an electric double layer at the interface between the solid particle and the liquid phase. When the two phases have a relative movement in an electrical field or are affected by other forces, the system can exhibit certain electrical properties, called *electrokinetic properties*.

Electrokinetic properties of soils are the overall reflection of the distribution of various kinds of ions in the electric double layer of a soil-water system. They are related to both the characteristics of the soil and the nature of ions. For variable charge soils, because they adsorb anions as well as cations and during the adsorption both electrostatic force and specific force are involved, their electrokinetic properties frequently manifest themselves in a complex manner. As shall be seen in the present chapter, the electrokinetic properties of variable charge soils exhibit certain characteristics different from those of constant charge soils, and these characteristics are of significance for further distinguishing soil types among these soils.

7.1 ELECTROKINETIC PROPERTIES AND ELECTRIC DOUBLE LAYER

All the electrokinetic phenomena occurring in any colloid system result from the existence of the electric double layer. The same holds true for soils. Therefore, in this section the theory of the electric double layer along with its relation to various electrokinetic properties will be introduced first, and then the complexities in soil systems in this respect will be examined.

7.1.1 Classical Theory of Electric Double Layer

When two phases are in contact, owing to the difference in properties, a redistribution of electric charge will occur at the interface between the two phases, leading to the formation of two layers with charges equal in quantity but opposite in sign between the two sides of the interface. This pair of charged layers is called *electric double layer*. It is a microscopically charged system present in the interfacial region between the two phases. The electric potential may vary at different positions within the system, but the system as a whole is electrically neutral.

The term electric double layer was first introduced by Helmholtz in 1853. He considered that the structure of the electric double layer was analogous to that of a parallel plate capacitor, and that the potential varied linearly with the distance from the surface of a given phase. However, Helmholtz's theory is only a simplified description of the electric double layer. Gouy (1910) and Chapman (1913) independently proposed the theory of diffuse double layer at a solid–liquid interface. This theory has become the basis of a variety of other theories about the electric double layer. The electric double layer theory was first introduced to soil chemistry by Schofield, which was also based on the Gouy–Chapman diffuse double layer theory (Bolt, 1979).

In the diffuse double layer theory, several assumptions are made: (1) Charged surfaces are dimensionally infinite planes with electric charges distributed evenly on them; (2) ions are point charges and interact electrostatically with the surface; (3) the medium of the solution is structureless, continuous, and uniform; (4) the Poisson equation can be used as a basic electrostatic equation to describe the whole double–layer system:

$$\nabla^2 \psi = -\frac{\rho}{\varepsilon_0 D} \qquad (7\text{-}1)$$

where ψ is the potential of the double layer, ρ the volume charge density, D the dielectric constant of the medium ($D=\varepsilon/\varepsilon_0$), and ε_0 the permittivity in vacuum; and (5) since ions in the double layer are also affected by their thermal movement, their distribution complies with the Boltzmann equation:

$$n_i = n_i^0 \exp\left(-z_i e \psi / kT \right) \qquad (7\text{-}2)$$

where n_i is the number of ion species i in a unit volume in the double layer, n_i^0 the number of ion species i in a unit volume of bulk solution, k the Boltzmann constant, and T the absolute temperature.

Based on the above assumptions, a basic equation for describing the diffuse double layer (i.e., the Gouy–Chapman equation) can be obtained:

$$\nabla^2\psi = \frac{d^2\psi}{dx^2} = -\frac{1}{\varepsilon_0 D}\sum_i n_i^0 z_i e\, \exp(-z_i e\psi/kT) \qquad (7\text{-}3)$$

In the following sections, several of the important parameters in the Gouy–Chapman theory shall be explained.

7.1.1.1 *Potential Distribution*

In the Gouy–Chapman theory, the distribution of potential in the double layer is given by

$$\tanh(z\psi'/4) = \tanh(z\psi_0'/4)\exp(-\kappa x) \qquad (7\text{-}4)$$

where $\psi' = e\psi/\kappa T$, $\kappa = (e^2\Sigma n_i^0 z_i^2/\varepsilon kT)^{1/2}$, which is called the Debye–Hückel parameter, and ψ the potential at position x in the double layer.

At a certain position far away from the double layer, the potential is extremely low, so it can be assumed that $\tanh P \approx P$. Then

$$z\psi' = 4re^{-\kappa x} \qquad (7\text{-}5)$$

where

$$r = \tanh(z\psi_0/4)$$
$$= \frac{\exp(z\psi_0'/2) - 1}{\exp(z\psi_0'/2) + 1}$$

When ψ is small enough, the Debye–Hückel approximation can be made, that is, $z_i e\psi << kT$. Therefore,

$$\psi = \psi_0\exp(-\kappa x) \qquad (7\text{-}6)$$

This means that the potential decreases exponentially with distance.

7.1.1.2 *Surface Charge Density*

For nonsymmetrical electrolytes, one obtains

$$\sigma_0 = \left\{ 2\varepsilon kT \sum_i n_i^0[\exp(-z_i e\psi_0/kT) - 1]\right\}^{1/2} \qquad (7\text{-}7)$$

and for symmetrical electrolytes, one has

$$\sigma_0 = \frac{4n^0 ze}{\kappa} \sinh(z\psi_0'/2) \tag{7-8}$$

When the potential value is very small, we can have

$$\sigma_0 = \varepsilon_0 D\kappa\psi_0 \tag{7-9}$$

7.1.1.3 Distribution of Ions in the Double Layer

The relationship is

$$n_i = n_i^0 \exp(-z_i e\psi/kT) \tag{7-10}$$

and when the potential value is very small,

$$n_+ = n^0(1 - ze\psi/kT) \tag{7-11}$$

$$n_- = n^0(1 + ze\psi/kT) \tag{7-12}$$

7.1.1.4 Electrokinetic Charge Density

It can be known from the above discussions that, when $\psi = \zeta$, for nonsymmetrical electrolytes, one has

$$\sigma_e = \left\{ 2\varepsilon kT \sum_i n_i^0 [\exp(-z_i e\zeta/kT] - 1 \right\}^{1/2} \tag{7-13}$$

and for symmetrical electrolytes, one obtains

$$\sigma_e = \frac{4n^0 ze}{\kappa} \sinh(z\zeta'/2) \tag{7-14}$$

where $\zeta' = e\zeta/kT$. When the potential value is very small, we obtain

$$\sigma_e = \varepsilon_0 D\kappa\zeta \tag{7-15}$$

7.1.1.5 κ

The reciprocal of κ mentioned above ($1/\kappa$) represents the thickness of the double layer. $\kappa = f(I^{1/2})$, where I is the ionic strength.

However, experiments showed that actually the second and the third assumptions cannot hold. For these reasons, Stern (1924) reconsidered the assumptions in the Gouy–Chapman theory and developed the Gouy–Chapman–Stern theory by making a significant modification.

In the Gouy–Chapman–Stern theory, ions distributed in the double layer are divided into two sublayers. The layer adjacent to the surface is called the *Stern layer* or *compact layer*, which is analogous to the Helmhotz layer. The outer layer is a diffuse layer, analogous to the diffuse layer in the Gouy–Chapman theory. In the Stern layer, ions are adsorbed electrostatically or specifically on the surface of the solid phase, with a thickness determined by the geometrical radius of the adsorbed ions. The potential in the layer varies linearly with the distance. The distributions of potential and ions in the outer layer obey the Gouy–Chapman equation.

According to this model, one has (Wiese et al., 1976)

$$\sigma_0 + \sigma_\delta + \sigma_d = 0 \tag{7-16}$$

$$\sigma_\delta = \frac{z_+ e N_s}{1 + \dfrac{N}{n_+ M} \exp\left(\dfrac{z_+ e \psi_\delta + \phi_+}{kT}\right)}$$
$$+ \frac{z_- e N_s}{1 + \dfrac{N}{n_- M} \exp\left(\dfrac{z_- e \psi_\delta + \phi_-}{kT}\right)} \tag{7-17}$$

where N_s is the number of effective adsorption sites, N the Avogadro constant, M the molecular weight of the solvent, and ϕ_+ and ϕ_- the specific adsorption potential, with a negative value favorable to the occurrence of adsorption, z_+ and z_- are the valence of the cations and anions (including the sign), respectively, and n_+ and n_- are the bulk concentrations of the cations and anions, respectively.

One also has

$$\sigma_d = -(8n^0 \varepsilon \varepsilon_0 kT)^{1/2} \sinh(ze\psi_d / 2kT) \tag{7-18}$$

The relationship among the potentials is expressed by

$$\psi_0 - \psi_\delta = \frac{\delta \sigma_0}{\varepsilon_1 \varepsilon_0} \tag{7-19}$$

where ε is the permittivity of the medium in the inner layer. According to Stern's assumption, $\psi_\delta = \psi_d$, for the flat double layer we can get

$$\psi_0 = \psi_d + \frac{\delta}{\varepsilon_1 \varepsilon_0} \Bigg[(8\varepsilon \varepsilon_0 kTn_0)^{1/2} \sinh(ze\psi_d/2kT)$$

$$- \frac{zeN_s}{1 + \frac{N}{n^0 M} \exp\left(\frac{ze\psi_d + \phi}{kT}\right)} \Bigg] \tag{7-20}$$

When ψ_d is very small, for a spherical double layer, one can have

$$\psi_0 = \psi_d + \frac{\delta}{\varepsilon_1 \varepsilon_0} \Bigg[\frac{\varepsilon \varepsilon_0}{(r + \delta)} \psi_d (1 + \kappa(r + \delta)$$

$$- \frac{zeN_s}{1 + \frac{N}{n^0 M} \exp\left(\frac{ze\psi_d + \phi}{kT}\right)} \Bigg] \tag{7-21}$$

where r is the radius of the particles, and δ is the distance from the Stern plane to the colloid surface.

Grahame (1947) further developed Stern's theory. He considered that the Stern layer is composed of two sub-layers. The inner layer is called the *inner Helmholtz phase* (IHP), which is composed of unhydrated ions specifically adsorbed on the solid surface. The other layer is called the *outer Helmholtz phase* (OHP), which is composed of hydrated ions. The layer outside of these two layers is the diffuse layer.

The modified electric double layer theory developed by Stern and modified by Grahame can explain some phenomena that the Gouy–Chapman theory cannot, such as the change in sign of electrokinetic potential, and provides a more reasonable and complete description of the structure and nature of the electric double layer. This theory has become a generally accepted theoretical model for describing the electric double layer. As a matter of fact, both Helmholtz's model and Gouy–Chapman's model can be regarded as the approximation of the Gouy–Chapman–Stern–Grahame (GCSG) model under different specified conditions. When the concentration of ions in the liquid phase is very high, the GCSG model is close to Helmholtz's model, whereas when the concentration is very low, on the other hand, the GCSG model turns out to be similar to Gouy–Chapman's model (Westall and Hohl, 1980).

7.1.2 Electrokinetic Phenomena and Electrokinetic Potential

Electrokinetic phenomena refer to those physical phenomena that occur at the interface between two phases in contact with each other when they have a relative movement. Such phenomena can occur under two circumstances.

One is the movement of a certain phase in an externally applied electrical field, and the other is the movement of one phase relative to another phase caused by an external mechanical force. Four kinds of electrokinetic phenomena can be distinguished:

1. *Electrophoresis.* The liquid phase remains static, while the solid phase migrates when affected by an applied electrical field,

2. *Electro-osmosis.* The solid phase remains static, and the liquid phase moves in an applied electrical field,

3. *Streaming potential.* This is the potential difference occurring at the two ends of a solid when a liquid in contact with it passes through it,

4. *Sedimentation potential.* This is the potential difference produced between the upper part and the lower part of a liquid when solid particles sink in the liquid caused by a gravitational force.

Electrokinetic phenomena can be observed experimentally. However, in order to interpret them, it is necessary to use the double layer theory. It is also necessary to introduce a new concept, the plane of shear (slip surface), and a new parameter, electrokinetic potential (also called ζ potential, zeta potential).

The slip surface is an imaginary plane in the double layer, and it exists close to the solid surface. Inside the slip surface, the solid phase and the liquid phase remain in a relative static state. When an electrokinetic phenomenon occurs, the solid phase together with its neighboring liquid phase inside the slip surface will move relative to the liquid phase outside the slip surface. Therefore, electrokinetic phenomena can reflect a series of electrochemical properties at the slip surface in the double layer. The potential at this slip surface is the electrokinetic potential, and the charge density is the electrokinetic charge density (σ_e).

Two points are worth mentioning here with respect to the concept of slip surface. First, the exact location of this surface in the double layer cannot be ascertained. It is generally considered that it is located near the interface between the Stern layer and the diffuse layer, with the exact location varying under different conditions. Second, while it is generally held that the viscosity and permittivity of the liquid medium can change abruptly at the slip surface, actually this view may deviate from real situations. It is quite possible that these properties change gradually near this surface, and therefore the surface is actually a layer of shear with a thickness equivalent to the radius of a molecule (Hiemenz, 1977). In colloid chemistry, electrokinetic potential is generally regarded as approximately equal to the potential of the Stern layer ψ_d. This assumption is reasonable and has been supported by experimental results (Hunter, 1981). The important significance of such assumption is that it relates the electrokinetic potential with the important parameter ψ_d in the double layer, thus making the physical meaning of electrokinetic potential more clear. Consequently, the introduction of electrokinetic potential and slip surface provides an effective experimental approach for studying the structure and properties of the double layer at the solid–liquid interface through determination of electrokinetic parameters.

In colloid chemistry and soil chemistry, electrophoretic determination is the most useful technique among the four electrokinetic determinations. In the following, the quantitative relationship between zeta potential and electrophoretic mobility will be presented briefly.

There are two basic equations for describing the relationship between electrophoretic mobility (u) and zeta potential (ζ). Both of them are related to the value κr, where κ is the Debye–Hückel parameter and r the real radius of the particle (Hiemenz, 1977). When $\kappa r > 100$, the equation is

$$u = \varepsilon \zeta / 4\pi \eta \qquad (7\text{-}22)$$

where ε is the permittivity of the medium and η is the viscosity of the liquid phase.

This equation is called the *Helmholtz–Smoluchowski equation* and is applicable to solid particles of any geometry.

When $\kappa r < 0.1$, the equation becomes

$$u = \varepsilon \zeta / 6\pi \eta \qquad (7\text{-}23)$$

This equation is called the *Hückel equation*.

The above two equations are applicable only to two extreme cases where the value κr is either very large or very small. For cases where the κr lies between the two extremes, Henry derived the following equation:

$$
\begin{aligned}
u = \frac{\varepsilon \eta}{6\pi \eta} \Big\{ 1 &+ \frac{1}{16}(\kappa r)^2 - \frac{5}{48}(\kappa r)^3 \\
&- \frac{1}{96}(\kappa r)^4 + \frac{1}{96}(\kappa r)^5 \\
&- \left[\frac{1}{8}(\kappa r)^4 - \frac{1}{96}(\kappa r)^6 \right] \exp(\kappa r) \int_{\infty}^{\kappa r} \frac{e^{-t}dt}{t} \Big\} \qquad (7\text{-}24)
\end{aligned}
$$

This equation is applicable to any value of κr.

In all the above equations for describing the relationship between u and ζ, the following assumptions are made: The surface of the colloid does not have electric conductivity; the conductivity, dielectric constant, and viscosity in the electric double layer are the same as in the bulk solution; and the relaxation effect of the colloid system is negligible so that the distortion of the electric double layer caused by electrokinetic phenomena can be ignored. Therefore, the zeta potentials calculated by these equations are reasonable only under certain conditions. For complex systems, the results are only of semiquantitative significance. Under such circumstances, it would be more advisable to use electrophoretic mobility to describe the electrokinetic phenomena (Van Olphen, 1963).

There are a number of factors that can affect the zeta potential. They include the following:

1. *Concentration of electrolytes*. As seen in Fig. 7.1, the higher the concentration of nonspecifically adsorbed ions in the liquid phase, the smaller the thickness of the electric double layer and thus the lower the zeta potential.

2. *Valence of ions*. The higher the valence of the counterions in the diffuse layer, the lower the zeta potential.

3. *Specific adsorption*. As shown in Fig. 7.2, specific adsorption of ions can lead to a remarkable change in zeta potential and can even make the sign

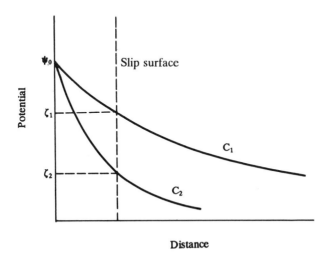

Fig. 7.1. Effect of electrolyte concentration on zeta potential ζ $(C_2 > C_1)$.

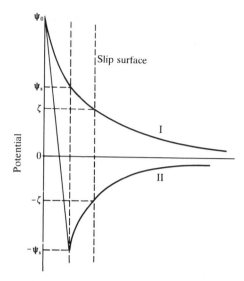

Fig. 7.2. Effect of specific adsorption of ions on zeta potential.

of the potential reversed (curve II). This is quite different from the curve without specific adsorption (curve I).

4. *Shape and size of colloid particles.* The situation here is rather complex. In general, for spherical ad rod–shaped particles, an identical equation can be used to calculate the zeta potential. For particles with irregular or random geometry, however, the required equations are different.

5. *Surface geometry of particles.* Strictly speaking, all the equations for describing the relationship between zeta potential and other electrokinetic parameters are applicable only to colloid particles with smooth surfaces. Actually, however, the surface of many colloid particles are far from smooth. Owing to the difference in surface geometry, the microregional structure of the electrical double layer may differ considerably, which in turn can result in the difference in zeta potential.

In summary, all the equations discussed above for describing the relationship between zeta potential and other electrokinetic parameters refer to ideal situations, and the calculated zeta potentials are only an approximate average of the zeta potential of a real system.

Zeta potential and other electrokinetic parameters are of important theoretical significance (Harsh and Xu, 1990; Sposito, 1984; Wiese et al., 1976). Zeta potential can provide much valuable information about the structure and properties of the electric double layer, even though the exact location of the slip surface cannot be ascertained. In the absence of specific adsorption in the system, $\zeta = 0$ when $\psi_0 = 0$. Therefore, the isoelectric point (IEP) can be obtained through determination of electrokinetic parameters. In this case the IEP is identical to the zero point of charge (ZPC). As mentioned before, under certain conditions it may be considered that $\zeta = \psi_d$, and thus $\psi_d = 0$ when $\zeta = 0$. Besides theoretical significance, zeta potential and other electrokinetic parameters may also have some practical applications. For instance, they determine the dispersion, flocculation, and sedimentation of colloids, including soil colloids. This point has been reviewed by Marshall (1964) and by Yu (1976).

7.1.3 Electric Double Layer at Metal Oxide-Water Interface

The classical theory of electric double layer can give a good description of the structure and properties of the double layer of reversible $AgX-H_2O$ (such as $AgI-H_2O$) interfaces. The Nernst equation is completely applicable to such interfaces. Both Ag^+ ions and X^- ions are potential-determining ions in the $AgX-H_2O$ system. Potential-determining ions are those ions that are present in both the solid phase and the liquid phase. When an equilibrium of these ions is established between the two phases, the electrical potential at the interface is determined by the activity of such ions in the bulk solution (James, 1981). Hence, one has

$$d\psi_0 = (kT/e)d\ln a_{Ag^+}. \qquad (7-25)$$

where a_{Ag^+} is the activity of Ag^+ ions in the bulk solution.

The surface of metal oxides is amphoteric. Their surface reaction is

$$\overset{\displaystyle H^+ \qquad\quad OH^-}{MOH_2^+ \; \leftarrow \; MOH \; \rightarrow \; MO^- + H_2O} \qquad (7-26)$$

In metal oxide-aqueous solution systems, H^+ ions and OH^- ions play the same role as Ag^+ ions and X^- ions in the $AgX-H_2O$ system, also functioning as potential-determining ions. For instance, the sign and magnitude of the zeta potential are determined by the pH of the solution, and the intersection point among isotherms of H^+ adsorption and OH^- adsorption in electrolyte solutions of different concentrations without the presence of specifically adsorbed ions is the ZPC. However, H^+ and OH^- ions differ from Ag^+ and X^- ions in that, while Ag^+ and X^- ions may be the constitutional part of the solid phase, H^+ and OH^- ions are present only on the surface of the metal oxides. Because of this, Bolt (1979) called Ag^+ and X^- ions the primary potential-determining ions and referred to H^+ and OH^- ions as the secondary potential-determining ions.

Since H^+ ions and OH^- ions are the potential-determining ions, the

Nernst equation should be applicable to metal oxide–aqueous solution systems:

$$\psi_0 = 0.059\,(\,pH_{zpc} - pH\,) \tag{7-27}$$

However, experimental results have shown that this treatment is not in conformity with the reality (Hunter, 1981; Wiese et al., 1976). Therefore, it is necessary to formulate more accurate models to describe the relationship between potential–determining ions and surface potential. A number of models have been developed in this respect (Healy and White, 1978; Sposito, 1984; Westall and Hohl, 1980). The basic points of these models are almost identical. Material equilibrium equation, mass action law, and statistic mechanics are used to describe the chemical equilibria of those ions that cause the occurrence of surface potential and the contribution of these ions to surface charge, so that interrelations between surface potential and potential–determining ions can be formulated through thermodynamic considerations. Then, the corresponding relationship between surface potential and surface charge density is established. Westall and Hohl (1980) classified these models into five types: constant capacitance model of Stumm and Schindler, diffuse layer model of Stumm, Huang, and Jenkins, VSC–VSP model adopted by Bowden, Posner, and Quirk, three–layer model proposed by Yates, Levine, and Healy and developed by Davis, James and Leckie, and Stern's model. The differences among these models lie mainly in (a) the ascertainment of those ions contributing to surface potential and their position at the surface and (b) the establishment of the relationship between surface potential and surface charge. Most of the experimental verifications of these models are made with the potentiometric titration method. The direct use of electrokinetic methods is less common. Only in a few cases the experimental results with both kinds of methods can be interpreted satisfactorily (Westall and Hohl, 1980). The main reason for this is that, owing to the complexities of the system in question, difficulties may be encountered in the calculation of zeta potential from electrokinetically determined results.

7.1.4 Electrokinetic Properties of Variable Charge Soils

There are two kinds of components in variable charge soils. One is a constant–charge component (i.e., clay minerals) which carries permanent negative surface charges originating from isomorphic substitutions. The other one is a variable charge component or a constant–potential component, including various kinds of crystalline and noncrystalline metal oxides, organic matter, and the surface of broken edges of some clay minerals. The quantity of surface charge of these variable charge components vary with the pH of the solution and other environmental factors. Consequently, variable charge soils are mixtures of both a constant charge component and a variable charge component, and their surface electrochemical properties are different from those of metal oxides as pure variable charge colloids and are also different from those of constant charge soils mainly carrying permanent negative surface charge. They can carry both negative surface charge and positive surface charge. Because metal oxides and organic matter carrying variable charges are present mainly on the surface of soil particles, the surface electrochemical properties of these soils are determined to a large extent by metal oxides and organic matter. Since metal oxides are the major carriers of positive charges while organic matter is the major carrier of variable negative charges, the contributions of these two components to surface electrochemical properties of the soil are different. Besides, variable charge soils possess a peculiar and important property, that is, they can adsorb some anions and cations, particularly polyvalent ions, on their surface specifically and then change their surface electrochemical properties. Because of these characteristics of variable charge soils, their electrokinetic properties may exhibit some special features.

Figure 7.3 shows the change in zeta potential of two variable charge soils (Rhodic Ferralsol and Ali–Haplic Acrisol) with the change in pH. For comparison, the zeta potentials of a phaeozem and a Cambisol are also shown. Phaeozem can be taken as the representative of constant charge soils. For this soil the zeta potentials were all negative in sign within the pH range studied, and the magnitude was less affected by pH, varying between –45 mV and –27 mV. By contrast, the negative zeta potentials of the

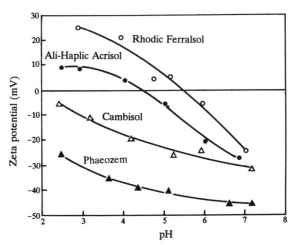

Fig. 7.3. Zeta potentials of variable charge soils and constant charge soils (0.001 M NaCl).

Rhodic Ferralsol and the Ali–Haplic Acrisol decreased gradually with the decrease in pH and then turned to be positive in sign. The variation ranges of zeta potential for these two soils were -23 mV to 25 mV and -25 mV to 10 mV, respectively. An IEP occurred for both soils. The IEPs were 5.5 and 4.5, respectively. The electrokinetic behavior of these variable charge soils is quite similar to that of iron oxides such as hematite as the typical variable charge colloid. It can also be seen from the figure that the electrokinetic properties may differ to some extent among different types of variable charge soils. As shall be seen in Section 7.5, this point is of significance in soil classification.

The electrokinetic characteristics of variable charge soils are related to their composition and are also affected by environmental factors, especially the presence of specific ions. These will be discussed in the following sections.

7.2 CONTRIBUTION OF SOIL COMPONENTS TO ELECTROKINETIC PROPERTIES

7.2.1 General Description

The electrokinetic properties of a mixture are related to the electrokinetic properties of its constitutional components, with each component contributing its part to the overall property. In order to quantify this principle, Parks (1967) established the relationship between the overall ZPC of a mixture and the individual IEP of each component:

$$\text{ZPC} = \sum_i f_i \text{IEP}_i \qquad (7\text{-}28)$$

where f_i is the fraction of i component, and IEP_i is the IEP of that component. This equation indicates that the ZPC of a mixture reflects the geometrical average of the IEP of each component. This theory has been widely accepted (Elliott and Sparks, 1981). Based on this theory, Hendershot and Lavkulich (1978) studied soil development and changes in ZPC as related to $\Delta\text{pH}_{\text{ZPC}}$, that is, $\text{ZPC} - \text{pH}_{\text{KCl}}$. Escuday and Galindo (1983) calculated the theoretical ZPC of clays composed mainly of gibbsite with the aid of equation (7-28) and compared with actually measured results, showing the important effect of iron oxides coated on the surface of clay particles on their electrokinetic properties.

Variable charge soils are composed of clay minerals different in structure and properties as well as of metal oxides, particularly iron and aluminum oxides. From the viewpoint of surface chemical structure, the surface of variable charge soils can be broadly distinguished into two categories: siloxane surface and hydrous oxide surface (Greenland and Hayes, 1978).

They belong to constant charge surface and variable charge surface, respectively. Soil organic matter carrying variable charges can also influence the surface properties of variable charge soils. Soil organic matter differs from hydrous oxides in that, while the latter are distinctly amphoteric in behavior, the former generally does not possess such a property. On the edges of clay minerals there also may be the presence of some variable charge surfaces, especially for kaolinite, although most of the surfaces are of siloxane type. All the surfaces of these components can affect the electrokinetic properties of a variable charge soil as an entity.

Various components of variable charge soils are not intermingled mechanically, but are combined together in different forms and in different ways. For instance, iron oxides can be present either as single crystalline particles or as amorphous substances coated on the surface of soil particles. Some aluminum oxides can even exist in the interlayers of clay minerals. Consequently, the relationship between electrokinetic properties of variable charge soils and those of their components may be quite complicated.

Nevertheless, it is still possible to examine the contribution of individual component of variable charge soils to the electrokinetic properties of the soil, despite the complexities mentioned above. Such studies can help us to understand some affecting factors of electrokinetic properties of the soil and the cause of the difference in these properties among different types of soils, and they can also help us to inquire into the relationship between electrokinetic properties of variable charge soils and their surface charge characteristics.

7.2.2 Iron Oxides

Iron oxides are one of the most important components in inducing the variability in surface properties of variable charge soils. With a variety of forms, iron oxides may behave quite differently with respect to surface charge characteristics and other surface chemical properties. Therefore, the electrokinetic properties of different kinds of iron oxides and the effect of these oxides on electrokinetic properties of the soil may vary to a certain extent.

According to Parks (1965), the IEP of various iron oxides generally ranges from 5.2 to 8.6, but sometimes may be higher. Van Schuylenborgh (1950) has reported low values of 3.2 for α-FeOOH and 2.1 for α-Fe$_2$O$_3$.

Figure 7.4 shows the change in zeta potential with the change in pH for a synthesized goethite treated at 70°C and a natural goethite. It can be seen from the figure that the zeta potentials of the synthesized goethite are positive within most of the pH range studied, and the potentials are quite high. The variation range is from 38 mV to −15 mV, with an IEP of 8.8. This is the typical electrokinetic behavior of iron oxides. By contrast, the

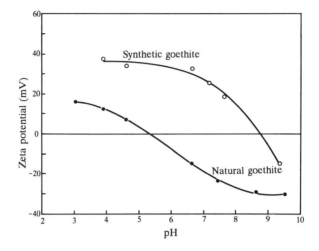

Fig. 7.4. Zeta potential of goethites (10^{-4} M NaCl).

electrokinetic behavior of natural goethite differs from that of the synthesized one remarkably. The zeta potentials of this mineral vary from 17 mV to -29 mV, with an IEP of 5.4, lower than that of the synthesized oxide by 3.4 pH unit. Parks (1965) thought that the extent of hydration, impurities, and structural defect may all have an effect on the electrokinetic behavior of iron oxides. Another possible factor may be the crystallinity and the degree of aging of the iron oxides. When studying the effect of heating treatment at different temperatures on the IEP of Fe_2O_3, Al_2O_3, and Cr_2O_3, Kittaka (1974) found that the IEP decreased with the rise in temperature. He considered that this effect was related to the release of OH^- groups on the surfaces of the oxides, which in turn was associated with the distance between the metal ions and their neighboring oxygen ions and with the coordination number of the metal ions.

In order to study the contribution of iron oxides to electrokinetic properties of variable charge soils, one can treat the soil by coating with iron oxides or removing iron oxides originally contained in the soil, and then observe the change in electrokinetic properties.

Figs. 7.5 and 7.6 show the change in zeta potential after the removal of iron oxides for the Hyper–Rhodic Ferralsol and Ferrali–Haplic Acrisol and the Ali–Haplic Acrisol, respectively. It can be seen from the figures that the zeta potential of the Hyper–Rhodic Ferralsol decreased markedly after the removal. The magnitude of decrease was about 20 mV within the pH range of 2.5–5.5. The decrement decreased with the rise in pH, and there was no remarkable difference in zeta potential from the original soil when the pH rose to 7.5. The IEP decreased from 6.4 to 5.3 after removal. The zeta potential of the Ali–Haplic Acrisol also decreased after the removal,

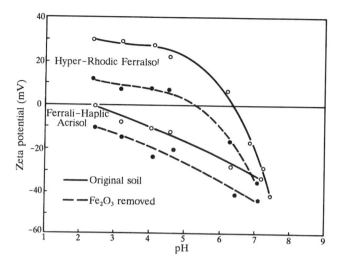

Fig. 7.5. Change in zeta potential of Hyper-Rhodic Ferralsol and Ferrali-Haplic Acrisol after removal of iron oxides.

although the decrement was not as large as for the Hyper-Rhodic Ferralsol. The Ferrali-Haplic Acrisol behaved similar to the Ali-Haplic Acrisol. These results demonstrate that the contribution of iron oxides to electrokinetic properties of variable charge soils is remarkable. Escudey and Galindo (1983) also reported similar changes in zeta potential and IEP of the clay fraction of Andosols composed mainly of noncrystalline allophane and iron

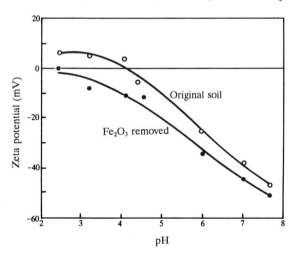

Fig. 7.6. Change in zeta potential of Ali-Haplic Acrisol (Jinhua) after removal of iron oxides (Zhang and Zhang, 1992a).

oxides after the removal of iron oxides.

The material presented above showed that the change in electrokinetic properties after the removal of iron oxides differs among various soil types. The marked decrease in zeta potential of the Hyper–Rhodic Ferralsol should be attributed to the large amount of iron oxides contained by this soil, which is as high as 23%. By contrast, the Ali–Haplic Acrisol contains 8.65% of iron oxides, and therefore the change in zeta potential caused by the removal of Fe_2O_3 is not as remarkable as that of the Hyper–Rhodic Ferralsol.

The IEP of iron oxides generally ranges from 6.5 to 8.5. Hence, iron oxides chiefly carry positive surface charges under ordinary conditions. Besides, these oxides in variable charge soils can mask permanent negative surface charges. Therefore, it can be suggested that decreases in zeta potential and IEP of these soils are the combined results of the decrease in positive surface charge and the increase in negative surface charge.

Escudey and Galindo (1983) used the difference between theoretical IEP calculated from Equation (7-28) and the actually measured value to account for the extent of coating of iron oxides on the surface of clay particles and the effect of such coating on surface charge of soils. The contents of Fe_2O_3, Al_2O_3, and SiO_2 of the Hyper–Rhodic Ferralsol are 23.12%, 31.14%, and 25.04%, respectively. Thus, its theoretical IEP should be 5.6, lower by 0.8 than the actually measured value 5.6. This indicates that there is a part of iron oxides coated on the surface of soil particles indeed, and this exerts a significant effect on the charge characteristics of the soil.

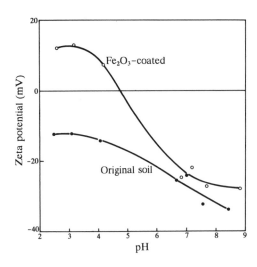

Fig. 7.7. Change in zeta potential of kaolinite after coating with iron oxides (Zhang and Zhang, 1992a).

Another approach for studying the contribution of iron oxides to electrokinetic properties of soils is to coat iron oxides on some pure clay minerals, such as kaolinite. Figure 7.7 shows the change in zeta potential of a kaolinite sample after coating with iron oxides. It can be seen that the zeta potentials of kaolinite were all negative before treatment. After coating, the potential rose markedly and might became positive in sign, resulting in an IEP of 4.75. Within the lower pH range, the zeta potential was more positive by approximately 25 mV than that of the original sample. Studies made by Rengasamy and Oades (1977) also showed a marked increase in zeta potential when kaolinite reacted with polymerized [Fe(III)–OH] and Fe^{3+}, and the more the amount of iron, the greater the effect. They also found that the change in electrophoretic mobility with the change in pH for the kaolinite after reaction with iron paralleled the change in net surface charge. These results demonstrate from another perspective that iron oxides possess a great effect on electrokinetic properties of variable charge soils and that one of the main reasons for the marked variability in surface charge of these soils is associated with the high content of iron oxides.

7.2.3 Aluminum Oxides

Aluminum oxides present in soils include crystalline gibbsite and amorphous aluminum oxides. There may also be the existence of some interlayer hydroxyl aluminum polymers in soils containing a large quantity of swelling minerals.

The electrokinetic properties of aluminum oxides are similar to those of

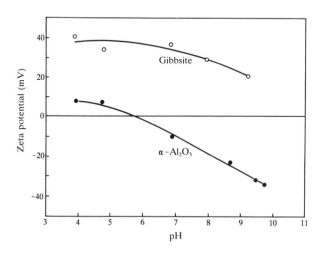

Fig. 7.8. Zeta potentials of gibbsite and α-Al_2O_3.

iron oxides. The IEP are quite high, and the kind, form, degree of aging, and extent of hydration of these oxides can all influence the electrokinetic properties.

Figure 7.8 shows the change in zeta potential of a synthetic gibbsite and an α-Al_2O_3 with the change in pH. The zeta potentials of the gibbsite were all positive within the whole pH range studied, whereas those of the α-Al_2O_3 were comparatively low, ranging from 9 mV to -34 mV, with an IEP of 5.9. In the literature, there have been numerous reports about the electrokinetic properties of various kinds of aluminum oxides. According to the review of Parks (1965), their IEP generally ranges from 5.0 to 9.3. But the IEP of corundum is quite low, and sometimes it may be as low as 2-3.

The contribution of aluminum oxides to electrokinetic properties of soils can be investigated through coating with aluminum oxides. Figure 7.9 shows the change in zeta potential of a Xanthic-Haplic Acrisol after coating with aluminum oxides. The zeta potentials increased remarkably within the pH range of 4-7, with an increment of as large as 15-20 mV. The IEP rose to 5.3 from the original value 4.3. This indicates that aluminum oxides precipitated on the surface of soil particles have an obvious effect on electrokinetic properties of the soil. Harsh et al. (1988) obtained similar results when studying the effect of coating with aluminum oxides on electrokinetic properties of Li-montmorillonite.

The change in zeta potential of a Hyper-Rhodic Ferralsol after coating with aluminum oxides is shown in Fig. 7.10. After coating, the zeta potential increased markedly below pH 7. However, the IEP only rose from 6.2 to 6.8. This soil contains more than 20% of free iron oxides and a large

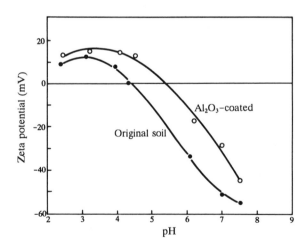

Fig. 7.9. Effect of coating with aluminum oxides on zeta potential of Xanthic-Haplic Acrisol (Zhang and Zhang, 1992a).

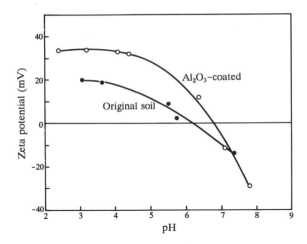

Fig. 7.10. Effect of coating with aluminum oxides on zeta potential of Hyper-Rhodic Ferralsol (Zhang and Zhang, 1992a).

amount of gibbsite. Consequently, the IEP of the soil is quite high and is less affected by coating with aluminum oxides.

It can be seen from the above two examples that the effect of aluminum oxides on electrokinetic properties of soils differs with the soil type. The effect of gibbsite in the Xanthic–Haplic Acrisol is more remarkable in the high pH range, while it is more remarkable in the low pH range for the Hyper-Rhodic Ferralsol. Presumably this is caused by the difference in composition of the two soils.

7.2.4 Silicon Oxides

Silicon oxides in soils chiefly exist in forms of amorphous opal, silicon oxide gel, secondary quartz, and phosphorous quartz (Chen, 1983; Wada and Harward, 1974). Well-crystallized silicon oxides are generally found in coarse-size particles. Similar to aluminum, silicon is one of the major elements of the solid part of soils. On the other hand, its electrokinetic properties are distinctly different from those of aluminum.

According to the data of Parks (1965), the IEP of natural quartz lies generally at about 2. For silicon oxide gel it is about 1.8. Therefore, among soil components capable of influencing electrokinetic properties of variable charge soils, silicon is one component causing the IEP to be low. In this respect, free silicon oxides behave differently from the siloxane surface in the crystal lattice of soil solid phase. The latter is affected by the aluminum in the lattice, while free silicon oxides not. These oxides can combine with other components in a variety of ways.

Studies of Li and Bruyn (1966) showed that the adsorption densities of Na^+, Cl^- and acetate ions at different pH and different concentrations of electrolytes coincided quite well with zeta potentials of quartz. Studies by Michael and Williams (1984) indicated that the GCSG model failed to interpret the experimental data obtained by two kinds of methods, namely, the electrokinetic method (electrophoresis) and the electrostatic method (potentiometric titration), possibly because of microregional differences at the surface of quartz particles.

7.2.5 Allophane

As an amorphous substance, allophane is one important component of Andosols. It has a large surface area and carries predominantly variable surface charges. The investigation of Escudey and Galindo (1983) showed that the zeta potentials of the clay fraction of Andosols composed mainly of allophane ranged from 50 mV to -30 mV within the pH range of 3–11, and the IEP varied from 8.7 to 10. This means that the IEP of allophane corresponds to that of iron and aluminum oxides. Thus, it may be postulated that the effect of allophane on electrokinetic properties of variable charge soils may be similar to that of iron and aluminum oxides.

7.2.6 Clay Minerals

The amount of clay minerals in variable charge soils is larger than that of iron and aluminum oxides in absolute quantity and may have a considerable contribution to electrokinetic properties of the soil. In terms of relative importance, however, their effect is smaller than that of the oxides. The contents of 1:1-type and 2:1-type clay minerals in variable charge soils varies with soil type. For instance, Rhodic Ferralsol is dominated by 1:1-type mineral kaolinite, while Ali–Haplic Acrisol contains a certain amount of 2:1-type minerals such as illite and vermiculite. Owing to the difference in structure, the two types of minerals have different surface properties. While both types carry some permanent negative surface charges originated from isomorphic substitution, the 2:1-type mineral has more charges. On the other hand, in addition to permanent charges, 1:1-type minerals also have some variable charges on their edge face. Therefore, the electrokinetic properties of these two types of clay minerals differ to a certain extent. Bergna (1950) reported that the negative value of zeta potential of kaolinite was markedly smaller than that of montmorillonite and illite. Delgado et al. (1986) have studied the electrokinetic properties and surface charge density of montmorillonite. Friend and Hunter (1970) suggested that the electrokinetic properties of clay minerals are closely related to their chemical composition and mineralogical structure as well as

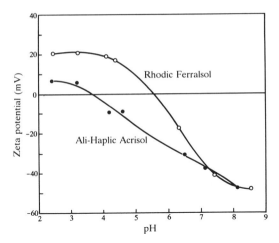

Fig. 7.11. Zeta potentials of variable charge soils with different mineralogical compositions.

composition of the medium.

The zeta potentials of a Rhodic Ferralsol derived from basalt and an Ali-Haplic Acrisol derived from phyllite are shown in Fig. 7.11. The contents of Fe_2O_3 of the two soils are 19.3% and 19.7%, respectively, very close to each other. Hence, the difference in electrokinetic properties between the two soils should be related to differences in their mineralogical composition. The Rhodic Ferralsol is dominated by kaolinite and gibbsite, while the Ali-Haplic Acrisol also contains a considerable amount of vermiculite and interstratified minerals. The higher zeta potential and higher IEP of Rhodic Ferralsols are probably caused by their higher contents of kaolinite and especially of gibbsite. This example implies that, in addition to iron and aluminum oxides, clay minerals may also have an important effect on the electrokinetic properties of variable charge soils.

7.2.7 Organic Matter

In addition to iron and aluminum oxides, humus is another major component in causing marked variability in surface charge of variable charge soils. Humic substances themselves are variable charge colloids with a huge surface area. Many authors have investigated the effect of organic matter on the electrostatic properties of variable charge soils (Hendershot and Lavkulich, 1978; Morais et al., 1976; Perrott et al., 1976). These substances should also exert influences on the electrokinetic properties of these soils.

Figure 7.12 shows the change in zeta potential for a Rhodic Ferralsol after the removal of organic matter. This sample contains 1.51% of organic

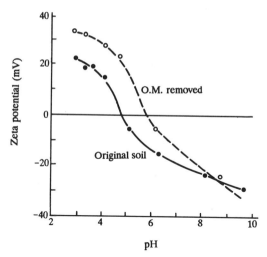

Fig. 7.12. Effect of removal of organic matter on zeta potential of Rhodic Ferralsol (Zhang and Jiang, 1964).

matter originally, and the content decreased to 0.38% after the removal. It can be seen from the figure that the zeta potential increased markedly and the IEP rose to 5.6 from the original value 4.7. The change in potential became less distinct when the pH was higher than 8. Jiang and Shen (1962) obtained similar results when studying the effect of organic matter on isoelectric properties of colloids of variable charge soils.

The content of organic matter in the soil profile generally decreases from the surface layer downwards. This can be reflected in electrokinetic properties, as shall be seen in Section 7.5.3.

7.2.8 Effect of Particle Size

Generally, the chemical compositions of soil particles with different sizes are different. Hence, the electrokinetic behaviors of different size fractions of a same soil may have a certain difference.

Figure 7.13 shows the zeta potentials of different size fractions of an Ali–Haplic Acrisol derived from Quaternary red clay. The zeta potentials of particles with sizes smaller than 1 μm are higher than those with particles 1–5 μm in size. Analyses show that the latter fraction contains less Fe_2O_3 and more quartz than the former fraction. Since the zeta potential of quartz is generally very low, the potentials of coarse particles should also be low.

Different from the case mentioned above, the difference in zeta potential between colloids of two size fractions of a Rhodic Ferralsol is quite small, owing to a similar chemical composition for the two fractions.

Thus, the effect of particle size on electrokinetic properties of soils is in

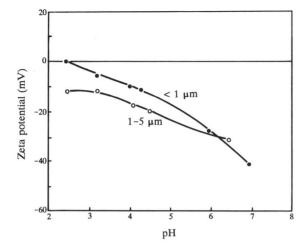

Fig. 7.13. Zeta potentials of clay fractions with different sizes of Ali–Haplic Acrisol (Zhang and Zhang, 1992a).

reality a reflection of the effect of chemical composition.

7.3 INFLUENCE OF ENVIRONMENTAL FACTORS

In the previous section, the contribution of the solid phase of variable charge soils to electrokinetic properties has been discussed. However, variable charge soils are not present in solid phase alone, but are in equilibrium with the surrounding liquid and gas phases. Therefore, environmental conditions surrounding the solid phase should also influence the electrokinetic properties. Actually, electrokinetic property is a reflection of electrochemical properties at the solid–liquid interface, and is related to both the solid phase and the liquid phase. For a variable charge soil system, H^+ ions and OH^- ions are potential–determining ions, and the production of variable charges is caused by the adsorption and dissociation of H^+ ions or OH^- ions on the surface of soil particles. Therefore, the pH of the liquid phase will have direct effects on surface charge and surface potential of the soil. Monovalent inorganic ions such as Cl^-, NO_3^-, Na^+, and K^+ are generally regarded as nonspecific ions. Their concentration can influence the thickness and structure of the electric double layer as well as the ionic strength of the liquid phase, which in turn can also influence the electrokinetic properties of soils. Specific ions can enter the Stern layer and can even change the sign of surface charge of soils, and thus they may have great effect on electrokinetic properties. Besides, physical factors such as water content and temperature of the soil can also affect its electrokinetic properties. In actual soil systems, these factors can interact mutually as well, and therefore their effects on electrokinetic properties of the soil may intermingle with one another. In the following sections, the effects of pH, kind of ions and concentration of neutral salts on electrokinetic properties of variable charge soils will be discussed. The effect of specific adsorption of ions will be treated in detail in Section 7.4.

7.3.1 pH

It has been seen in the preceding sections that the zeta potential of variable charge soils varies remarkably with the change pH, reflecting the variability of electrokinetic charge of these soils. The production of variable charges may be expressed by equation (7-26):

$$\overset{H^+ \qquad OH^-}{MOH_2^+ \; \leftarrow \; MOH \; \rightarrow \; MO^- + H_2O} \qquad (7\text{-}26)$$

At low pH, the concentration of H^+ ions in solution is high and the surface of variable charge soils adsorb excessive amount of H^+ ions. Hence, these soils carry positive charges and show a positive zeta potential. At high pH, the concentration of OH^- ions is high and H^+ ions will dissociate from the surface, and hence the soils carry negative charges and show a negative zeta potential. At a certain pH, IEP will occur.

According to the site–binding model (Hunter, 1981) for describing the electrical double layer of metal oxide–water systems, the relationship between surface potential and pH is

$$\psi_0 = 2.303 \frac{kT}{e} [pH_z - pH] - \frac{kT}{ze} \ln \frac{[AH_2^+]}{[A^-]} \qquad (7\text{-}29)$$

where pH_z is the pH at which $\psi_0 = 0$, and A is the solid phase. In the absence of specifically adsorbed ions in the system, zeta potential and ψ_0 change in the same direction. Hence, the above equation can also indirectly reflect the relationship between zeta potential and pH. Of course, since variable charge soils are mixed systems composed of both variable charge components and permanent–charge components, the relationship between zeta potential and pH is actually rather complex. Dixit (1982) discussed the complexity of the effect of pH on electrokinetic properties of soil colloids. Nevertheless, pH is the most important factor in influencing the electrokinetic property of variable charge soils. Since pH itself is apt to change in soil systems, the zeta potential of variable charge soils is liable to change under natural conditions.

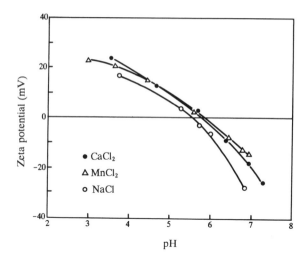

Fig. 7.14. Zeta potentials of Rhodic Ferralsol in solutions of different cations (Zhang and Zhang, 1991).

7.3.2 Kind of Ions

Because of differences in size, degree of hydration, and electric charge, various ions locate in the electric double layer differently. This would affect electrokinetic properties of the soils.

The zeta potentials of a Rhodic Ferralsol in solutions composed of different cation species are shown in Fig. 7.14. It can be seen that, when compared with Na^+ ions, the presence of Ca^{2+} ions and Mn^{2+} ions causes a more positive zeta potential of the Rhodic Ferralsol. This difference is more remarkable at high pH. The effects of two divalent cations Ca^{2+} and Mn^{2+} are nearly identical. The presence of divalent cations also leads to a higher IEP than in the presence of monovalent cations. For the Ali–Haplic Acrisol, similar trends can be observed (Fig. 7.15).

The electrokinetic properties of variable charge soils in solutions composed of different anion species are also different. In this case a rather complex specific adsorption may frequently be involved. This will be discussed in Section 7.4.

7.3.3 Concentration of Neutral Salt

Indifferent ions of neutral salts are generally distributed within the diffuse layer of the electric double layer. This is different from the case for potential-determining ions and specifically adsorbed ions. When the concentration of this kind of ions is increased, the electric double layer will

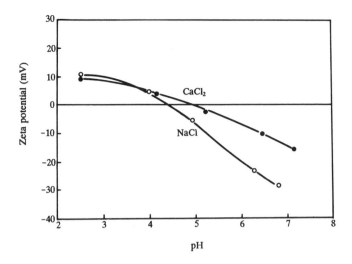

Fig. 7.15. Zeta potentials of Ali–Haplic Acrisol in NaCl solution and in CaCl₂ solution.

be compressed and hence the zeta potential will decrease. This has been demonstrated by experimental results obtained with pure oxides including hematite (Hunter, 1981), γ-Al_2O_3 (Sprycha, 1989), TiO_2 (Sprycha, 1986), quartz (Michael and Willams, 1984), and SnO_2 (Houchin and Warren, 1984). In this case, the zeta potential-pH curves in solutions of different concentrations intersect at the IEP of the oxides.

For various kinds of clay minerals, the experimental results are very inconsistent, and some of them may even be contrary to the rule mentioned above. The results obtained by Low (1987) and Miller (1984) showed that the zeta potential of montmorillonite did not vary with the concentration of neutral electrolytes. Harsh et al. (1988) found that the zeta potential of Li–montmorillonite coated with aluminum oxides even rose with the increase in ionic strength within the pH range 4-5. Studies with kaolinite (Williams and Williams, 1978) showed that the effect of neutral electrolytes was related to pH. When examining whether vermiculite can be used as a simulating system for the test of the electric double layer theory, Friend and Hunter (1970) observed that some particular effects could occur if the electrolyte concentration was high, and these effects might influence the coagulation of clay particles. Different explanations by various authors have been made for these results.

The changes in zeta potential for a Rhodic Ferralsol and an Ali–Haplic Acrisol with the change in pH in NaCl solutions of two concentrations are shown in Fig. 7.16. It can be seen that the change in NaCl concentration did not lead to an appreciable change in electrokinetic properties of the two

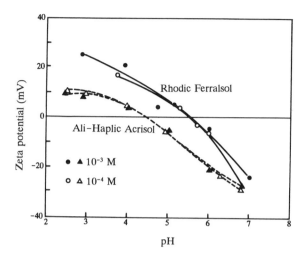

Fig. 7.16. Zeta potentials of Rhodic Ferralsol and Ali–Haplic Acrisol in NaCl solutions of different concentrations.

soils. When the neutral electrolyte was KNO_3 (Fig. 7.17), the zeta potential rose slightly with the increase in electrolyte concentration at high pH. These results indicate that the relationship between electrokinetic properties of variable charge soils and concentration of neutral electrolytes is different not only from that of pure oxides, but also from that of clay minerals. Because variable charge soils consist of both constant charge components

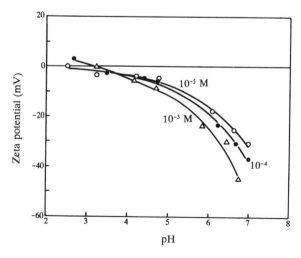

Fig. 7.17. Zeta potentials of Ali–Haplic Acrisol in KNO_3 solutions of different concentrations.

and variable charge components and can adsorb both cations and anions of the neutral electrolyte in which specific adsorption may also be involved, at present no unequivocal generalization can be made for this complex problem.

7.4 EFFECT OF SPECIFIC ADSORPTION OF IONS

The specific adsorption of some cations and anions by variable charge soils has been referred to many times in the preceding chapters. The mechanisms of such adsorption differ with the ion species. From the viewpoint of electrokinetic properties of soils, since the dominant factor is the distribution of ions within the electric double layer and since the common characteristic feature of specifically adsorbed ions is that they can enter the Stern layer or even become a constituent part of the soil solid phase, thus functioning as potential–determining ions, the effect of specifically adsorbed ions on electrokinetic properties of soils would be much stronger than that of indifferent ions. In particular, since variable charge soils can adsorb ions specifically more strongly than do constant charge soils, the effect of specific adsorption of ions on electrokinetic properties of these soils is of important significance. One of the important consequences of specific adsorption is the shift of both ZPC and IEP. After specific adsorption of anions, the ZPC usually shifts towards higher pH and the IEP shifts towards lower pH, while after specific adsorption of cations the ZPC shifts towards lower pH and the IEP shifts towards higher pH (Arnold, 1978). However, there are different reports and views regarding the direction of the shift in ZPC (Parker et al., 1979; Sposito, 1981).

In soils there are a variety of ions that can be adsorbed specifically on the surface of soil particles. Both the form and the valence of these ions may be different. Two types of ions are of special significance. One type is specific anions, such as F^-, SO_4^{2-}, PO_4^{3-}, SiO_4^{2-}, $C_2O_4^{2-}$, MoO_4^{2-}, and BO_3^{2-}. The other is hydrolytic metal cations, including some heavy metal ions and transitional metal ions. This latter type of ions has been treated in Chapter 4. In the following, the effects of the former type of ions on electrokinetic properties of variable charge soils when these ion species exist alone or in combination with the latter type of ions will be discussed.

7.4.1 Specific Adsorption of Anions

The change in zeta potential of a Rhodic Ferralsol in solutions containing NaCl, Na_2SO_4, or NaF with the change in pH is shown in Fig. 7.18. It can be seen that the zeta potentials are lower in the Na_2SO_4 solution and especially in the NaF solution than in the NaCl solution. This difference vanishes when the pH is higher than about 7. The IEPs in the two solutions

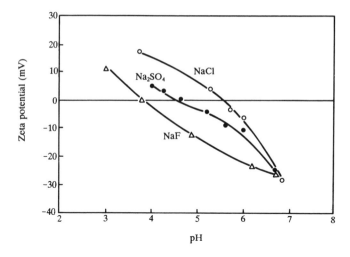

Fig. 7.18. Zeta potentials of Rhodic Ferralsol in NaCl, Na$_2$SO$_4$, and NaF solutions (Zhang and Zhang, 1991).

are 4.6 and 3.8, respectively, lower by 0.9 and 1.7 pH unit, respectively, than in the NaCl solution. It has been shown in Chapter 6 that this type of soil can adsorb sulfate coordinately, and the coordination adsorption of fluoride ions is even stronger. Obviously, the decrease in zeta potential of the soil in the presence of the two ion species is caused by the coordination adsorption of these ions.

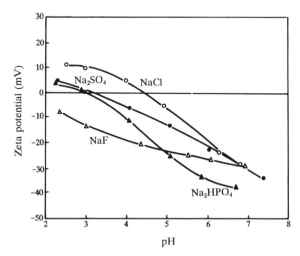

Fig. 7.19.   Effect of specific adsorption of anions on electrokinetic properties of Ali–Haplic Acrisol (Zhang and Zhang, 1991).

Figure 7.19 shows the effects of SO_4^{2-}, F^-, and HPO_4^{2-} ions on electrokinetic properties of the colloids of an Ali–Haplic Acrisol. In the figure, the zeta potential–pH curves of the soil in the presence specifically adsorbed anions lie lower than in presence of chloride ions. The effects of various anions are different within different pH ranges. In the presence of HPO_4^{2-} ions, the zeta potential–pH curve almost parallels that in the presence of NaCl within the whole pH range studied, whereas in the presence of SO_4^{2-} or F^- ions the magnitude of decrease in zeta potential is large only within the lower pH range.

Figure 7.20 shows the effects of sulfate, oxalate and citrate ions on electrokinetic properties of a Rhodic Ferralsol of Brasilia. The specific adsorption of these three kinds of anions causes the zeta potential to be lower by maximally 20, 45, and 50 mV, respectively, than the potential when chloride ions are present, with the difference in extent of influence varying with pH. At pH lower than 4.2, the order of effect is oxalate > citrate > sulfate, while within the pH range above 4.2 the order is citrate > oxalate > sulfate.

Specific adsorption of anions leads to a distinct change in IEP of the Rhodic Ferralsol. For sulfate, citrate, and oxalate systems, the IEPs are 6.1, 3.4, and 2.7, respectively, whereas it is 6.6 for the chloride system.

The examples presented above showed that one of the important characteristics in the effect of specific adsorption of anions on electrokinetic properties of variable charge soils is that the effect may differ within different pH ranges. This should be related to the difference in adsorption, because the effect of anions on surface potential is associated with the

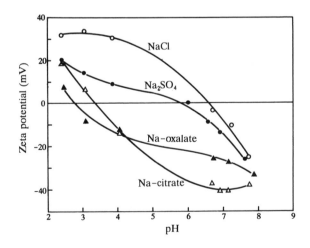

Fig. 7.20. Effect of specific adsorption of anions on electrokinetic properties of Rhodic Ferralsol of Brasilia.

number of those ions combined with the surface, the extent of binding, and their location in the electric double layer. Regarding the effect of pH, in the literature there have been many reports (Barrow, 1984; Barrow and Ellis, 1986; Bar–Yosef et al., 1988; Hingston et al., 1972; Kafkafi and Bar–Yosef, 1969; Muligadi et al., 1966; Nodvin et al., 1986; Perrott et al., 1976; Singh, 1984; Su and Harsh, 1993; White and Taylor, 1977; Zhang et al., 1987). It can thus be imagined that the weakening in the effect of sulfate on zeta potential of the three soils with the increase in pH shown in Figs. 7.18, 7.19 and 7.20 should be directly related to the decrease in adsorption of these ions (Zhang et al., 1987; cf. Chapter 6). For oxalate and citrate ions, the cause should be similar.

When comparisons are made among various anion species, it can be found that the influence by fluoride ions and phosphate ions on zeta potential and IEP is greater than that by sulfate ions, with the relative extent of influence between the former two ion species depending on pH. When the pH is low, the effect of fluoride ions is greater. When pH is high, the effect of phosphate ions is stronger (Fig. 7.19). The greater dependence of the effect of phosphate ions on pH as compared to fluoride ions should be related to its pattern of change in valence. Of course, fluoride ions also have a chemical equilibrium $HF \rightleftharpoons F^-$ which is influenced by pH. However, the influence of pH on the equilibrium of phosphate ions, $H_2PO_4^- \rightleftharpoons HPO_4^{2-} \rightleftharpoons PO_4^{3-}$, is much greater. Likely, the effect of oxalate on zeta potential is greater than that of citrate at low pH while the reverse is true at high pH (Fig. 7.20). This can also be explained by the greater influence of pH on the valence of trivalent citrate ions than on oxalate ions.

With respect to soil type, it can be noticed that the effect of specific adsorption of anions on electrokinetic properties of Rhodic Ferralsols is distinctly greater than that on Haplic Acrisols. This should be attributed to the higher contents of iron and aluminum oxides and thus the stronger adsorption of anions by the former type of soil.

For the same anion species, its effect on electrokinetic properties of soils is also related to its concentration.

Figure 7.21 shows the effect of SO_4^{2-} ions of different concentrations on electrokinetic properties of the Rhodic Ferralsol, in which the higher the concentration of the anions, the larger the decrement in zeta potential and the more the IEP shift towards lower pH. In solutions of 5×10^{-6} and 2.5×10^{-4} SO_4^{2-} mol L^{-1}, the IEPs are 5.0 and 3.4, respectively. The IEP of this soil in Cl^- is 5.6. The difference in the effect of concentration vanishes gradually with the increase in pH, because the zeta potential–pH curve tends to close to the curve in Cl^- solution. This trend in the change of zeta potential is in conformity with the change in adsorption of sulfate with the change in pH (Nodvin et al., 1986; Zhang and Zhang, 1989). The effect of phosphate ions on electrokinetic properties of the Ali–Haplic Acrisol also

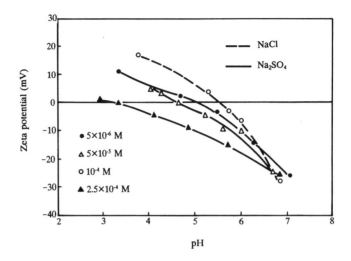

Fig. 7.21. Effect of concentration of Na_2SO_4 on electrokinetic properties of Rhodic Ferralsol (Zhang and Zhang, 1991).

becomes more remarkable with the increase in concentration (Fig.7.22). Different from the case for sulfate, the zeta potential–pH curves for several concentrations almost parallel one another, and the difference among concentrations remains visible even at high pH. This reflects the difference in characteristics of adsorption of the two kinds of ions.

The effect of concentration for fluoride ions is even more striking. Such

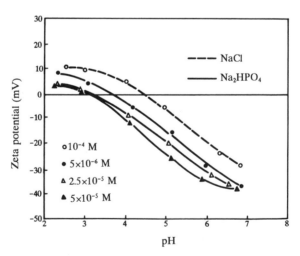

Fig. 7.22. Effect of concentration of Na_2HPO_4 on electrokinetic properties of Ali–Haplic Acrisol.

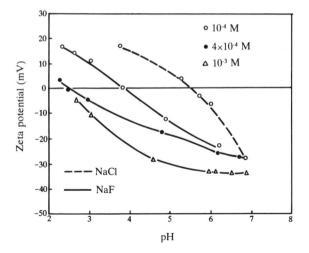

Fig. 7.23. Effect of concentration of NaF on electrokinetic properties of Rhodic Ferralsol (Zhang and Zhang, 1991).

effects for a Rhodic Ferralsol and an Ali–Haplic Acrisol are shown in Figs. 7.23 and 7.24, respectively. For the Rhodic Ferralsol, fluoride ions of 10^{-4} M and 4×10^{-4} M cause the IEP to drop to 3.8 and 2.4, respectively. For the Ali–Haplic Acrisol, fluoride ions of 10^{-5} M and 5×10^{-5} M cause the IEP to decrease to 3.4 and 2.8, respectively. When the concentration is further higher, both of the soils exhibit negative zeta potential within the whole pH

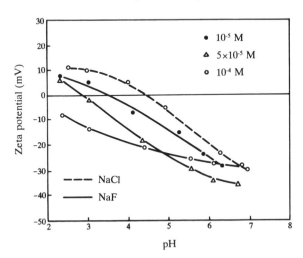

Fig. 7.24. Effect of concentration of NaF on electrokinetic properties of Ali–Haplic Acrisol.

range studied. This demonstrates the strong effect of fluoride ions on electrokinetic properties of variable charge soils.

7.4.2 Competitive Adsorption of Anions

When two or more species of specific anions coexist, a competitive adsorption will occur. The effect of such competition on surface characteristics of the soil can also be reflected in electrokinetic properties.

Figure 7.25 shows the change in zeta potential of a Rhodic Ferralsol with the change in pH in systems containing both sodium phosphate and sodium arsenate with different ratios. The zeta potentials are lowest when the ratio is 4:1 and highest when the ratio is 1:4. The IEPs are 3.1 and 4.8, respectively, in the two cases, differing by 1.7 pH units. The zeta potential in the two cases differs by about 20 mV, with the difference almost independent of pH. The IEP is 4.1 when the ratio is 1:1. These results indicate that the effect of phosphate ions is greater than that of arsenate ions, consistent with the order of adsorption by soils between the two anion species (Parfitt, 1978). Livesey and Huang (1981) showed that the presence of a certain amount of phosphate ions led to the decrease in adsorption of arsenate ions to a large extent. Obviously, the difference in zeta potential shown in Fig. 7.25 reflects the stronger ability of phosphate ions for competitive adsorption.

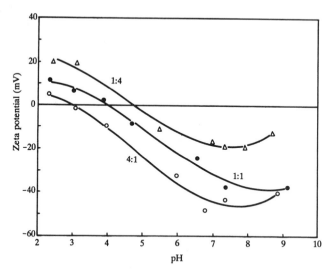

Fig. 7.25. Effect of competitive adsorption of phosphate ions with arsenate ions on electrokinetic properties of Rhodic Ferralsol (total concentration, 10^{-4} M; numerals in the figure are phosphate–to–arsenate ratios).

7.4.3 Combined Adsorption of Specific Anions and Cations

When specific anions and cations coexist in the system, variable charge soils may adsorb both of them. The two kinds of ions will distribute in different regions of the electrical double layers and exert different effects. The zeta potential at the slip surface presents a comprehensive reflection of the effect of adsorption of these two kinds of specific ions.

A comparison between the electrokinetic properties of a Rhodic Ferralsol in $ZnSO_4$ solution and those when the two ion species Zn^{2+} and SO_4^{2-} are present separately is shown in Fig. 7.26. It can be seen that within different pH ranges the influence of $ZnSO_4$ on electrokinetic properties of the soil is different. Within the low pH range, the zeta potential–pH curve is similar to that in Na_2SO_4 solution, with the exception that the potentials are comparatively lower. This indicates the dominant effect of specific anions SO_4^{2-}. Within the high pH range, the curve becomes close to that in $ZnCl_2$ solution and the potential becomes much higher than that in $NaCl$ solution. This implies that in this case specific cations Zn^{2+} play the leading role. At the vicinity of IEP, the zeta potential–pH curve lies between that in Na_2SO_4 solution and that in $ZnCl_2$ solution. This means that in this pH range both Zn^{2+} ions and SO_4^{2-} ions play a role simultaneously. It can also be seen from the figure that after adsorption of Zn^{2+} ions and SO_4^{2-} ions the IEP of the Rhodic Ferralsol is 4.6, close to the IEP in Na_2SO_4 solution. This implies that when the pH

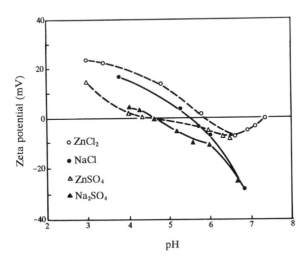

Fig. 7.26. Electrokinetic properties of Rhodic Ferralsol in $ZnSO_4$ solution as compared to those when the two ion species are present separately (Zhang and Zhang, 1991).

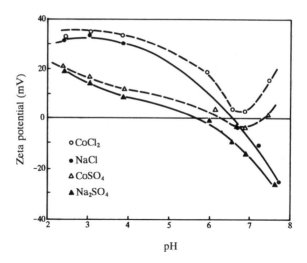

Fig. 7.27. Electrokinetic properties of Rhodic Ferralsol of Brasilia in CoSO$_4$ solution as compared to those when the two ion species are present separately.

is low, the effect of SO$_4^{2-}$ ions is greater than that of Zn^{2+} ions. It has been seen in the preceding sections that when specifically adsorbed ions influence electrokinetic properties of the soil separately, the influence of sulfate ions is greater within low pH range while that of Zn^{2+} ions is greater within high pH range. It can be seen from the tendency shown in Fig. 7.26 that such characteristic features of anions and cations can also be reflected when they coexist.

Figure 7.27 shows a comparison between electrokinetic properties of the Rhodic Ferralsol from Brasilia in the presence of CoSO$_4$ and those when the two ion species are present separately. It can be seen that within the low pH range the zeta potential–pH curve is close to but slightly higher than that when only SO$_4^{2-}$ ions are adsorbed. This suggests that at this time the effect of SO$_4^{2-}$ ions is the major one but Co^{2+} ions also play a certain role. Within the high pH range, the curve is similar to that in the presence of only Co^{2+} ions but lies slightly lower than the latter curve. This implies that at this time Co^{2+} ions play the dominant role but SO$_4^{2-}$ also exert a certain influence. In this case, two IEPs occur on the curve, with the first one lying between the two IEPs when cations and anions are adsorbed separately.

In summary, the situation in electrokinetic behavior of variable charge soils in systems containing both specifically adsorbed anions and specifically adsorbed cations is more complex than when these ions are present separately. Since zeta potential is the overall reflection of electrochemical properties of the slip surface in the electric double layer, and since it displays the influence of both anions and cations, the overall influence is determined by the relative reaction intensities of the two ion species with the soil. Thus, the electrokinetic properties are related to the nature of the two ion species as well as the composition of the soil, and they are also affected by pH and concentration of the two kinds of ions. Since pH can affect surface charge characteristics of the soil as well as the forms of the two ion species, it would be an extremely important factor in determining electrokinetic properties of variable charge soils, just like when either of the two kinds of ions is present alone.

7.5 ELECTROKINETIC PROPERTIES AND SOIL TYPE

Soils are formed through pedogenesis from parent materials as affected by environmental factors. Because of the differences in pedogenic factors, soils of different types vary in chemical and mineralogical compositions. It was shown in Section 7.2 that various soil components have different contributions to electrokinetic properties. Soil as an entity composed of these components should also manifest itself in this respect. For variable charge soils with a great variability in surface charge, this point is of more important significance. As early as 60 years ago Mattson (1932) proposed the well-known isoelectric weathering theory. In recent years, with the further development in research on variable charge soils, the relationship between electrokinetic properties and soil type received renewed interest (Elliott and Sparks, 1981; Schulthess and Sparks, 1988; Zhang and Zhang, 1991b).

Variable charge soils distribute widely in the world, including China. Because of the complexities in pedogenic factors, the types of these soils are numerous. Therefore, electrokinetic properties may be different among various types of soils and even among different horizons within a profile. These differences are related to the chemical composition of the soil, and hence they are of certain significance in soil classification.

7.5.1 Difference Among Soil Types

It was seen in Fig. 7.3 that the electrokinetic properties of variable charge soils are remarkably different from those of constant charge soils. Such differences can also be reflected among various types of variable charge soils. Figure 7.28 shows zeta potential-pH curves for four representative types of such soils. The Rhodic Ferralsol derived from basalt was collected from Hainan Island. The Ferrali-Haplic Acrisol derived from granite was collected from Guangdong Province. The Ali-Haplic Acrisol derived from Quaternary red clay was collected from Zhejiang Province. The Xanthic

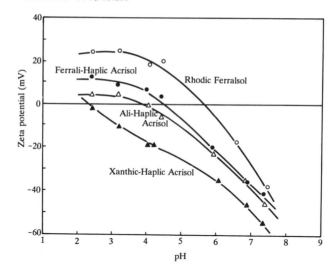

Fig. 7.28. Electrokinetic properties of four variable charge soils (Zhang and Zhang, 1992b).

Haplic Acrisol, collected from Guangdong Province, was derived from schist located at an altitude of 930 m.

The figure shows that, following the increase in pH, the zeta potential of the former three soils changes progressively from positive to negative in sign, and that of the Xanthic–Haplic Acrisol increases in negative value, although for this latter soil no IEP occurs. This indicates that the four soils all have a great variability in surface charge. From the slopes of the zeta potential–pH curves it can be judged that the variability in surface charge is the greatest for the Rhodic Ferralsol, intermediate for the Ferrali–Haplic Acrisol and Ali–Haplic Acrisol, and the smallest for the Xanthic–Haplic Acrisol. At a given pH the magnitude of zeta potential follows the same order. For instance, at pH 3 the zeta potentials of the four soils are 24 mV, 10 mV, 4 mV, and –8 mV, respectively, differing by 32 mV between the extremes. The IEPs of the former three soils are 5.7, 4.7, and 4.1, respectively. For the Xanthic–Haplic Acrisol, the estimated IEP is about 2. The materials presented above are representative for the four principal types of variable charge soils. A large number of determinations with other samples made by the author indicate that the IEPs of the four types of soils follow the same order shown in Fig. 7.28.

The difference in electrokinetic properties among the four types of soils is caused mainly by differences in mineralogical composition, quantity, and form of iron and aluminum oxides, as well as by the manner of combination of clay minerals with iron and aluminum oxides. The clay of Rhodic Ferralsol is dominated by kaolinite carrying little permanent charge. It contains large amounts of gibbsite and iron oxides. The content of Fe_2O_3 is 20% (Zhang and Li, 1958). The presence of large amounts of iron oxides and gibbsite with high IEP causes the high IEP of the soil. The predominant clay mineral of the Ferrali-Haplic Acrisol is also kaolinite. However, for this soil the content of iron oxides is much lower than that of the Rhodic Ferralsol. Besides, this soil generally does not contain gibbsite. This is the reason why its zeta potential and IEP are lower than those of the Rhodic Ferralsol. For the Ali-Haplic Acrisol, in addition to kaolinite, there is also a certain amount of 2:1-type clay minerals carrying permanent negative surface charge. Therefore, its zeta potential and IEP are even lower. The zeta potential and IEP of the Xanthic-Haplic Acrisol are the lowest among the four types of soils, due to the presence of large amounts of 2:1-type clay minerals.

If examined on a global scale, it can be found that the electrokinetic properties of the same type of soil in different regions may be quite different. Figure 7.29 shows such a comparison among Rhodic Ferralsols of Brasilia, Hawaii, and Hainan, China. The content of free iron oxides of the Rhodic Ferralsol of Hawaii is 15%. The zeta potential is relatively negative and the IEP is relatively low. The zeta potential of the Rhodic Ferralsol of Brasilia is more positive and the IEP is the highest, although the content of free iron oxides is only 9%. Since the predominant clay minerals of the three soils are all kaolinite, the difference in electrokinetic properties among them should be related to the differences in the form of iron oxides and the manner of combination of these oxides with clay minerals. It was

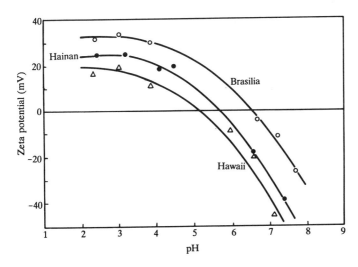

Fig. 7.29. Comparison of electrokinetic properties among three Rhodic Ferralsols (Zhang and Zhang, 1992b).

seen in Chapter 6 that the ability for adsorbing sulfate ions among the three soils is also different, and, interestingly, this difference coincides with the trend in difference shown in Fig. 7.29.

The results presented above imply that variable charge soils are different in electrokinetic properties markedly not only among different types, but also within the same type to a certain extent, owing to a variety of reasons.

7.5.2 Microregional Difference

Within a small area, owing to the differences in topography and other factors, microregional hydrological, thermal, and biological conditions may be different. These would influence the formation of soils. For instance, it can frequently be observed that on different slopes of a hill the color of soils may differ considerably. The two samples of Ferrali–Haplic Acrisols shown in Fig. 7.30 were taken from the north slope and the south slope of a small hill, respectively. The color of the former one is brownish yellow, while that of the latter is red, although both of them are derived from granite. It can be seen from the figure that their electrokinetic properties differ markedly. The zeta potential may differ by 25 mV. For the latter sample there occurs an IEP at pH 4.6, whereas for the former sample no IEP can be observed. Chemical analyses show that the chief difference in composition between the two samples is the contents of free Fe_2O_3, which are 3.7% and 10.2%, respectively (Zhang and Li, 1958). It can be postulated that the difference in electrokinetic properties between the two soil samples mainly reflects differences in the content and form of Fe_2O_3.

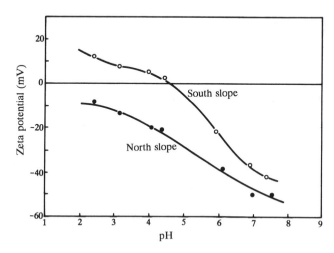

Fig. 7.30. Electrokinetic properties of Ferrali–Haplic Acrisols at north slope and south slope of a hill (Zhang and Zhang, 1992b).

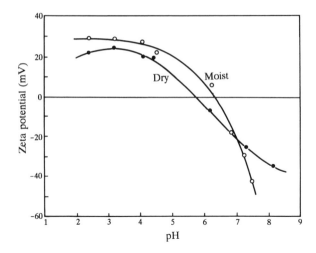

Fig. 7.31. Electrokinetic properties of Rhodic Ferralsols developed under different moisture conditions (Zhang and Zhang, 1992b).

Local difference in moisture regime can also cause differences in chemical properties of the soils. For example, Rhodic Ferralsols widely distributed in Leizhou peninsula are all derived from basalt. However, soils in the central part are dark red in color due to the occurrence of a distinctly dry season during the year, while those in the coastal area with a humid climate are brownish red in color. Figure 7.31 shows that the zeta potential of the latter soil is more positive and the IEP higher than the former soil. The mineralogical compositions of these two samples are similar, both predominated by kaolinite, gibbsite, and iron–containing minerals. The only difference is that the Fe_2O_3 content of the latter soil is 23% chiefly in the form of goethite, while that of the former soil is 18% dominated by hematite (Zhang and Li, 1958). It seems that the difference in electrokinetic properties between the two samples is caused by differences in content and form of the Fe_2O_3.

7.5.3 Differentiation Within a Profile

Having been highly weathered, variable charge soils generally are very deep, and the mineral part differs little throughout the whole profile when compared with other types of soils. However, two important factors can cause a remarkable difference in electrokinetic properties among various horizons within the profile. One factor is the effect of organic matter. It was shown in Section 7.2.7 that organic matter causes the zeta potential to become more negative. Although the content of organic matter of variable charge soils is much lower than that of the mineral part, its effect on

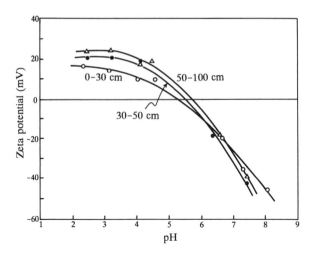

Fig. 7.32. Electrokinetic properties of soil samples of different horizons of a Rhodic Ferralsol profile (Zhang and Zhang, 1992b).

electrokinetic properties of the soil cannot be ignored, because it possesses a large surface area. Figure 7.32 shows the electrokinetic properties of different horizons of a Rhodic Ferralsol profile. It can be seen that at pH 4 the zeta potentials for the 0- to 30, 30- to 50 and 50-to 100-cm horizons are 12, 18, and 22 mV, respectively, and at pH 5 they are 3, 8 and 13 mV, respectively. The IEPs are 5.3, 5.5, and 5.7, respectively. The chemical composition and mineralogical composition of various horizons of this highly weathered Rhodic Ferralsol are quite similar. The Fe_2O_3 contents of the clay fractions are all within the range $19.0 \pm 0.4\%$, and the SiO_2/Al_2O_3 ratios are all about 1.5. However, the contents of organic matter are 3.94%, 1.02%, and 0.64%, respectively, from the surface to lower horizons (Zhang and Li, 1958). It can thus be imagined that the difference in electrokinetic properties among the three horizons should be related to the higher content of organic matter of the surface horizon.

Another factor is the difference in redox conditions caused by different water regimes within a soil profile. The principal consequence of this difference is the segregation of iron in different horizons and within different parts of the same horizon. The most obvious morphological reflection in this respect is the formation of plinthitic horizon. It can frequently be observed that in the lower part of many variable charge soils there is the presence of such a horizon. In this horizon, red and white spots and veins are interwoven with one another, exhibiting the characteristic plinthitic pattern. This is the morphological reflection of the local segregation of iron in the soil. The mechanisms of the formation of plinthitic horizon have been studied from the standpoints of mineralogy and

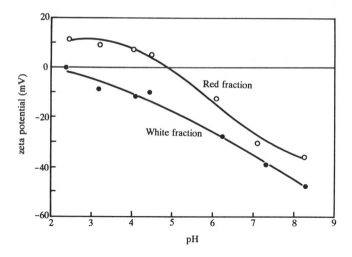

Fig. 7.33. Electrokinetic properties of red fraction and white fraction of plinthite (Zhang and Zhang, 1992b).

micromorphology (Ambrost et al., 1986; Gu, 1989; Muller and Bocquier, 1986).

Figure 7.33 shows the difference in electrokinetic properties between the red fraction and the white fraction of the plinthitic horizon of a Ferrali–Haplic Acrisol. The zeta potentials of the red fraction are distinctly higher than those of the white fraction, especially within the low pH range. The IEP of the red fraction is 4.9, while it is as low as 2.4 for the white fraction, differing by 2.5 pH units. Chemical analyses show that the SiO_2/Al_2O_3 ratios for the red fraction and the white fraction are 2.23 and 2.46, respectively, differing only slightly, but the contents of iron oxides are 14.42% and 4.77%, respectively (Zhang, 1981). Most of the iron oxides contained by the red fraction are hematite and meghemite, and a part of the oxides is coated on clay particles as amorphous gels. Apparently, the large quantity of iron oxides in the red fraction is the major cause of the distinct difference in electrokinetic properties of this fraction from those of the white fraction.

Some variable charge soils have been cultivated for rice. In paddy soils, the water regime changes more intensely and more frequently. The consequence of such change is the development of some characteristic horizons, including cultivated horizon, plowpan, illuvial horizon, glei horizon, and substratum. The difference in electrokinetic properties among different horizons of the profile of a paddy soil derived from Quaternary red clay is shown in Fig. 7.34. As seen in the figure, at pH below 5.5 the zeta potentials of the substratum are the highest. The IEP of this horizon is 3.5, essentially retaining the characteristics of the parent material, the Ali-

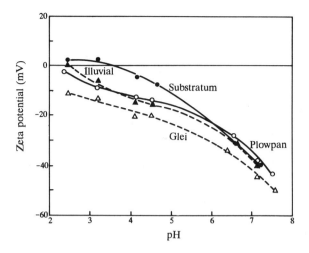

Fig. 7.34. Electrokinetic properties of soil samples of different horizons of a paddy soil profile.

Haplic Acrisol. The IEPs of the plowpan and the illuvial horizon, about 2.2 and 2.4, respectively, are much lower than that of the substratum, indicating the result of eluviation of iron oxides, despite the fact that in the illuvial horizon iron has accumulated again to a certain extent. The most striking phenomenon is that the electrokinetic properties of the gley horizon have changed to be similar to those of constant charge soils. It has been known that under long-term submergence the major part of iron oxides in the gley horizon are lost, leaving only 1-3% in content, which corresponds to one-fourth of that of the parent material (Yu, 1985). It can thus be suggested that the characteristics in electrokinetic properties of the gley horizon are a direct reflection of the reduction and eluviation of iron oxides from that horizon.

7.5.4 Soil Classification

In the preceding sections, the characteristics and their causes in electrokinetic properties of variable charge soils have been discussed. In the present section, an attempt shall be made to quantitatively relate electrokinetic property to soil classification, using the principal parameter IEP as a reference index. In Section 7.2 the contributions of various components to electrokinetic properties of the soil have been presented. Studies with mixtures of various oxides showed that their IEP is determined by the IEP of each oxide per se and their relative quantities. Variable charge soils contain, in addition to aluminosilicate minerals, large amounts of iron and aluminum oxides. Therefore, their IEP should be closely related to the kind

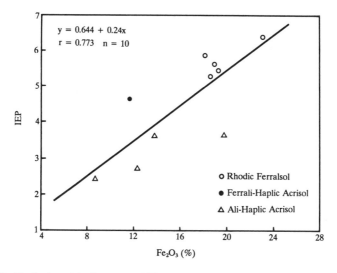

Fig. 7.35. Relationship between IEP and Fe_2O_3 content of variable charge soils (Zhang and Zhang, 1992b).

and quantity of these oxides.

Iron oxides exist in variable charge soils chiefly in free form. And, because of their large quantity, they should have a marked effect on the IEP of the soils. The relationship between IEP of 10 variable charge soils and their Fe_2O_3 content is shown in Fig. 7.35. These soils include Rhodic Ferralsols, Ferrali–Haplic Acrisols, and Ali–Haplic Acrisols. Despite the difference in mineralogical composition, the IEP correlates with the Fe_2O_3 content fairly well, with a correlation coefficient of 0.773. It can frequently be observed in electrophoretic measurements that, for a given soil sample, the electrophoretic mobility of the majority of particles falls within a certain range. It can thus be postulated that most of the iron oxides in the clay fraction are combined with aluminosilicate minerals rather than mixed mechanically with the minerals as single particles. This is the basic cause for the important effect of iron oxides on zeta potential of soils.

On the surface of aluminosilicates there are Al–OH groups as well as Si–OH groups. Aluminum oxides also have Al–OH groups. Si–OH groups can release protons, producing negative surface charges. Al–OH groups can accept protons or release OH^- ions, causing the surface to carry positive charges. Hence, the existence of Si–OH groups causes the ZPC of colloids to decrease, while the existence of Al–OH groups causes the ZPC to rise. The ZPC of the whole soil should be related to the ratio of Si–OH to Al–OH groups. Perrott (1977) and Gonzales–Batista (1982) have shown, using amorphous aluminosilicates and allophane, that the ZPC correlated

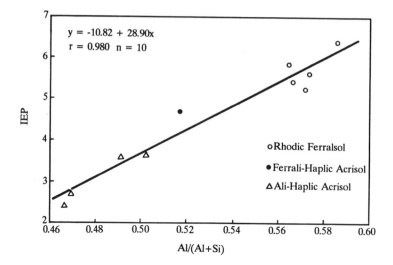

Fig. 7.36. Relationship between IEP and Al/(Al+Si) ratio of variable charge soils (Zhang and Zhang, 1992b).

with the ratio Al/(Al+Si) quite well. This correlation should also be reflected in electrokinetic properties. If the IEPs of the 10 variable charge soils are plotted against the corresponding molar ratios Al/(Al+Si) (Fig. 7.36), it can be found that the correlation coefficient is 0.980. This means that the IEP of variable charge soils is also closely related to the

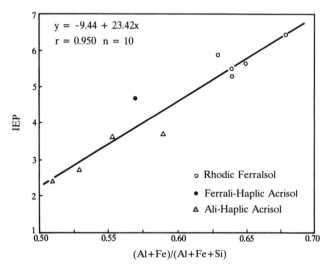

Fig. 7.37. Relationship between IEP and (Al+Fe)/(Al+Fe+Si) ratio of variable charge soils (Zhang and Zhang, 1992b).

Al/(Al+Si) ratio.

In addition to Al–OH and Si–OH groups, on the surface of variable charge soils there are also Fe–OH groups, which are similar to Al–OH groups in electrokinetic characteristics. Hence, the IEP of soils should also be related to the ratio (Al+Si)/(Al+Fe+Si). This is the actual case. As shown in Fig. 7.37, the correlation coefficient attains 0.950.

Hendershot and Lavkulich (1978) and Hendershot et al. (1979) suggested that the difference between ZPC and pH_{KCl} can be used as a comprehensive index for characterizing the degree of soil development and that this difference becomes smaller following the progress in pedogenic development. The significance of pH_{KCl} in soil chemistry will be discussed in Chapter 11. Here, the IEP-pH_{KCl} values of 10 samples may be plotted against the corresponding ratios (Fe+Al)/(Fe+Al+Si). As shown in Fig. 7.38, the value increases with the increase in the ratio, that is, with the progress in pedogenic development. It can also be found that the IEP-pH_{KCl} values for Rhodic Ferralsols and Ferrali–Haplic Acrisols are all positive, while those for Ali–Haplic Acrisols are all negative, which implies that the degree of weathering of Rhodic Ferralsols and Ferrali–Haplic Acrisols is higher than that of Ali–Haplic Acrisols.

From the materials present above it can be concluded that the IEP of variable charge soils is a fairly good index in reflecting the chemical composition of these soils. Apparently, chemical composition and the related mineralogical composition of soils are the basis of soil classification. When examining the materials about the electrokinetic characteristics of

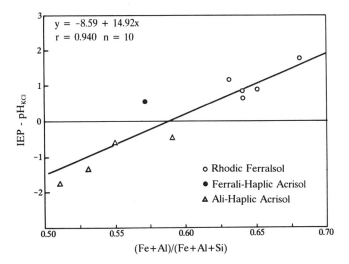

Fig. 7.38. IEP-pH_{KCl} of variable charge soils in relation to chemical composition of the soil (Zhang and Zhang, 1992b).

variable charge soils of China it may be found that the approximate IEP ranges are >5.5 for Rhodic Ferralsols, 4.0–5.5 for Ferrali–Haplic Acrisols, and <4.0 for Ali–Haplic Acrisols. The IEP of Xanthic–Haplic Acrisols is generally lower than that of Ali–Haplic Acrisols. As has been seen in Fig.7.3, for constant–charge soils such as Cambisols and Phaeozems, no IEP can be observed under ordinary conditions. Gallez et al. (1976) used ZPC as an index for classification of tropical soils and obtained the following approximate ranges: 5.5 for Oxisols, 4.0 for Ultisols, and 3.5 for Alfisols.

In addition to IEP, the sign and magnitude of zeta potential of the soil at a certain pH (particularly within the lower pH range) as well as the changing patterns of zeta potential with the change in pH may frequently reflect the characteristics of various kinds of variable charge soils. As was seen in Fig. 7.28, the zeta potential curves of Rhodic Ferralsols, Ferrali–Haplic Acrisols, Ali–Haplic Acrisols and Xanthic–Haplic Acrisols are markedly different. Thus, the electrokinetic property is a valuable parameter for characterizing soil type. Here, the pioneering concepts of amphoteric properties of soils, isoelectric weathering, and electrochemistry of soil formation raised by Mattson 60 years ago merit recollection, although they are too simplified as judged from our present knowledge of soil chemistry. Of course, for the same soil the electrokinetic properties are intimately related to ionic environment. It is for this reason that these properties bear their practical significance in soil management, such as the formation of soil structure.

BIBLIOGRAPHY

Ambrost, J. P., Nahon, D., and Herbillon, A. J. (1986) *Geoderma*, 37:283–294.

Arnold, P. W. in *The Chemistry of Soil Constituents* (D. J. Greenland and M. H. B. Hayes, eds.). John Wiley & Sons, New York., pp. 355-404.

Barrow, N. J. (1984) *J. Soil Sci.*, 35:283–297.

Barrow, N. J. and Ellis, A. S. (1986) *J. Soil Sci.*, 37:287–293.

Bar–Yosef, B., Kafkafi, U., Rosenberg, R., and Sposito, G. (1988) *Soil Sci. Soc. Am. J.*, 52:1580–1585.

Bergna, H. E. (1950) *Trans. 4th Intern. Congr. Soil Sci.*, III:75–80.

Bleam, W. F. and McBride, M. B. (1986) *J. Colloid Interface Sci.*, 110:335-346.

Bolt, G. H. (1979) *Soil Chemistry. B. Physical–Chemical Models*. Elsevier, Amsterdam.

Chen, J. F. (1983) in *Soil Colloids*, Vol. I (Y. Xiong, ed.). Science Press, Beijing, pp. 132–275.

Davidtz, J. C. and Sumner, M. E. (1965) *J. Soil Sci.*, 16:270–274.

Delgado, A., Gonzalez–Caballero, F. and Bruque, J. M. (1986) *J. Colloid*

Interface Sci., 113:203–211.

Dixit, S. P. (1982) *Soil Sci.*, 133:144–149.

Elliott, H. A. and Sparks, D. L. (1981) *Soil Sci.*, 132:402–409.

Escudey, M. and Galindo, G. (1983) *J. Colloid Interface Sci.*, 93:78–83.

Escudey, M. and Gil-Llambias, G. (1985) *J. Colloid Interface Sci.*, 107:272–275.

Farrah, H., Slavek, J., and Pickering, W. F. (1987) *Aust. J. Soil Res.*, 25:55–69.

Friend, J. P. and Hunter, R. J. (1970) *Clays Clay Miner.*, 18:275–283.

Gallez, A., Juo, A. S. R., and Herbillon, A. J. (1976) *Soil Sci. Soc. Am. J.*, 40:601–608.

Gonzales-Batista, A., Hernandez-Moreno, J. M., Fernandez-Caldas, E., and Herbillon, A. J. (1982) *Clays Clay Miner.*, 30:103–110.

Greenland, D. J. and Hayes, M. H. B. (eds.) (1978) *The Chemistry of Soil Constituents*. John Wiley & Sons, New York.

Gu, X. Y. (1989) *Soil Bull.*, 43:57–66.

Harsh, J. B. and Xu, S. H. (1990) *Adv. Soil Sci.*, 14:131–165.

Harsh, J. B., Doner, H. E., and Fuerstenau, D. W. (1988) *Soil Sci. Soc. Am. J.*, 52:1589–1592.

Healy, T. W. and White, L. R. (1978) *Adv. Colloid Interface Sci.*, 9:303–345.

Hendershot, W. H. and Lavkulich, L. M. (1978) *Soil Sci.*, 128:136–141.

Hendershot, W. H., Singleton, G. A., and Lavkulich, L. M. (1979) *Soil Sci. Soc. Am. J.*, 43:387–389.

Hiemenz, P. C. (1977) *Principles of Colloid and Surface Chemistry*. Marcel Dekker, New York.

Hingston, F. J., Posner, A. M., and Quirk, J. P. (1971) *Discuss. Faraday Soc.*, 52:334–342.

Hingston, F. J., Posner, A. M., and Quirk, J. P. (1972) *J. Soil Sci.*, 23:177–192.

Houchin, M. R. and Warren, L. J. (1984) *J. Colloid Interface Sci.*, 100:278–286.

Huang, C. P. and Rhoads, E. A. (1989) *J. Colloid Interface Sci.*, 131:289–306.

Huang, C. P. and Stumm, W. (1973) *J. Colloid Interface Sci.*, 43:409–420.

Hunter, R. J. (1981) *Zeta Potential in Colloid Science*. Academic Press, London.

Hunter, R. J. and James, M. (1992) *Clays Clay Miner.*, 40:644–649.

James, R. O. (1981) in *Adsorption of Inorganics at Solid-Liquid Interfaces* (M. A. Anderson and A. J. Rubin, eds.). Ann Arbor Science, Ann Arbor, MI.

James, R. O. and Healy, T. W. (1972a) *J. Colloid Interf. Sci.*, 40:42–52.

James, R. O. and Healy, T. W. (1972b) *J. Colloid Interf. Sci.*, 40:53–64.

James, R. O. and Healy, T. W. (1972c) *J. Colloid Interf. Sci.*, 40:65-81.

Jiang, J. M. and Shen, R. S. (1962) *Acta Pedol. Sinica*, 10:354-360.

Kafkafi, U. and Bar-Yosef, B. (1969). *Int. Clay Conf. Tokyo*, I:691-696.

Kinniburgh, D. G. (1981) in *Adsorption of Inorganics at Solid-Liquid Interfaces* (M. A. Anderson and A. J. Rubin, eds.). Ann Arbor Science, Ann Arbor, MI, pp. 91-160.

Kittaka, S. (1974) *J. Colloid Interface Sci.*, 48:327-333.

Kuo, S. and Baker, A. S. (1980) *Soil Sci. Soc. Am. J.*, 44:969-974.

Leckie, J. O. and James, R. O. (1974)   in *Aqueous-Environmental Chemistry of Metals* (A. J. Rubin et al., eds.). Ann Arbor Science, Ann Arbor, MI.

Li, C. K. and Zhang, X. N. (1957) *Acta Pedol. Sinica*, 5:78-94.

Li, H. C. and Bruyn, P. L. D. (1966) *Surface Sci.*, 5:203-220.

Livesey, N. T. and Huang, P. M. (1981) *Soil Sci.*, 131:88-94.

Loganathan, P. and Burau, R. G. (1973) *Geochim. Cosmochim. Acta*, 37:1277-1293.

Low, P. F. (1987) in *Proceedings of the International Clay Conference, Denver, 1985* (L. G. Schultz et al., eds.). Clay Mineral Society, Bloomington, MN.

Macnaughton, M. G. and James, R. O. (1974) *J. Colloid Interface Sci.*, 47:431-440.

Marsh, K. B., Tillman, R. W., and Syers, J. K. (1987) *Soil Sci. Soc. Am. J.*, 51:318-323.

Marshall, C. E. (1964) *The Physical Chemistry and Mineralogy of Soils*. vol. I. John Wiley & Sons, New York.

Mattson, S. (1932) *Soil Sci.*, 34:209-240.

McBride, M. B. (1989) *Adv. Soil Sci.*, 10:1-56.

McBride, M. B. and Blasiak, J. J. (1979) *Soil Sci. Soc. Am. J.*, 43:866-870.

Michael, H. L. and Williams, D. J. A. (1984) *J. Electroanal. Chem.*, 179:131-139.

Miller, S. E. (1984) *Diss. Abstr.*, 85:07733.

Morais, F. I., Page, A. L. and Lund, L. J. (1976) *Soil Sci. Soc. Am. J.*, 40:521-527.

Muljadi, D., Posner, A. M., and Quirk, J. P. (1966) *J. Soil Sci.*, 17:212-229.

Muller, J. P. and Bocquier, G. (1986) *Geoderma*, 37:113-136.

Nodvin, S. C. (1986) *Soil Sci.*, 142:69-75.

Overbeek, J. Th. G. (1952) in *Colloid Science* (H. R. Kruyt, ed.). Vol.1, Elsevier, Amsterdam.

Parfitt, R. L. (1978) *Adv. Agron.*, 30:1-50.

Parker, J. C., Zelazny, L. W., Sampath, S. and Harris, W. G. (1979) *Soil Sci. Soc. Am. J.*, 43:668-674.

Parks, G. A. (1965) *Chem. Rev.*, 65:177-198.

Parks, G. A. (1967)   in *Equilibrium Concepts in Natural Water Systems,*

Advances in Chemistry Series 67. American Chemical Society, Washington, DC.

Perrott, K. W. (1977) *Clays Clay Miner.*, 25:417–421.

Perrott, K. W., Smith, B. F. L., and Mitchell, B. D. (1976) *J. Soil Sci.*, 27:348–356.

Rengasamy, P. and Oades, J. M. (1977) *Aust. J. Soil Res.*, 15:235–242.

Roy, W. R. et al. (1986) *Soil Sci. Soc. Am. J.*, 50:1176–1182.

Schulthess, C. P. and Sparks, D. L. (1987) *Soil Sci. Soc. Am. J.*, 51:1136–1144.

Schulthess, C. P. and Sparks, D. L. (1988) *Soil Sci. Soc. Am. J.*, 52:92–97.

Singh, B. R. (1984) *Soil Sci.*, 138:346–353.

Sposito, G. (1981) *Soil Sci. Soc. Am. J.*, 45:292–297

Sposito, G. (1984) *The Surface Chemistry of Soils*. Oxford University Press, New York.

Sprycha, R. (1986) *J. Colloid Interface Sci.*, 110:278–281.

Sprycha, R. (1989) *J. Colloid Interface Sci.*, 127:1–11.

Su, C. M. and Harsh, J. B. (1993) *Clays Clay Miner.*, 41:463–473.

Tewari, P. H. and Lee, W. (1975) *J. Colloid Interface Sci.*, 52:77–88.

Van Olphen, H. (1963) *An Introduction to Clay Colloid Chemistry*. Interscience, New York.

Van Schuylenborgh, J. (1950) *Trans. 4th Intern. Congr. Soil Sci.*, I:89–92.

Wada, K. and Harward, M. E. (1974) *Adv. Agron.*, 26:211–260.

Westall, J. and Hohl, H. (1980) *Adv. Colloid Interface Sci.*, 12:265–294.

White, R. E and Taylor, A. W. (1977) *J. Soil Sci.*, 28:48–61.

Wiese, G. R., James, R. O., Yates, D. E., and Healy, T. W. (1976) in *International Review of Science. Physical Chemistry Series Two, Vol. 6, Electrochemistry* (J. O'M Bockris, ed.). Butterworth, London.

Williams, D. J. A. and Williams, K. P. (1978) *J. Colloid Interface Sci.*, 65:79–87.

Yu, T. R. (ed.) (1976) *Electrochemical Properties of Soils and Their Research Methods*. Science Press, Beijing.

Yu, T. R. (ed.) (1985) *Physical Chemistry of Paddy Soils*. Science Press/Springer Verlag, Beijing/Berlin.

Zhang, G. Y. and Zhang, X. N. (1989) *Soils*, 21:239–242.

Zhang, G. Y., Zhang, G. Y., and Yu, T. R. (1987) *J. Soil Sci.*, 38:29–38.

Zhang, H. and Zhang, X. N. (1991) *Pedosphere*, 1:41–60.

Zhang, H. and Zhang, X. N. (1992a) *Pedosphere*, 2:31–42.

Zhang, H. and Zhang, X. N. (1992b) *Geoderma*, 54:173–188.

Zhang, X. N. (1981) in *Proceedings of International Symposium on Paddy Soil*. Science Press/Springer Verlag, Beijing/Berlin, pp. 475–479.

Zhang, X. N. and Jiang, N. H. (1964) *Acta Pedol. Sinica*, 12:120–131.

Zhang, X. N. and Li. C. K. (1958) *Acta Pedol. Sinica*, 6:178–192.

Zhang, X. N. and Zhao, A. Z. (1988) *Acta Pedol. Sinica*, 25:164–174.

8

ELECTRIC CONDUCTANCE

C. B. Li

The migration of colloidal soil particles in an applied electric field has been discussed in Chapter 7. Soil particles carrying electric charges invariably adsorb equivalent amounts of ions of the opposite charge. Generally there is a certain amount of free ions present in soil solution. When an electric field is applied to a soil system, a phenomenon known as *electric conductance* occurs. As in the case for electrolyte solutions, soil particles and various ions interact with one another during their migration, and these interactions can affect the electric conductance of the system. Variable charge soils carry both positive and negative surface charges, and it can be expected that their interactions with various ions would be rather complicated during conductance. On the other hand, this makes the measurement of electric conductance an effective means in elucidating the mechanisms of interactions between variable charge soils and ions.

Both direct-current (DC) electric fields and alternating-current (AC) electric fields can induce the migration of charged particles. In the latter case, the migration of these particles should be related to the frequency of the applied AC electric field. Therefore, in this chapter, after describing the principles of electric conductance of ions and colloids and the factors that affect the conductance of a soil, emphasis shall be placed on the interaction between variable charge soils and various ions as reflected by the frequency effect in electric conductance.

8.1 ELECTRIC CONDUCTANCE OF IONS AND COLLOIDS

8.1.1 Definition

For a colloidal suspension, the electric conductance may be regarded as the contribution of conductances of both charged colloidal particles and ions. These two parts may be called the *electric conductance of colloidal particles* and the *electric conductance of ions*, respectively. However, in actual cases it is difficult to distinguish between these two parts. Therefore, it is a general practice to distinguish the electric conductance as that caused by colloidal particles plus their counterions from that caused by ions of the free solution. These may be called *electric conductance of the colloid* and *electric conductance of the free solution*. The former conductance is the difference between the electric conductance of the suspension and that of the free solution. The electric conductance of the colloid system minus that of the colloidal particles is called *surface conductance*, which is the conductance caused by ions of the electric double layer.

8.1.2 Electric Conductance of Ions

The electric conductivity of electrolyte solutions obeys Ohm's law. This conductivity is generally expressed by the reciprocal of the electric resistance (R), called electric conductance L. According to Ohm's law, the electric resistance of a conductor is proportional to the length l and inversely proportional to the cross-sectional area A of the conductor:

$$R = \rho \frac{l}{A} \qquad (8\text{-}1)$$

where ρ is a proportional constant, called *specific resistance* or *resistivity*, with ohm cm as the unit.

Electric conductance is the reciprocal of electric resistance; therefore,

$$L = \frac{1}{\rho} \cdot \frac{A}{l} = \frac{\kappa}{K} \qquad (8\text{-}2)$$

where L is the electric conductance. The unit of electric conductance is siemens. κ is the reciprocal of ρ, called *specific conductance* or *conductivity*, which is the conductance of the conductor with a length of 1 cm and a cross-sectional area of 1 cm². The unit of specific conductance is siemens cm^{-1}. K is called *cell constant*.

In order to compare the conducting ability of different electrolytes, the concept of equivalent conductance is introduced. Equivalent conductance is defined as the electric conductance of the solution containing 1 gram equivalent of electrolyte between two electrodes 1 cm apart. The equivalent conductance is related to the practically measured specific conductance as

$$\lambda = \frac{1000\kappa}{C} \qquad (8\text{-}3)$$

where C is the concentration of the electrolyte in equivalent and λ is the equivalent conductance. The unit of equivalent conductance is siemens cm² gram-equivalent^{-1}.

Due to interactions among ions in solution, equivalent conductance decreases with the increase in electrolyte concentration. For dilute solutions of strong electrolytes, Kohlrausch suggested an empirical equation:

$$\lambda = \lambda_0 - K\sqrt{C} \tag{8-4}$$

where K is a constant, and λ_0 is the equivalent conductance of the solution at infinitesimal dilution, called *limiting equivalent conductance*. The electric conductance of a solution at infinitesimal dilution is the sum of the electric conductances of the cations and the anions. This can be expressed as

$$\lambda_0 = \lambda_{0,+} + \lambda_{0,-} \tag{8-5}$$

where $\lambda_{0,+}$ and $\lambda_{0,-}$ are the equivalent conductances of the cations and the anions at infinitesimal dilution, respectively. The limiting equivalent conductances of cations and anions commonly found in soils are given in Table 8.1. As shall be seen later, the difference in equivalent conductance among various ions is of practical significance in soils. In particular, the equivalent conductances of hydrogen ions and hydroxyl ions are especially large. This is the basis of conductometric titration of soils.

The equivalent conductance of an ion species is related to its migration rate. In a DC electric field, the migration rate is proportional to the potential gradient of the electric field:

$$V_+ = U_+ \frac{dE}{dl} \tag{8-6a}$$

$$V_- = U_- \frac{dE}{dl} \tag{8-6b}$$

where V_+ and V_- are the migration rate of the cations and the anions, respectively, and proportional constants U_+ and U_- are the migration rate of the ions under unit potential gradient, called *mobility*. The unit of mobility is cm^2 s^{-1} V^{-1}. In an infinitesimally diluted solution the mobility attains a limiting value, called *absolute mobility U_0*, and the migration rate of the ion is called *absolute migration rate*, because in this case there is no interaction among ions. Thus, equation (8-6) can be rewritten as

$$V_{0,+} = U_{0,+} \frac{dE}{dl} \tag{8-6c}$$

Table 8.1 Limiting Equivalent Conductance of Ions in Water Solution
(25°C)

Cation	$\lambda_{0,+}$	Anion	$\lambda_{0,-}$
H^+	349.8	OH^-	198.0
Na^+	50.11	Cl^-	76.34
K^+	73.52	NO_3^-	71.44
NH_4^+	73.4	HCO_3^-	44.5
$\frac{1}{2}Mg^{2+}$	53.06	$\frac{1}{2}SO_4^{2-}$	80.0
$\frac{1}{2}Ca^{2+}$	59.50		
$\frac{1}{2}Ba^{2+}$	63.64		

$$V_{0,-} = U_{0,-} \frac{dE}{dl} \tag{8-6d}$$

For completely dissociated electrolytes, the limiting equivalent conductance λ_0 is related to the absolute mobility of the ion as

$$\lambda_0 = (U_{0,+} + U_{0,-}) F \tag{8-7}$$

where F is the Faraday constant.

Thus, the limiting equivalent conductance is related to the absolute mobility of the ion species as:

$$\lambda_{0,+} = U_{0,+} F \tag{8-8a}$$

$$\lambda_{0,-} = U_{0,-} F \tag{8-8b}$$

This means that the absolute mobility of an ion is equal to the limiting equivalent conductance divided by Faraday constant (96494 coulomb).

8.1.3 Interactions Among Ions in Electric Conductance

According to the Debye–Hückel theory, because of electrostatic attraction, in an electrolyte solution one central ion is surrounded by a global ionic atmosphere of oppositely charged ions. Under steady conditions the distribution of ions of the ionic atmosphere is symmetrical with the central ion as the center of the globe. In this case the charge density is the highest at locations closely adjacent to the central ion and diminishes with the increase in distance from the central ion. In an infinitesimally diluted solution, because the distance among ions is extremely large, there should be no effect exerted by other ions when an ion migrates. On the other hand, when a central ion with an ionic atmosphere migrates in an applied electric field, the migration rate is decreased by the following retarding forces.

8.1.3.1 *Relaxation Force (Relaxation Effect)*

If the central ion is a cation and the surrounding ions are anions, the central ion would move to the negative electrode in an electric field and the surrounding ionic atmosphere will show a tendency to move toward the positive electrode. Thus, the symmetry of the ionic atmosphere is disturbed and the charge density of the ionic atmosphere behind the central ion becomes larger than that ahead of the central ion. This results in the presence of surplus opposite charges behind the central ion. As long as the central ion is moving, this asymmetry of ionic atmosphere will exist. Due to the effect of electrostatic attraction, this asymmetrical ionic atmosphere would exert an electrostatic retarding force on the movement of the central ion. This is called *relaxation effect*. This effect induces a reduction in the migration rate of the central ion, resulting in a decrease in electric conductance. However, the asymmetrical ionic atmosphere has a tendency to restore its global symmetry. If the movement of the central ion is stopped, the ionic atmosphere will rapidly establish its symmetry again. The time required for the disturbance and restoration of the symmetry of the ionic atmosphere is called *relaxation time*.

8.1.3.2 *Electrophoretic Force (Electrophoretic Effect)*

Under the influence of an applied electric field, central cations together with their solvent molecules move to the negative electrode, whereas the atmospheric anions together with their solvent molecules show a tendency toward moving to the positive electrode. This is another factor for retarding the movement of the central ion to the negative electrode. Because this effect is equivalent to the increase in viscosity of the solvent and is similar to the retarding force of the solvent when a colloidal particle migrates in an electric field, it is called *electrophoretic force* or *electrophoretic effect*. This effect also results in the reduction of migration rate of the ion which decreases its equivalent conductance.

When an alternating voltage is applied, if the frequency is so high that the vibration period of the central ion is smaller than the relaxation time of the ionic atmosphere, there would not be sufficient time for a big change in the symmetrical distribution of electric charge surrounding the central ion. Therefore, the retarding force caused by the relaxation effect decreases with the increase in frequency of the applied field. The increase in electric conductance in this way is called *conductivity dispersion*. When the vibration frequency is higher than about 10^6 circles per second, the central ion seems stationary and the ionic atmosphere would be symmetrical. Thus, the electric conductance attains a limiting value. However, in this case the electrophoretic effect still exists. Therefore, the electric conductance cannot attain the maximum value at infinitesimal dilution.

The higher the valency of the ion and the concentration of the solution, the higher the frequency required for inducing the conductivity dispersion.

If the applied electric field attains a voltage gradient of 20000 V cm^{-1}, the ion would move with a speed of about 1 m per second. In this case, within the relaxation time the pass–length of the ion may exceed the effective thickness of the ionic atmosphere by many times, and, therefore, there would not be sufficient time for the moving ion to induce the formation of its ionic atmosphere. Under such a condition both relaxation effect and electrophoretic effect would be reduced greatly or even vanish, resulting in the increase of electric conductance to the maximum value at infinitesimal dilution. The phenomenon of the increase in electric conductance of electrolyte solution in a high strength field is called the *Wien effect*.

8.1.4 Electric Conductance of Colloids

The colloidal particle can be regarded as one kind of polyvalent ion. For example, for a colloid particle carrying negative charges, the electric conductance should be (Overbeek, 1952)

$$L = NQU \qquad (8\text{-}9)$$

where Q is the amount of electric charge carried by one colloidal particle, N is the number of colloidal particles per cubic centimeter, and U is the migration rate of the colloidal particle when the field gradient is 1 V cm^{-1}.

For clay minerals carrying negative charges, such as montmorillonite, if the adsorbed cations dissociate completely, NQ would correspond to the cation exchange capacity. It is then possible to calculate the contribution of clay particles to the total electric conductance of the clay suspension from the experimentally measured electrophoretic velocity. This was conducted when studying the electric conductance of clays by Deshpande and Marshall (1959, 1961).

For colloids not containing free electrolytes, the difference between total electric conductance and the conductance of colloidal particles is called *surface conductance*. This surface conductance is the contribution of ions of the electric double layer to the electric conductance of the colloidal system.

In principle, surface conductance can be calculated with an equation similar to equation (8-9), assuming that this conductance is equal to the product of electric charges of all the ions in the electric double layer times the mobilities of the relevant ions. However, because the distribution of ions in the electric double layer is uneven and because the mobility of ions varies with the position in the double layer, the practical utility of equations that have been proposed based on some assumptions (Yu, 1976) is limited. At present it is generally assumed that, for colloids carrying negative charges, unhydrated cations directly positioned on the surface of the colloid do not participate in conduction appreciably and the mobility of cations in another layer adjacent to this inner layer is only several tenths of the limiting conductance, while cations in the diffuse layer have a normal mobility. Therefore, the surface conductance as calculated by subtraction method is an apparent average value of the conductances of these ions.

8.2 FACTORS AFFECTING ELECTRIC CONDUCTANCE OF SOILS

8.2.1 General Discussion

Soil is a polyphase system consisted of solid particles, solution, and air. When charged particles, including colloids and ions, migrate under the influence of an applied electric field, many physical factors of the soil, including texture, structure, and water content, can affect the electric conductance through affecting the interrelationship between colloidal particles and ions. In order to reduce the effect of these complicating factors, one possible way is to study the separated colloid fraction because these colloids are the chief charge–carriers of the solid part of the soil. These colloids can still have a volume effect on electric conductance of the system, which is not simply proportional to the volume fraction of the colloid (Cremers and Laudelout, 1965; Dakshinumurti, 1960; Fricke and Curtis, 1936; Yu, 1976). This point is the principal difference in behavior between soils and solutions with respect to electric conductance. Therefore, when examining chemical factors in affecting the electric conductance of the soil, it is necessary to consider the effect of these physical factors at the same time or to make comparisons under the same physical conditions.

For the colloids of variable charge soils carrying both negative charges and positive charges and adsorbed with both cations and anions, they may be regarded as a mixture consisting of acidoids with their salts and basoids with their salts. Some of the chemical properties of these acidoid and basoids are similar to those of weak acids or weak bases. Because most of the variable charge soils carry more negative charges than positive charges under ordinary pH conditions, their behavior with respect to electric conductance is more like that of weak acids. As in the case of electric conductance of a weak acid with its salt, the most important factors that can affect the electric conductance of soil colloids are the kind of ion species, the type of the soil, and the colloid concentration in the soil-water system. These will be discussed in the following sections.

8.2.2 Kind of Ions

It has been seen in Chapters 3 and 4 that the affinities of various cations and anions for the surface of soil colloids differ markedly. Thus, their distribution between the electric double layer and the free solution and within the double layer would be different. Besides, as seen in Table 8.1, the conductivity of various ions is different. These two factors combined together induce the dependence of electric conductance of a soil on the kind of the ions. For clay minerals and constant charge soils, this point has been reviewed by Marshall (1964) and by Yu (1976). In the following, the effect of cations for variable charge soils will be examined.

For the convenience in discussion, the conductometric titration curve of the colloidal suspension of variable charge soils, that is, the change in electric conductance with the increase in amount of alkali added, is examined first. One example for the titration of the colloidal suspension of an Ali-Haplic Acrisol with NaOH is shown in Fig. 8.1. Most of the exchangeable cations of the electrodialyzed soil colloid are aluminum ions. Because of the extraordinarily high affinity of aluminum ions for the colloidal surface, their degree of dissociation in water is extremely small. Hence, the electric conductance of the whole suspension is very low. After the addition of NaOH, hydroxyl ions are consumed in the neutralization of aluminum ions, while the exchange sites originally occupied by aluminum ions are occupied by sodium ions. Because of their weak affinity for soil colloids, a considerable portion of sodium ions are in the diffuse double layer; therefore, the electric conductance increases. This is the reason why in the first part of the titration curve the electric conductance increases with the increase of NaOH. After the occupation of all the exchange sites by sodium ions, the additional sodium ions can only exist in free solution. In this case most of the additional hydroxyl ions, which have a high mobility, are also present in solution. These two factors make the electric conductance of the suspension increase markedly. The whole titration curve can be distinguished as two straight lines. The quantity of hydroxyl ions added to the system at the intersection point of the two extrapolated lines should in principle correspond to the apparent cation-exchange capacity of the soil colloid. Of course, at this point the number of cation-exchange sites has increased considerably as compared to the original colloid, because the pH of the suspension has increased.

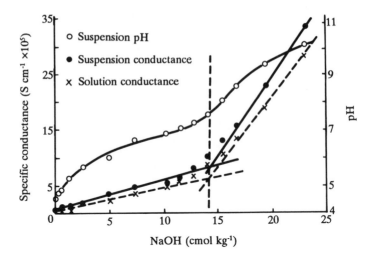

Fig 8.1. Titration curves of the colloid fraction of Ali–Haplic Acrisol (Yu, 1976).

In Fig. 8.1, the titration curve of the centrifugalized solution is similar to that of the suspension, with the exception that the curve lies below the latter curve. According to definition, the difference between the two conductances would be the electric conductance of the colloid, that is, the conductance contributed by both colloidal particles and ions in the electric double layer. The figure indicates that the electric conductance of the soil colloid shows a tendency to increase with the increase in the amount of NaOH. This increase reflects a change in the distribution of sodium ions in the electric double layer.

Among cations commonly found in soils, aluminum and sodium represent two extremes as far as the affinity for the surface of soil colloid is concerned. This is because the affinity is extraordinarily strong for aluminum ions and very weak for sodium ions. It can be expected that the effect of other cations on electric conductance lies intermediately between these two extremes. The changes in electric conductance when the colloidal suspensions of a Ferrali–Haplic Acrisol and a Rhodic Ferralsol are titrated with NaOH, KOH, and $Ca(OH)_2$, are shown in Figs. 8.2 and 8.3, respectively. The patterns of change in electric conductance when titrated with NaOH or KOH are similar to those of the Ali–Haplic Acrisol shown in Fig. 8.1. Note that at the first part of the curves the electric conductance of the sodium system is higher than that of the potassium system. This could be caused by the higher proportion of sodium ions in participating in conduction. On the other hand, when the quantity of cations exceeds a certain limit, the electric conductance of the potassium system becomes

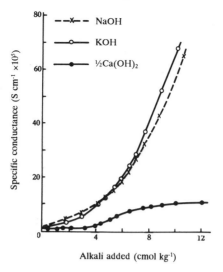

Fig. 8.2.   Conductometric titration curves of the colloid fraction of Ferrali–Haplic Acrisol.

higher than that of the sodium system. This would reflect the greater effect of the higher mobility of potassium ions as compared to sodium ions than the effect of the degree of dissociation of adsorbed cations. As seen in Table 8.1, the limiting equivalent conductance of potassium ions is about 1.5 times that of sodium ions.

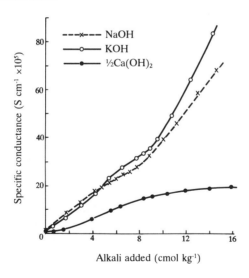

Fig. 8.3. Conductometric titration curve of the colloid fraction of Rhodic Ferralsol.

The electric conductance of the calcium system is much lower than that of alkali metal systems. This is understandable because most of the divalent cations distribute in the inner part of the electric double layer.

In the conductometric titration curves shown in Figs. 8.1 to 8.3 there is a complicating factor involved. When alkalies are added to variable charge soils, the change in surface charge caused by the rise in pH would affect the interrelationships between cations and the colloidal surface and, thus, the surface conductance. The effect of different alkalies on pH is dependent on both the kind of the cations and the type of the soil. This complicating factor can be excluded to a certain extent, if the comparison is made at the same pH. The electric conductances of the colloidal suspension of three variable charge soils with the same colloid concentration under different pH conditions are shown in Figs. 8.4 to 8.6. Note that at pH higher than about 6.5 where a large proportion of cations is present in solution, the effect of the difference in limiting equivalent conductance between potassium ions and sodium ions shows itself clearly.

From the above discussions it can be concluded that both the conductivity of cations and the affinity of these ions with the surface of soil colloid can affect the electric conductance of variable charge soils. The relative importance of these two factors is determined chiefly by the quantity of cations present in the soil system.

8.2.3 Type of Soil

When comparing Figs. 8.1 to 8.3 it can be found that there are two

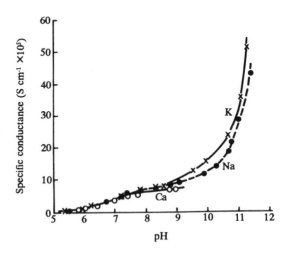

Fig. 8.4. Electric conductance of the colloid fraction of Ali–Haplic Acrisol at different pH.

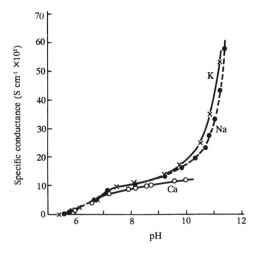

Fig. 8.5. Electric conductance of the colloid fraction of Ferrali–Haplic Acrisol at different pH.

important differences among the three representative variable charge soils: The quantities of cations required for a marked increase in electric conductance are of the order Rhodic Ferralsol < Ferrali–Haplic Acrisol < Ali–Haplic Acrisol, and conversely, the electric conductances of the soil at the same quantity of cations are of the order Rhodic Ferralsol > Ferrali-Haplic Acrisol > Ali–Haplic Acrisol. The first phenomenon is in reality the reflection of the difference in the quantity of cation exchange sites of the

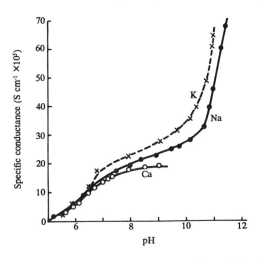

Fig. 8.6. Electric conductance of the colloid fraction of Rhodic Ferralsol at different pH.

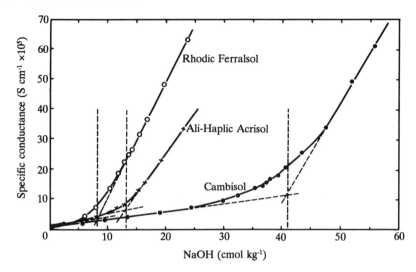

Fig. 8.7. Estimation of apparent cation–exchange capacity of the colloid fraction of soils by conductometric titration method (Yu and Zhang, 1986).

soil colloid, while the latter phenomenon reflects the difference in the fraction of cations participating in conduction, which is determined by the negative surface charge density of the soil colloid.

It was mentioned in the last section that the quantity of cations at which the two extrapolated straight lines intersect on the conductometric titration curve equals the apparent cation–exchange capacity, that is, the number of negative surface charges of the soil colloid. In Chapter 2 it was shown that the quantity of negative surface charge carried by different soil colloids differs markedly. For constant charge soils, it is possible to estimate the cation–exchange capacity from the conductometric titration curve. For variable charge soils, one can also make the estimation in a similar way. Such an example for the colloid fractions of an Ali-Haplic Acrisol and a Rhodic Ferralsol is shown in Fig. 8.7. In the figure the behavior of the colloid fraction of a constant charge soil (Cambisol) is also shown for comparison. It can be seen that for the colloid fraction of the three soils the apparent cation–exchange capacities, as estimated from the intersection point of the two extrapolated straight lines, correspond to 8.2, 13.5, and 41.0 cmol kg^{-1}, respectively. The apparent cation–exchange capacity estimated in this way frequently coincides approximately with the value estimated by the potentiometric titration method. However, as shown in Chapter 2, since the quantity of negative surface charge carried by variable charge soils varies with environmental conditions, particularly pH, the so–called cation-exchange capacity determined by any method is of relative significance only.

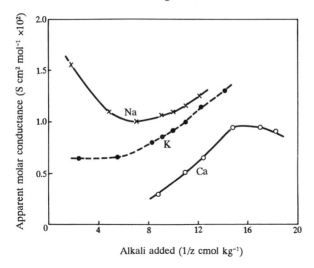

Fig. 8.8. Apparent molar conductance of cations in suspension of the colloid fraction of Ali–Haplic Acrisol.

The phenomenon that the electric conductance for the same ion species varies with the kind of the soil is more interesting because, based on such a phenomenon, it is possible to examine the interrelationships between ions and the surface of soil colloid.

For the convenience of comparison, one can calculate the apparent molar conductance of relevant ion species from the electric conductance of the colloidal suspension. Colloidal particles may contribute to the total electric conductance to some extent. However, if the determination is carried out with the DC method and if the voltage gradient applied to the two electrodes of the conductance cell is very small, the variation range of the contribution of colloidal particles to total electric conductance would not be large. It is then possible to examine the difference in the relationship between the same ion species and different soil colloids from the variation patterns of the apparent molar conductance of relevant ions.

The results for three soil colloids are shown graphically in Figs. 8.8 to 8.10. Note that the apparent molar conductance of various cations in different soils differs considerably. For example, when the quantity of sodium ions is 4 cmol kg^{-1} colloid, the molar conductances for the colloid suspension of Ali–Haplic Acrisol, Ferrali–Haplic Acrisol and Rhodic Ferralsol are 1.2, 2.9, and 4.8×10^{-2} S m^{-1} mol^{-1}, respectively. This means that in the colloid suspension of Rhodic Ferralsol either a larger proportion of sodium ions participates in conduction or the mobility of sodium ions is large as compared to other soils, or both of these phenomena exist.

This is not difficult to understand. The electric conductance of ions in a

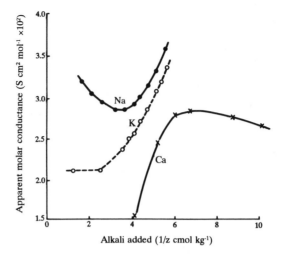

Fig. 8.9. Apparent molar conductance of cations in suspension of the colloid fraction of Ferrali–Haplic Acrisol.

colloidal system is determined by the distribution of these ions between the electric double layer and the free solution and their distribution within the double layer. These distributions are dependent mainly on the surface charge density of the soil colloid. The negative surface charge density is the highest for the colloid of the Ali–Haplic Acrisol and the lowest for the

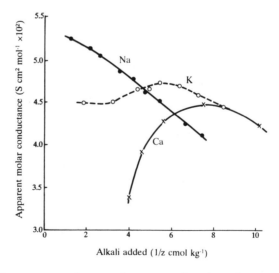

Fig. 8.10. Apparent molar conductance of cations in suspension of the colloid fraction of Rhodic Ferralsol.

Rhodic Ferralsol. Hence, at the same quantity of the same cation species, the electric conductance differs with the kind of the soil.

The pattern of change in apparent molar conductance with the change in quantity of ions for various cations also differs with the type of the soil. This is caused by the difference in variation pattern of surface charge for different soil colloids.

For potassium ions, when the quantity is small, the apparent molar conductance hardly changes with the change in quantity of ions in the three soils. This indicates that at this time the effect of the increase in degree of dissociation caused by the increase in quantity of potassium ions is approximately equal to the effect of the increase in negative surface charge density caused by the rise in pH, allowing the two effects to compensate one another. With a further increase in quantity of potassium ions, the former effect becomes larger than the latter effect, and hence the electric conductance increases. When dissociation increases to a certain extent, it will not increase any more, whereas the negative surface charge density may increase continuously with the rise in pH, particularly for the Rhodic Ferralsol. Therefore, eventually the apparent molar conductance shows a tendency to decrease (Fig. 8.10).

The behavior of sodium ions differs from that of potassium ions in that the degree of dissociation is very high. Hence, the degree of dissociation varies within a narrower range with the change in the quantity of the ions. In this case, the change in surface charge density of the soil caused by the change in pH becomes the dominant factor in determining the apparent molar conductance. In Figs. 8.8 to 8.10, the first part of the apparent molar conductance curve for sodium ions shows a tendency to decrease in the three soil colloids. This is a reflection of the increase in negative surface charge density of the soil colloid with the rise in pH. When the quantity of sodium ions increases to a certain extent, these ions would dissociate in solution and become less affected by the colloid, and therefore the apparent molar conductance may increase. This is the case for the Ali–Haplic Acrisol and the Ferrali–Haplic Acrisol. On the other hand, for the Rhodic Ferralsol with a large variability in surface charge, the apparent molar conductance of sodium ions still decreases, which is caused by the effect of the increased negative surface charge. It can be expected that for this latter soil the molar conductance would also increase when the quantity of sodium ions exceeds a certain limit.

The change in molar conductance of calcium ions is more distinct than that of sodium ions or potassium ions. This indicates that the distribution of divalent calcium ions in the electric double layer is more accessible to the effect of the quantity of these ions. When the quantity of the ions exceeds a certain limit, the apparent molar conductance decreases slightly. At this time a specific adsorption might occur.

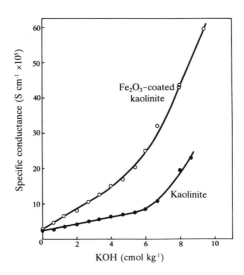

Fig. 8.11. Effect of iron oxides on electric conductance of kaolinite (Yu, 1985).

Because the surface charge of soil colloids is the principal factor in inducing the difference in electric conductance of ions in different soils, it can be expected that, in addition to the change in pH, the change in surface conditions of the soil caused by other means can also affect the electric conductance. It has been seen in previous chapters that iron oxide is an important material in affecting the surface charge of variable charge soils. The addition of iron oxide coatings to an iron–free clay mineral, such as kaolinite, would result in an increase in electric conductance, caused by the weakening of affinity of the mineral with cations. This is just the case. As shown in Fig. 8.11, the higher the pH, the more pronounced the effect of the addition of 2–3% iron oxide coatings.

From the discussions made in this section it can be concluded that the quantity of electric charge at the surface of soil colloids determines the quantity of cations adsorbed by the colloid, whereas the surface charge density determines the distribution of these ions within the electric double layer. The trends of effect of these two factors on electric conductance are not always of the same direction. Both of these factors are affected by environmental conditions, particularly by pH. Therefore, complicated situations may frequently be encountered when examining the electric conductance of variable charge soils. This is one of the important characteristics of these soils as compared to constant charge soils.

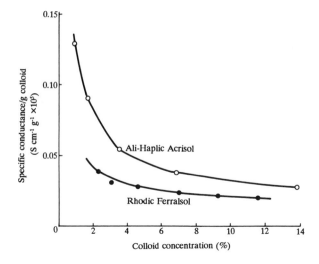

Fig. 8.12. Effect of colloid concentration on electric conductance of soils (Yu, 1976).

8.2.4 Concentration of Colloid

The degree of dissociation of all weak electrolytes decreases with the increase in solution concentration. As a consequence, the equivalent conductance will decrease with the increase in concentration, although the electric conductance of the whole system increases. For soil colloid as one kind of polyvalent weak electrolyte, the electric conductance will also change with the colloid concentration, as shown in Fig. 8.12 for two variable charge soils. If the electric conductance is expressed as the specific conductance per gram of soil colloid, it can be seen that it decreases with the increase in colloid concentration. The effect of colloid concentration is especially pronounced within the low concentration range, similar to the case for weak electrolytes.

According to the theory of electrolyte solution, the logarithm of the equivalent conductance of a weak electrolyte is linearly related to the logarithm of the electrolyte concentration with a slope of −0.5. If the behavior of soil colloids is similar to that of weak electrolytes, there should be a linear relationship between the logarithm of "weight conductance" (i.e., the electric conductance per 1 g of colloid in 100 ml of suspension) and the logarithm of the colloid concentration expressed in percent. This kind of relationship for the colloid fraction of two variable charge soils is shown in Fig. 8.13, taking one constant charge soil (Cambisol) for comparison. It can be seen that for the H,Al−saturated colloid of Rhodic Ferralsol, Ali−Haplic Acrisol and Cambisol, the slopes of the straight lines are −0.45, −0.60, and

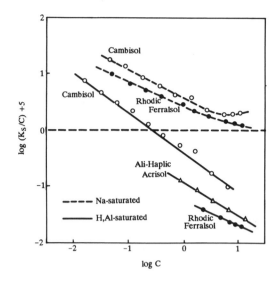

Fig. 8.13. Weight conductance (K_s/C) in relation to colloid concentration (C) of soils (Yu and Zhang, 1986).

-0.70, respectively. For the Na-saturated Rhodic Ferralsol, the slope of the straight line is -0.45. For the Na-saturated Cambisol, the slope of the straight line at low colloid concentrations is -0.50, but the line becomes curved at high colloid concentrations. Thus, if regarded as an electrolyte, variable charge soils are relatively weak, while the constant charge soil Cambisol behaves like a weak electrolyte only under certain conditions.

The H,Al-saturated soil colloids can be regarded as acidoids, and the Na-saturated colloids can be regarded as the salt of the acidoids. According to the theory of weak electrolyte solution, at the same concentration the larger the difference in electric conductance between the weak acid and its salt, the weaker this acid. As can be seen in Fig. 8.13, this difference for the colloid of Rhodic Ferralsol is much larger than that of Cambisol, reflecting the difference in acidoid strength between the variable charge soil and the constant charge soil.

8.3 FREQUENCY EFFECT IN ELECTRIC CONDUCTANCE OF SOILS

8.3.1 Significance

The phenomenon of conductivity dispersion of ions has been mentioned in Section 8.1.3. Conductivity dispersion is caused by the weakening or vanishing of retarding effect of the ionic atmosphere on the migration of the central ion when the changing frequency of applied alternating electric field is sufficiently high. For a soil colloid with its adsorbed ions, this phenomenon should also exist. However, the detailed situation may be different from that of the solution. In a soil colloidal system, the distribution of ions is not only heterogeneous microscopically but also heterogeneous at a relatively macroscopic level. Microscopically, the distribution of ions in the electric double layer on the surface of the colloid is even more complex. In the interactions of colloids with ions, both electrostatic force and specific force may be involved. Thus, various forces that can retard the movement of ions are intermingled together, and the overall expression may not be continuous. On the other hand, these complexities also make it possible to distinguish different forces between soil colloids and ions and to deduce the distribution patterns of ions in the system based on the conductivity dispersion phenomenon. As early as in the 1950s Deshpande and Marshall (1959, 1961) suggested that adsorbed ions can be distinguished into three portions: One portion adheres to the surface of the colloid and does not participate in conduction; one portion located in the Stern layer does not conduct electricity in DC field or in AC field of low frequency (60 Hz), but can show normal conductance in AC field of high frequency (10,000 Hz); and another portion of ions of the Gouy layer conducts electricity in a normal manner. Arulanandan and Mitchell (1968) and Mehran and Arulanandan (1977) observed the conductivity dispersion phenomenon of some clay suspensions at more higher frequencies.

In the interactions of variable charge soils with ions, in addition to electrostatic adsorption, the specific adsorption of both cations and anions may be involved. It can then be expected that the frequency effect in electric conductance would be more complex. These complexities shall be discussed in the following sections.

8.3.2 Effect of Anions

Similar to the case for the potentiometric titration of soil colloids by the addition of different amounts of alkali, one can apply an AC voltage of different frequencies and observe the variation in electric conductance. After subtracting the electric conductance of the free solution from the total conductance of the suspension and taking account of the volume effect of colloidal particles, the electric conductance of the colloid together with the adsorbed ions can be obtained.

The changes in electric conductance with the change in applied frequency for the colloid of Ali–Haplic Acrisol, Ferrali–Haplic Acrisol, and Rhodic Ferralsol in different dilute acids are shown in Figs. 8.14 to 8.16, respectively. In the figure, the unit is expressed as relative conductivity, taking the conductivity at 300 Hz as 1.

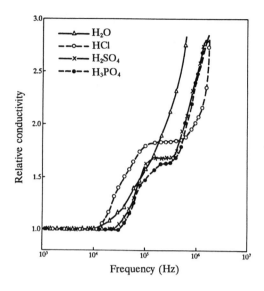

Fig. 8.14. Change in relative conductivity with applied frequency for the colloid fraction of Ali–Haplic Acrisol in different acids (concentration of acid, 10^{-4} N).

It can be seen that all the soils show a conductivity dispersion phenomenon in all the acids. Note that the pattern of conductivity dispersion in acids

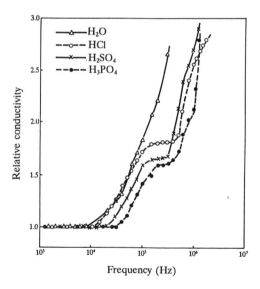

Fig. 8.15. Change in relative conductivity with applied frequency for the colloid fraction of Ferrali–Haplic Acrisol in different acids.

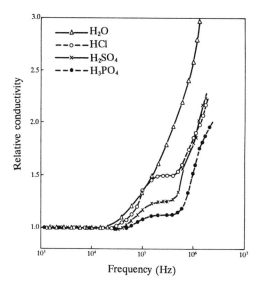

Fig. 8.16. Change in relative conductivity with applied frequency for the colloid fraction of Rhodic Ferralsol in different acids.

is markedly different from that in water. In water suspensions, the frequency at which conductivity dispersion begins to appear, about 10^4 Hz, is much lower than that in electrolyte solutions.

For soils adsorbed with chloride, sulfate, or phosphate ions, the electric conductance begins to increase when the frequency is increased to a certain limit, and then it reaches a plateau region. The relative conductivity at the plateau region differs with the kind of the anion species. Then, following a further increase in frequency, the electric conductance increases sharply again.

The pattern of change in electric conductance for different soil colloids adsorbed with the same anion species is similar. However, the magnitude of conductivity dispersion differs with the type of the soil. As can be seen in Figs. 8.17 and 8.18, for colloids adsorbed with either chloride ions or phosphate ions, the magnitude of the increase in relative conductivity at the plateau region is of the order Ali–Haplic Acrisol > Ferrali–Haplic Acrisol > Rhodic Ferralsol. The difference is especially remarkable between Rhodic Ferralsol and the former two soils.

Thus, it can be suggested that the effect of anions on the electric conductance of variable charge soils is related to both the nature of the anion species and the type of the soil.

Two meaningful parameters can be distinguished on the curve. The frequency at which the relative conductivity equals 1.1 may be called the *frequency of initial dispersion*, that is, the frequency at which conductivity

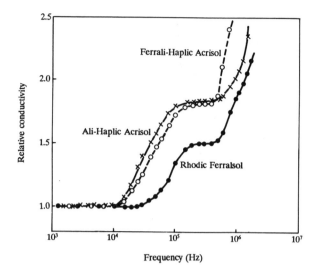

Fig. 8.17. Conductivity dispersion of the colloid fraction of three soils adsorbed with chloride ions (Li and Yu, 1990).

dispersion starts to appear. This frequency should be related to certain energy barriers that must be overcome in order to initiate the migration of some anions in the electric double layer under the influence of the applied

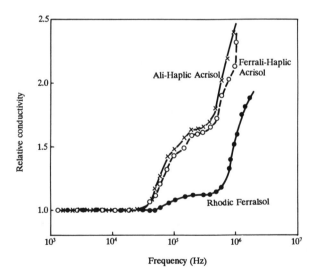

Fig. 8.18. Conductivity dispersion of the colloid fraction of three soils adsorbed with phosphate ions (Li and Yu, 1990).

Table 8.2 Initial Frequency and Plateau Range of Conductivity Dispersion of Soils (Li and Yu, 1990)

Soil	Initial Frequency (kHz)				Plateau Range (kHz)		
	H_2O	HCl	H_2SO_4	H_3PO_4	HCl	H_2SO_4	H_3PO_4
Ali–Haplic Acrisol	20	17	38	45	400	260	230
Ferrali–Haplic Acrisol	17	18	30	46	340	250	230
Rhodic Ferralsol	40	55	69	79	360	330	360

electric field. The difference in frequency at the two sides of the plateau region may be called the *plateau range*. In Table 8.2 both of these parameters are related to the kind of the anion species and the type of the soil. When adsorbed with chloride ions, the frequency of initial dispersion for the Rhodic Ferralsol is much higher than that in water suspension, although it does not differ from that in water suspension for the Ali–Haplic Acrisol and the Ferrali–Haplic Acrisol. When adsorbed with sulfate ions and particularly with phosphate ions, the frequency of initial dispersion increases for all of the three soil colloids. In this case, the plateau range also differs with the kind of the anion species and the type of the soil colloid. For Rhodic Ferralsol, the range in the three anion systems is relatively wide. For the other two soils, the plateau range is wider when chloride ions are adsorbed.

The numerical value of relative conductivity minus 1 represents the increase in electric conductance caused by conductivity dispersion. This value should be related to the quantity of those anions at a certain bonding energy level. Hence, the relative conductivity at two threshold frequencies may be chosen to compare the quantity of anions at the two respective bonding energy levels. The ratio of these two relative conductivities may reflect the relative magnitude of the two bonding forces. This kind of comparison for relative conductivity at plateau frequency and at 1.7 MHz is given in Table 8.3. The affinity of those anions that can participate in conduction only at 1.7 MHz with soil colloids could be stronger than those anions conducting electricity at the plateau frequency. The table shows that, in the weak affinity region, the increase in electric conductance associated with chloride ions is larger than that associated with sulfate ions or phosphate ions, whereas at the strong affinity region the reverse is true. Superficially, Rhodic Ferralsol adsorbed with phosphate ions is an exception. This is caused by the extremely strong affinity of this soil for phosphate ions, and therefore only a small portion of the anions can

Table 8.3 Relative Conductivity of Soil Colloids at Plateau Frequency
and at 1.7 MHz (Li and Yu, 1990)

Item	Treatment	Soil A[b]	Soil B	Soil C
Relative conductivity	HCl	1.84	1.81	1.50
at plateau region	H_2SO_4	1.68	1.65	1.25
	H_3PO_4	1.64	1.60	1.12
Relative conductivity	HCl	2.74	2.80	2.17
at 1.7 MHz	H_2SO_4	3.31	3.40	2.23
	H_3PO_4	3.00	3.08	1.88
Relative affinity[a]	HCl	1.07	1.22	1.34
	H_2SO_4	2.40	2.69	3.92
	H_3PO_4	2.13	2.47	6.33

[a]Relative affinity = (relative conductivity at 1.7 MHz
– relative conductivity at plateau frequency)
/(relative conductivity at plateau frequency – 1).
[b]Soil A, Ali–Haplic Acrisol; Soil B, Ferrali–Haplic Acrisol;
Soil C, Rhodic Ferralsol.

participate in electric conduction even at a frequency of 1.7 MHz.

The conductivity dispersion phenomenon of various anions in different types of variable charge soils shows that the affinity of Rhodic Ferralsol for anions is stronger than that of Ali–Haplic Acrisol and Ferrali–Haplic Acrisol. Among the three anion species, the order of affinity is phosphate > sulfate > chloride. On the other hand, even for chloride ions there is a specific affinity force with variable charge soils. These conclusions are consistent with the conclusions from previous chapters based on other phenomena.

Technical difficulties limit the further increase in frequency of applied AC field in the measurement of electric conductance. It can be expected that, like the case for solutions, the electric conductance would attain a limiting value when the frequency is increased continuously. When the frequency effect of soil colloid suspensions is compared with that of solutions, it may be noted that the required frequency for the initiation of conductivity dispersion is lower while the required frequency for inducing a maximum conductivity dispersion is higher in the former case than in the latter case. This indicates that in the interactions of the colloid of variable charge soils with anions a wide range of energy level is involved.

8.3.3 Effect of Cations

The effect of cations on the electric conductance of soil colloid of variable charge soils also shows a peculiar feature. It is generally considered that, among exchangeable cations commonly found in soils, the properties of sodium ions differ from other cations, especially from potassium ions, only in quantity but not in nature. Contrary to this assumption, Figs. 8.19 to 8.21 show that the change in electric conductance with the change in applied frequency for soil colloids adsorbed with sodium ions differs distinctly from that for soil colloids adsorbed with other cations. In the former case, conductivity dispersion begins to appear when the frequency attains several thousand hertz, and then the electric conductance increases with the increase in frequency in a nearly linear fashion. For soil colloids adsorbed with other cations on the conductance–frequency curve there is a plateau region, similar to the curve for anions. The distinctness of this plateau varies with the kind of cation species.

The extent of conductivity dispersion of different soil colloids also varies with the type of the soil. Figures 8.22 and 8.23 show that for both the potassium system and the calcium system the effect of frequency is of the order Ali–Haplic Acrisol > Ferrali–Haplic Acrisol > Rhodic Ferralsol. This seems to indicate that the stronger the adsorption of cations, the larger the proportion of adsorbed cations in the Stern layer and the greater the frequency effect.

Two parameters can be used to compare the frequency effect in electric

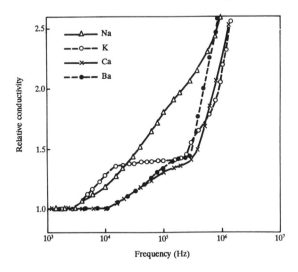

Fig. 8.19. Effect of cations on conductivity dispersion of the colloid fraction of Ali–Haplic Acrisol (conductivity at 1 kHz taken as 1).

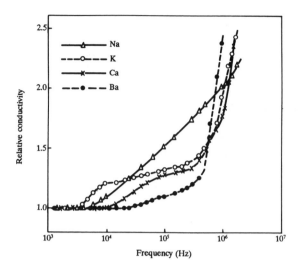

Fig. 8.20. Effect of cations on conductivity dispersion of the colloid fraction of Ferrali–Haplic Acrisol.

conductance of colloids of various soils adsorbed with different cations. In Table 8.4, the meaning of initial frequency of conductivity dispersion is the same as that in the anion system. Inflection frequency is the frequency at which the relative conductivity increases sharply for the second time. The

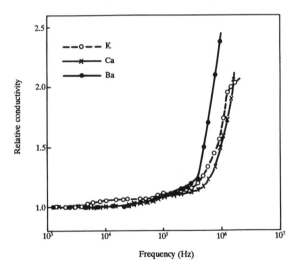

Fig. 8.21. Effect of cations on conductivity dispersion of the colloid fraction of Rhodic Ferralsol.

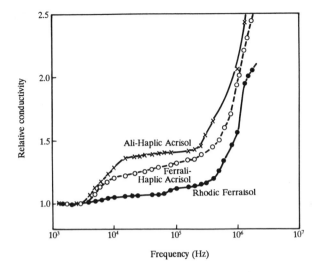

Fig. 8.22. Change in relative conductivity with applied frequency for the colloid fraction of soils (potassium system).

table shows that the initial frequency of conductivity dispersion for the potassium system is similar to that of the sodium system. By contrast, the initial frequencies of conductivity dispersion for systems of divalent calcium ions and especially of barium ions are much higher than those of

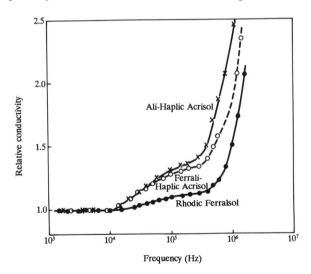

Fig. 8.23. Change in relative conductivity with applied frequency for the colloid fraction of soils (calcium system).

Table 8.4 Initial Frequency and Inflection Frequency of Conductivity
Dispersion of Soil Colloids in Different Cation Systems

Item	Cations	Soil A[a]	Soil B	Soil C
Initial frequency	Na	6	10	–
(kHz)	K	5	5	80
	Ca	23	23	100
	Ba	25	30	100
Inflection frequency	K	200	220	350
(kHz)	Ca	300	300	400
	Ba	300	350	400

[a]Soil A, Ali–Haplic Acrisol; Soil B, Ferrali–Haplic Acrisol;
Soil C, Rhodic Ferralsol.

monovalent cations. The inflection frequency for divalent cations is also higher than that for potassium ions markedly. Among the three soil colloids, the two parameters are particularly high for the Rhodic Ferralsol.

When the magnitudes of relativity conductivity at the two threshold frequencies are compared (Table 8.5), some interesting points can be observed. If the relative conductivity at the inflection frequency is a reflection of the ratio of the conductance of adsorbed ions with a relatively weaker bonding energy level to that of the ions in the diffuse layer, which is in turn a reflection of the ratio of the quantities of the two portions of ions, no remarkable difference can be found among potassium, calcium, and barium ions. On the other hand, for adsorbed ions with a stronger bonding energy level, that is, for ions that can show a big frequency effect at 1.7 MHz, the quantity in the barium system is much larger than that in other systems. If adsorbed with the same cation species, Rhodic Ferralsol differs distinctly from the other two types of soils in that the relative conductivities at both of the two threshold frequencies are relatively small, whereas the proportion of ions with a higher bonding energy level is relatively large.

At present, it is not possible to formulate quantitative relations between the two threshold frequencies or the magnitude of conductivity dispersion at the two frequencies and the distribution of ions within the electric double layer or the bonding strength of ions with the surface of soil colloids. Nevertheless, it is apparent that, qualitatively, the behaviors of potassium, calcium, and barium ions are markedly different from that of sodium ion, an ion species generally considered to be adsorbed solely by electrostatic force. It is noticeable that the Rhodic Ferralsol containing a large amount

Table 8.5 Relative Conductivity of Soil Colloids at Inflection Frequency and at 1.7 MHz in Different Cation Systems

Item	Cations	Soil A[a]	Soil B	Soil C
Relative conductivity at inflection frequency	K	1.43	1.35	1.17
	Ca	1.40	1.33	1.15
	Ba	1.45	1.40	1.25
Relative conductivity at 1.7 Mhz	K	2.72	2.43	2.04
	Ca	2.97	2.66	2.06
	Ba	5.53	5.28	4.50
Relative affinity[a]	K	3.00	3.09	5.12
	Ca	3.93	4.03	6.07
	Ba	9.09	9.70	13.00

[a]Relative affinity = (relative conductivity at 1.7 megHz
  - relative conductivity at inflexion frequency)
  /(relative conductivity at inflexion frequency - 1).
[b]Soil A, Ali–Haplic Acrisol; Soil B, Ferrali–Haplic Acrisol;
  Soil C, Rhodic Ferralsol.

of iron oxides behaves differently from the Ali–Haplic Acrisol and the Ferrali–Haplic Acrisol. In previous chapters, the specific adsorption of these cations by variable charge soil has been encountered several times. Here, the specific adsorption of these cations also manifests itself in frequency effect of conductance.

8.3.4 Effect of Electrolyte

In the presence of electrolytes, owing to the adsorption of both cations and anions, the frequency effect in electric conductance would be more complicated. In this case, both cations and anions can have an effect on conductivity dispersion, but the two effects may not be additive. The relative importance of the two effects would be dependent on both the kind of the ion species and the type of the soil.

The variations of relative conductivity of three soil colloids with the change in applied frequency in different electrolyte solutions are shown in Figs. 8.24 to 8.26, respectively. In the figure, the electric conductance at 300 Hz is taken as 1 in calculations.

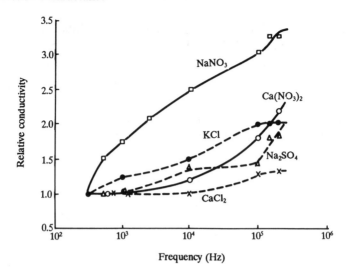

Fig. 8.24. Change in relative conductivity with applied frequency for the colloid fraction of Ali–Haplic Acrisol in different electrolyte solutions (10^{-4} N) (Li and Yu, 1987).

A common tendency for the three soil colloids is that the conductivity dispersion is most pronounced in $NaNO_3$ solution and least remarkable in $CaCl_2$ solution. The same is true when AC conductance is compared with

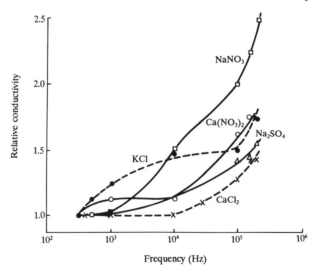

Fig. 8.25. Change in relative conductivity with applied frequency for the colloid fraction of Ferrali–Haplic Acrisol in different electrolyte solutions (Li and Yu, 1987).

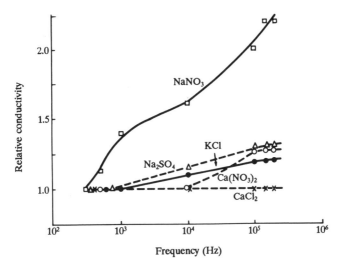

Fig. 8.26. Change in relative conductivity with applied frequency for the colloid fraction of Rhodic Ferralsol in different electrolyte solutions (Li and Yu, 1987).

DC conductance. When the three soil colloids are compared, the magnitude of conductivity dispersion is of the order Ali–Haplic Acrisol > Ferrali–Haplic Acrisol > Rhodic Ferralsol, the same order as when cations or anions are present separately.

In order to compare the function of various ion species in inducing conductivity dispersion, two parameters can be adopted: the ratio of electric conductance at 300 Hz to that at DC, L_{300}/L_{DC}, and the ratio of electric conductance at 200 kHz to that at 300 Hz, L_{200K}/L_{300}. The calculated results are given in Table 8.6.

The data in the table show that, within these low–frequency ranges, if compared based on the ability of inducing conductivity dispersion, the order for cations is $Na^+ > K^+ > Ca^{2+}$ and that for anions is $NO_3^- > Cl^- > SO_4^{2-}$. This is of the same order as when the two kinds of ions are present alone. Note that the effect of anions on conductivity dispersion is greater than that of cations. Two causes may be responsible for this difference. Electrodialyzed soil colloids contain a large amount of aluminum ions as counterions. It may be the conductivity dispersion caused by adsorbed aluminum ions that masks to a certain extent the effect of alkali metal ions and alkaline earth metal ions on conductivity dispersion. On the other hand, since the effect of polyvalent aluminum ions on conductivity dispersion within the range of low frequencies is not large, it is apparent that it is mainly the large difference in adsorption properties among anion species that induces a strong effect on conductivity dispersion. This also indicates

Table 8.6 Relative Conductivity of Soil Colloids in Different Electrolyte
Solutions (Li and Yu, 1987)

Item	Electrolyte	Soil A[a]	Soil B	Soil C	Mean
L_{300}/L_{DC}	$CaCl_2$	1.00	1.00	1.00	1.00
	Na_2SO_4	1.25	1.18	1.00	1.14
	KCl	1.33	1.33	1.00	1.22
	$Ca(NO_3)_2$	1.25	1.33	1.22	1.27
	$NaNO_3$	1.59	1.54	1.43	1.52
	Mean	1.28	1.28	1.13	1.23
L_{200K}/L_{300}	$CaCl_2$	1.32	1.43	1.00	1.25
	Na_2SO_4	1.83	1.55	1.31	1.56
	KCl	2.02	1.75	1.21	1.66
	$Ca(NO_3)_2$	2.20	1.82	1.28	1.77
	$NaNO_3$	3.26	2.48	2.20	2.65
	Mean	2.13	1.81	1.40	1.78

[a]Soil A, Ali–Haplic Acrisol; Soil B, Ferrali–Haplic Acrisol;
Soil C, Rhodic Ferralsol.

that, for variable charge soils, when the pH is sufficiently low, anions may
have very important effects on surface electrochemical properties of the
soil. This should be one of the important characteristics of variable charge
soils as compared to constant charge soils.

8.4 ELECTRIC CONDUCTANCE AND SOIL FERTILITY

Under natural conditions, most of the acid variable charge soils are chiefly
saturated by aluminum ions. Since the degree of dissociation of adsorbed
aluminum ions is extremely low, the electric conductance of these soils is
usually very low. The electric conductance increases after the application of
nutrient cations such as potassium, ammonium, magnesium, and calcium, as
well as accompanied anions. Despite the difference in function among these
ions in plant nutrition, there may exist a certain relationship between
electric conductance and the overall fertility level for the same type of soil.
For variable charge soil in particular, because differences in physical factors
among different fields within the same region and among different depths
within the same profile are not large, it can be generally observed that the

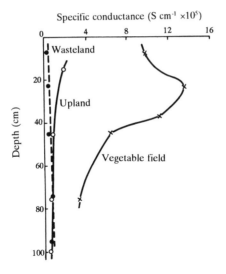

Fig. 8.27. Electric conductance of the profile of Ali–Haplic Acrisol under different utilization conditions (Sun et al., 1983).

electric conductance can reflect the fertility status of the soil. Such an example is shown in Fig. 8.27. The three soil profiles are all developed on Quaternary red clay. For the virginal wasteland, the electric conductance of the whole profile is extremely low. For the cultivated layer of the upland soil the conductance is a little higher, caused by fertilization. However, the electric conductance of low–lying layers remains low. For the cultivated layer of the vegetable soil, because of the effect of large amounts of fertilizers, the electric conductance is quite high. The high electric conductance in lower layers of this soil profile is caused by the eluviation of ions from the cultivated layer.

In Southeast Asia, many variable charge soils have been cultivated for rice. The fertility level of these paddy soils is determined by long–term fertilization practices, and it can also be reflected by the electric conductance of the soil profile. As seen in Fig. 8.28, for an infertile soil, except in the cultivated layer, the electric conductance of the whole profile differs from that of the parent soil only slightly. For well–developed soils, the electric conductance of the whole profile is high, due to long–term eluviation of ions applied to the cultivated layer through fertilization and irrigation.

During rice–growing season, the electric conductance of the cultivated layer is generally higher than that during dry–farming season. This is caused by the development of reduction processes and the decomposition of organic matter as well as the application of a large amount of fertilizers. In

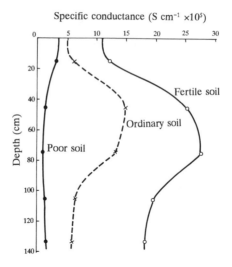

Fig. 8.28. Electric conductance of the profile of paddy soils (derived from Ali-Haplic Acrisol) under upland-crop conditions (Jinhua) (Sun et al., 1983).

this case, however, the electric conductance can still reflect the difference in fertility level among fields, as seen in Fig. 8.29.

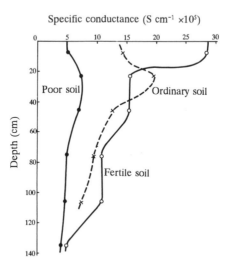

Fig. 8.29. Electric conductance of the profile of paddy soils (derived from Ali-Haplic Acrisol) under rice-growing conditions (Taihuo) (Sun et al, 1983).

The above-mentioned examples are quite representative for variable charge soils of China. This means that, for highly weathered variable charge soils, the electric conductance is intimately related to the fertility status of the soil. Therefore, as early as in the 1940s (Hardy and Rodrigues, 1949), electric conductance was used as a single-value routine estimation for tropical soils, and such kind of application was broadened considerably later (Yu, 1976; Yu and Ji, 1993).

BIBLIOGRAPHY

Arulanandan, K. and Mitchell, J. K. (1968) *Clays Clay Miner.*, 16:337–351.

Atkins, W. R. G. (1924) *J. Agric. Sci.*, 14:198–203.

Barrow, N. J. (1985) *Adv. Agron.*, 38:183–229.

Conkling, B. L. and Blanchar, R. W. (1986) *Soil Sci. Soc. Am. J.*, 50:14-55–1459.

Conkling, B. L. and Blanchar, R. W. (1989) *Soil Sci. Soc. Am. J.*, 53:16-85–1690.

Cremers, A. and Laudelout, H. (1965) *Soil Sci.*, 100:298–299.

Dakshinamurti, C. (1960) *Soil Sci.*, 90:302–306.

Davis, J. A. and Leckie, J. O. (1978) *J. Colloid Interface Sci.*, 67:90–107.

Davis, J. A., James, R. O., and Leckie, J. O. (1978) *J. Colloid Interface Sci.*, 63:480–499.

Deshpande, K. B. and Marshall, C. E. (1959) *J. Phys. Chem.*, 63:1659–1662.

Deshpande, K. B. and Marshall, C. E. (1961) *J. Phys. Chem.*, 65:33–36.

Fricke, H. and Curtis, H. J. (1936) *J. Phys. Chem.*, 40:715–722.

Hardy, F. and Rodrigues, G. (1949) *Proceedings of the First International Conference on Tropical and Subtropical Soils.* pp. 220–225.

Jurinak, J. J., Sandhu, S. S., and Duddley, L. M. (1987) *Soil Sci. Soc. Am. J.*, 51:625–630.

Kinniburgh, D. G., Syers, J. K., and Jackson, M. L. (1975) *Soil Sci. Soc. Am. Proc.*, 39:464–470.

Li, C. B. and Yu, T. R. (1987) *Soil Sci.*, 144:403–407.

Li, C. B. and Yu, T. R. (1990) *Soil Sci.*, 150:831–835.

Li, C. B. (1989a) *Acta Pedol. Sinica*, 26:16–21.

Li, C. B. (1989b) *Progr. Soil Sci.*, 17:1–8.

Low, P. F. (1958) *Soil Sci. Soc. Am. Proc.*, 22:395–398.

Marshall, C. E. (1949) *The Colloid Chemistry of the Silicate Minerals.* Academic Press, New York.

Marshall, C. E. (1964) *The Physical Chemistry and Mineralogy of Soils.* vol. 1. John Wiley & Sons, New York.

Mehran, M. and Arulanandan, K. (1977) *Clays Clay Miner.*, 25:39–48.

Overbeek, J. Th. G. (1952) in *Colloid Science* (H. R. Kruyt, ed.). Elsevier, Amsterdam, pp. 194–244.

Palmer, C. J. and Blanchar, R. W. (1980) *Soil Sci. Soc. Am. J.*, 44:925–929.

Parfitt, R. L. (1978) *Adv. Agron.*, 30:1–50.

Rhodes, J. D. and Ingvalson, R. D. (1971) *Soil Sci. Soc. Am. Proc.*, 35:54–60.

Rhodes, J. D. and van Schilfgaarde, J. (1976) *Soil Sci. Soc. Am. J.*, 40:647–651.

Rhue, D. (1992) *Soil Sci. Soc. Am. J.*, 56:683–689.

Shainberg, I. and Kemper, W. D. (1966) *Clays Clay Miner.*, 14:117–132.

Schwarz, G. (1962) *J. Phys. Chem.*, 66:2636–2642.

Shea, P. F. and Luthin, J. N. (1961) *Soil Sci.*, 92:331–339.

Sparks, D. L. (ed.) (1986) *Soil Physical Chemistry*. CRS Press, Boca Raton, FL.

Staunton, S. (1990) *J. Soil Sci.*, 41:643–653.

Sun, H. Z., Wu, J. and Yu, T. R. (1983) *Acta Pedol. Sinica*, 20:69–78.

Westall, J. and Hohl, H. (1980) *Adv. Colloid Interface Sci.*, 12:265–294.

Yu, T. R. (1976) *Electrochemical Properties of Soils and Their Research Methods*. Science Press, Beijing, pp. 258–301.

Yu, T. R. (1985) *Physical Chemistry of Paddy Soils*. Science Press/Springer Verlag, Beijing/Berlin, pp. 157–177.

Yu, T. R. and Zhang, X. N. (1986) in *Proceedings of the International Symposium on Red Soils*. Science Press/Elsevier, Beijing/Amsterdam, pp. 409–441.

Yu, T. R. and Ji, G. L. (1993) *Electrochemical Methods in Soil and Water Research*. Pergamon Press, Oxford, pp. 332–365.

Zhang, X. N. and Jiang, N. H. (1964) *Acta Pedol. Sinica*, 12:120–131.

9

ION DIFFUSION

T. R. Yu and E. J. Wang

The microregional transport of ions under an externally applied electric field has been discussed previously. When ions are distributed heterogeneously in soil on a macroscopic scale, because of the presence of concentration gradient (i.e., the difference in chemical potential), ions tend to migrate from a site of high concentration to a site of low concentration. Such a phenomenon is called *ion diffusion*. The diffusion rate of various ions in a soil is related to the nature of the ions and the interaction among them and is also affected by the chemical processes in the soil, such as adsorption, desorption, and repulsion. For variable charge soils carrying both positive and negative surface charges, the factors that affect ion diffusion are rather complex.

In the present chapter, after treatment of basic principles of ion diffusion, the characteristic features of ion diffusion in variable charge soils will be discussed, with the emphasis on diffusion of anions because this is one of the important means for elucidating the characteristics of variable charge soils.

9.1 PRINCIPLES OF ION DIFFUSION

9.1.1 Ion Diffusion in Solution

In a solution, if the ion concentration in point A is higher than that in point B, under static conditions, the number of ions moving from point A to point B will be larger than that moving in the opposite direction due to the random thermal motion of ions.

In order to express the net ion flux J within an unit time interval through an unit area, Fick introduced the first diffusion law:

$$J = -D \frac{dC}{dX} \qquad (9\text{-}1)$$

where dC/dX is the concentration gradient. The negative sign in the equation denotes that the flux is from high concentration to low concentration; that is, the direction of the flux is opposite to that of the concentration gradient. D is called the diffusion coefficient. It can be seen from the equation that the diffusion coefficient is the flux passing through an unit cross-sectional area within a unit time interval under a unit concentration gradient. D is the most important parameter in ion diffusion.

Fick's first diffusion law is applicable to both homogeneous and heterogeneous medium such as soil. The essential requirement for applying this law is that the diffusion is in a static state. Under a static state, dC/dX does not change with time. Usually, it needs a long time for the diffusion to reach a static state. If it is desired to conduct an experiment in a nonstatic state, the relationships in Fick's second diffusion law ($dC/dt = D(d^2C/dX^2)$) under different boundary conditions must be used.

In order to express the interaction energy between diffusing substances and the medium during diffusion, the concept of activation energy is introduced. The activation energy of diffusion includes two parts: the energy barrier that must be overcome for an ion during its diffusion, and the energy needed for the formation of a "hole" in the medium (water) to accommodate the diffusing ion. Activation energy of diffusion can be determined using the relationship between the diffusion coefficient and temperature:

$$D = A \exp\left(-\frac{E}{RT}\right) \qquad (9\text{-}2)$$

where E is the activation energy of diffusion and A is a constant. Taking the logarithmic form of equation (9-2), we can have

$$\ln D = \ln A - \frac{E}{RT} \qquad (9\text{-}3)$$

Determining D at different temperatures and plotting $\ln D$ against $1/T$, E can be obtained from the slope of the plot.

In an aqueous solution, ions diffuse through a medium that is formed by a hydrogen bond and has a loose structure. The charge of ions will break such a structure because water dipoles tend to orient around ions. If an ion and its surrounding water molecules are regarded as a spherical particle moving in a homogeneous medium, according to Stokes' law, the diffusion coefficient of the ion is related to its radius as

$$D = \frac{kT}{6\pi\eta r} \qquad (9\text{-}4)$$

where k is the Boltzmann constant and r is the effective radius of the hydrated ion. It has been found experimentally that the diffusion coefficient decreases with the decrease in the radius of the unhydrated ion. This is because a small ion has a stronger tendency of hydration and thus a larger effective radius.

It can be seen from equation (9-4) that the diffusion coefficient is inversely proportional to the viscosity η of the medium.

9.1.2 Ion Diffusion in Soils

Soil is a heterogeneous system consisting of gas, liquid, and solid phases. Therefore, the diffusion of ions in a soil is much more complex than that in solution. Ions in a soil can either diffuse to the interior of soil particles (diffusion within solid phase) or diffuse through the water film on the surface of soil particles. When the pore space of the soil is filled with water, ions can also diffuse in soil solution. In the solid phase the diffusion of ions requires an activation energy as high as 40–200 kJ mol^{-1}; therefore, the diffusion coefficient is only of the order of 10^{-16}–10^{-24} cm^2 s^{-1} (Haan et al., 1965), much smaller than the 10^{-5} cm^2 s^{-1} value for the diffusion in aqueous solutions where the activation energy of diffusion is only 16–18 kJ mol^{-1} (Yu, 1987). This type of diffusion will not be considered in the present discussion. The diffusion of ions at the soil–water interface is more complex. In this diffusion several important processes, such as interactions between ions and surface charges, adsorption–desorption processes of ions, exchange reactions among exchangeable ions, position exchange between ions and water molecules, and so on, are involved.

To derive a diffusion equation for heterogeneous soil systems, similar to the derivation of equation (9-1), it can be assumed that the total ion flux through an unit cross-sectional area of soil is the flux through the interstitial solution plus an additional flux. This additional flux is caused by the movement of exchangeable cations along the solid surface. Thus,

$$J = -D_l \theta f\left(\frac{dC_l}{dX}\right) + J_a \qquad (9\text{-}5)$$

where D_l is the diffusion coefficient of ions in free solution, θ is the volume fraction of the solution in soil which represents the cross section of diffusion, f is an impedance factor, C_l is the ion concentration in soil solution, and J_a is the additional flux caused by exchangeable ions. In the literature, reports on the contribution of J_a to J differ greatly (Conkling and Blanchar, 1989; Nye, 1979; Rhue, 1992; Staunton, 1986, 1990; Staunton and Nye, 1987; van Schaik et al., 1966). Apparently, this contribution depends on a variety of factors, including type of the soil, nature of the ion species, water content, and accompanying ions.

Combining equations (9-1) and (9-5), we can get

$$D = D_l \theta f \left(\frac{dC_l}{dC} \right) + D_a \qquad (9\text{-}6)$$

Here, D may be called the apparent diffusion coefficient of the ions in soil, or the diffusion coefficient. In the equation, θ is the volume fraction of water associated with ions in soil solution but does not include water associated with exchangeable ions. Thus, the product θC_l is the quantity of ions in solution in an unit volume of soil. dC_l/dC is the changing rate of ion concentration in solution with the change in total ion concentration. For adsorbed ions, dC_l/dC near the surface of clay particles decreases with the decrease in total ion concentration. Therefore, the diffusion coefficient of adsorbed ions increases with the increase in the quantity of the diffusing source. For repulsed ions, the reverse is true. However, because repulsion is usually much weaker than adsorption within the electric double layer of colloids, the effect of repulsion on ion diffusion is generally much smaller than that of adsorption. At a given condition, the relative intensities of adsorption and repulsion will affect the magnitude of the apparent diffusion coefficient.

Impedance factor f mainly depends on the tortuosity of the soil, that is, the effect caused by geometric factors. Because ions can only migrate along the surface of soil particles and through pore spaces when diffusing in a soil, the actual diffusion pathway is much longer than a straight pathway, whereas the effective cross-sectional area through which ions can pass is much smaller than the apparent cross-sectional area. Besides soil tortuosity, f also includes the effect of the increase in viscosity of water adjacent to particle surfaces on ion diffusion.

In equation (9-6), θ and f can be regarded as physical factors. Both of them are related to the water content of the soil. There have been many reports (Kemper, 1960; Low, 1962; Nye, 1979; Rhadoria et al., 1991; Sinha and Singh, 1977; Skogley and Schaff, 1985; Stigter, 1980; Wu, 1976) showing that the apparent diffusion coefficient of ions in soils is closely related to the water content. However, this relationship is quite complex and varies with soil type, ion species, and water content. Table 9.1 shows the effect of water content on the diffusion coefficient of four ion species in an Ali-Haplic Acrisol. All the data about diffusion coefficient of ions cited in this chapter were determined under similar water content conditions.

Table 9.1 Effect of Water Content on Diffusion Coefficient of Ions in Ali–Haplic Acrisol

Water	D (cm^2 s^{-1} $\times10^6$)			
(%)	Cl$^-$	NO$_3^-$	K$^+$	Ca^{2+}
25	0.30	–	–	–
30	0.82	–	–	–
35	2.2	–	0.21	0.08
40	–	3.0	0.84	–
49	4.4	4.1	1.8	1.0
55	7.3	8.1	7.0	1.6

The other two terms dC_1/dC and D_a in equation (9-6) are related to the surface properties of the soil and the nature of ions. These relationships may be very complex even for constant charge soils. On the other hand, research on these relationships may provide an important means for elucidating the interactions between ions and soils, particularly for variable charge soils carrying both charges and negative surface charges.

9.1.3 Determination of Diffusion Coefficient

As mentioned in the last section, diffusion coefficient is defined as the ion flux within a unit time interval through a unit cross–sectional area under a unit concentration gradient and is the most important parameter in characterizing the diffusion rate of ions. For a soil system, the diffusion coefficient is also an important parameter in characterizing the interactions between the diffusing ions and the soil.

In soil science, various methods for the determination of diffusion coefficients have been suggested (Nye, 1979; Wu, 1976). The most commonly used method is the thin–section method (Brown et al.,1964; Nye and Ameloko, 1986). This method can give accurate results but is tedious in its operation. The ion-exchange resin method (Barraclough and Tinker, 1981, 1982) and the electric conductance method (Conkling and Blanchar, 1989; Rhue, 1992) can only give an overall diffusion coefficient, but not the picture of ion distribution along the diffusing path. The present authors (Wang and Yu, 1989a) have developed a method for the determination of diffusion coefficient of ions with miniature ion–selective electrodes. The principle of this method is described below.

Based on the method proposed by Barrer (1951) for the study of ion diffusion, a certain quantity (Q) of a diffusing substance is placed at one end of the diffusion cell and allowed to diffuse for a certain period of time (t) at constant temperature, and then the concentration C at a distance x from the diffusing source is determined with a miniature ion-selective electrode. C is related to x as follows:

$$\log C = \log \frac{Q}{(\pi Dt)^{\frac{1}{2}}} - \frac{0.1086x^2}{Dt}$$
$$= A - \frac{0.1086x^2}{Dt} \tag{9-7}$$

where A is a constant and its numerical value depends on diffusion condition and D is the diffusion coefficient of the ions.

For ion diffusion in soils, if the contribution of adsorbed ions is neglected, the relationship between ion concentration in solution C_1 and x can be written as

$$\log C_1 = A' - \frac{0.1086x^2}{Dt} \tag{9-8}$$

According to the Nernst equation, the potential E_i of an ion-selective electrode is related to C_1 as follows:

$$E_i = E_i^0 \pm s \log \gamma C_1 \tag{9-9}$$

where E_i^0 is the "standard potential" of the electrode, s is a constant at a given temperature which is positive for cations and negative for anions, and γ is the activity coefficient of the ion species.

If the ionic strength in soil solution is low and thus variations in γ can be neglected, equation (9-9) becomes

$$E_i = E_i^{0'} \pm s \log C_1 \tag{9-10}$$

In the determination, an electrochemical cell is constructed with an ion-selective electrode and a reference electrode. The emf should be

$$E = E^0 \pm s \log C_1 + E_j \tag{9-11}$$

Because the position of the reference electrode is fixed, the liquid-junction potential (E_j) between the reference electrode and the soil can be regarded as constant. Because the diameter of the tip of the ion-selective electrode is less than 0.5 mm, the potential of the electrode at position x can be regarded as reflecting the ion concentration at this point. Combining equations (9-9) and (9-11) we have

$$A' - \frac{0.1086x^2}{Dt} = \pm \frac{(E - E^0 - E_j)}{s} \tag{9-12}$$

and

$$\pm \frac{E}{s} = B - \frac{0.1086x^2}{Dt} \tag{9-13}$$

where $B = A' + (E^0 + E_j)/s$.

Thus, after the measurement of E values of the electrode at various points, E/s is plotted against x^2 and the diffusion coefficient D can be obtained from the slope of the plot.

Figures 9.1 and 9.2 show the E vs. x^2 relationships for K^+ ions and NO_3^- ions in an Ali–Haplic Acrisol after diffusion for a given time. Note that within a certain distance range the relationship was linear. The procedure described above has been used to study ion diffusion in variable charge soils, and the results from the study will be presented in the next sections. Since the experiments were conducted under specified conditions, the D values obtained are apparent diffusion coefficients. For the sake of convenience, D will be called the *diffusion coefficient* in this chapter.

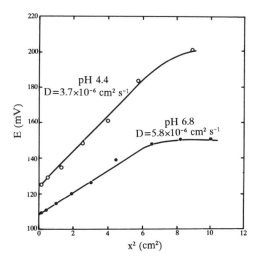

Fig. 9.1. E–x^2 relationship in the diffusion of nitrate ions in Ali–Haplic Acrisol (Wang and Yu, 1989a).

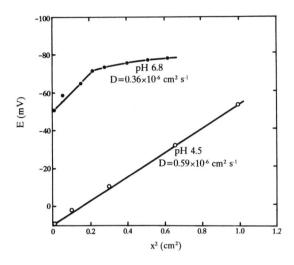

Fig. 9.2. $E-x^2$ relationship in the diffusion of potassium ions in Ali–Haplic Acrisol (Wang and Yu, 1989a).

9.2 DIFFUSION OF IONS IN SOILS

When various ions diffuse in variable charge soils, the diffusion rate will depend not only on the nature of the ions, but also on the surface charge properties of the soil. Other accompanying ions with the same or opposite electric charge will also affect the diffusion rate. These factors will be discussed in the following sections.

9.2.1 Relationship Between Diffusion Coefficient and Ion Species

Comparisons of the diffusion coefficients between two cation species and between two anion species in several soils are given in Table 9.2. The diffusion coefficients of anions are much larger than those of cations. This is because these soils either carry more negative surface charges than positive surface charges, or carry mainly negative surface charges. As a result, cations will encounter a greater electrostatic retarding force than anions during their diffusion. Because of electrostatic attraction exerted by soil particles, cations will diffuse mainly in the water film close to clay particles. Since the viscosity of water near the clay surface is larger than that in bulk solution, the retarding force for cation diffusion will be even greater. In contrast, the situation is different during the diffusion of anions. When the two kinds of cations are compared, it is seen that the diffusion coefficient of calcium ions is only one–half of that of potassium ions. This is because divalent calcium ions are attracted more strongly by clay particles. For anions, there are two phenomena deserving notice. First, the diffusion coefficients of the two anion species in the Cambisol are remarkably larger than those in other soils. This is because this soil chiefly carries negative surface charges, and therefore its negative surface charge density is higher than that of variable charge soils. Thus, this soil can have a repulsive action on anions. Second, a comparison between the two anion species can show that the mean value of diffusion coefficients of nitrate ions is larger than that of chloride ions. Since the reverse should be true for the two ion species in aqueous solution (Cussler, 1984), this phenomenon is difficult to explain. It can only be imagined that the adsorption of Cl^- ions by variable charge soils is stronger than that of NO_3^- ions. This point will be further discussed shortly.

9.2.2 Effect of pH

Soil pH affects ion diffusion in two ways. For certain kinds of ions, pH may affect their forms. This has been discussed in the previous chapters. It has been shown in research that even in constant charge soils the diffusion rate of some anions (Habib and Guennelon, 1983; Place et al., 1968) and cations (Cottenie and Verloo, 1984; Trefry and Metz, 1984; Melton et al., 1973)

Table 9.2 Diffusion Coefficients of Different Ions in Soils[a] (Wang and Yu, 1989b)

Soil	D (cm^2 s^{-1} $\times 10^6$)			
	Cl^-	NO_3^-	K^+	Ca^{2+}
Rhodic Ferralsol (Guangdong)	3.4	3.9	0.68	0.32
Ferrali–Haplic Acrisol (Guangdong)	2.6	3.2	0.78	0.54
Ali–Haplic Acrisol (Jiangxi)	2.5	2.6	0.47	0.27
Rhodic Ferralsol (Brasilia)	1.5	2.8	0.63	–
Rhodic Ferralsol (Australia)	3.1	3.6	–	–
Rhodic Ferralsol (Hawaii)	2.9	3.2	0.72	0.32
Xanthic Ferralsol (Manaus)	3.6	4.0	0.43	0.26
Cambisol (Nanjing)	6.0	6.1	–	–
Mean	3.2	3.7	0.62	0.34

[a]Electrodialyzed soil.
$Q_{Cl} = Q_{NO_3} = Q_K = 2Q_{Ca} = 1.1$ mmol cm^{-2}.

specifically adsorbed by the soil are affected by pH. In variable charge soils, as has been mentioned in previous chapters, pH can also affect the surface charge properties of the soil and its interactions with ions. This will directly affect the diffusion of ions, including those not participating in specific adsorption (Song and Ishiguro, 1992). This latter point is of concern in the present chapter.

The diffusion coefficients of Cl^-, NO_3^-, K^+, and Ca^{2+} ions in four soils as related to pH are shown in Figs. 9.3 to 9.6, respectively. It can be seen that the diffusion coefficients of Cl^- ions and NO_3^- ions increased with the increase in pH in the three variable charge soils. By contrast, in Cambisol carrying chiefly negative surface charges, the diffusion coefficient of NO_3^- ions was nearly independent of pH and that of Cl^- ions was only affected by pH to a small extent. When the two anion species are compared, it can be found that the slope of the D-pH curve for Cl^- ions was steeper than that for NO_3^- ions, showing that the trend was $D_{Cl} < D_{NO_3}$ at low pH but $D_{Cl} > D_{NO_3}$ at high pH. Since these two ion species are generally regarded as nonspecifically adsorbed ions, the increase in the diffusion coefficient with the rise in pH would be caused by the decrease in the retarding force exerted by the positive surface charges of the soil. These results also show that the diffusion of Cl^- ions is more strongly affected by soil pH than that of NO_3^- ions.

The diffusion coefficients of both potassium ions and calcium ions decreased steadily with the increase in pH. This is caused by the intensification of the effect of negative surface charge of soils.

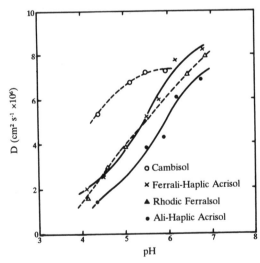

Fig. 9.3. Diffusion coefficient of chloride ions in relation to soil pH ($Q = 1.1$ mmol cm^{-2}) (Wang and Yu, 1989b).

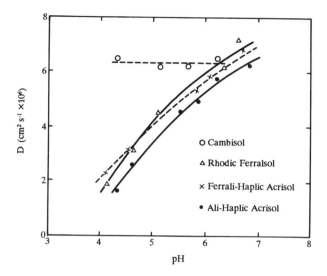

Fig. 9.4. Diffusion coefficient of nitrate ions in relation to soil pH ($Q = 1.1$ mmol cm^{-2}) (Wang and Yu, 1989b).

9.2.3 Effect of Iron Oxides

Because iron oxides are the principal carriers of the positive charge in soils, it should be expected that the diffusion coefficient of anions will decrease

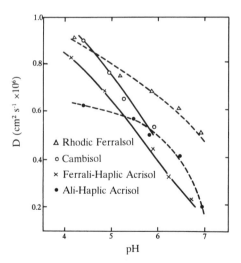

Fig. 9.5. Diffusion coefficient of potassium ions in relation to soil pH ($Q = 0.55$ mmol cm^{-2}) (Wang and Yu, 1989b).

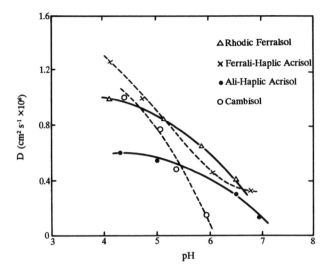

Fig. 9.6. Diffusion coefficient of calcium ions in relation to soil pH (Q = 1.1 mmol cm^{-1}) (Wang and Yu, 1989b).

when iron oxide coatings are added to soils. This is consistent with experimental results. The data in Table 9.3 show that the D_{Cl} and D_{NO_3} decreased in the three variable charge soils when iron oxides were added. The more iron oxide added, the smaller the diffusion coefficient. If a comparison between the two anion species is made, the diffusion of Cl$^-$ ions was affected more than that of NO$_3^-$ ions. This point is of theoretical interest and shall be discussed later.

9.2.4 Effect of Quantity of Diffusing Source

The quantity of the diffusing source determines the concentration gradient of ions during diffusion. Besides, for adsorptive ions, the lower the concentration, the larger the fraction adsorbed on the surface of clay particles. In this case, the ions will encounter a greater electrostatic attraction of clay particles and a larger viscous resistance of the medium during diffusion. As a result, the diffusion coefficient will be decreased. This is particularly true for strongly adsorbed ions such as copper, zinc, manganese (Ellis et al., 1970; Melton et al., 1973; Phillips et al., 1972), and phosphate (Habib and Guennedon, 1983; Hira and Singh, 1977, 1978).

It can be expected that in variable charge soils the effect of the quantity of diffusing source on the ion diffusion would be in a complicated manner.

The effects of the quantity of the diffusing source on diffusion coefficients of potassium and calcium ions in four soils are shown in Table 9.4.

Table 9.3 Effect of Iron Oxides on Diffusion Coefficient of Anions in Soils[a] (Wang and Yu, 1989b)

Soil	Treatment	pH	D (cm^2 s^{-1} ×10^6) Cl^-	NO_3^-	Relative Cl^-	NO_3^-
Rhodic Ferralsol	Original	6.7	7.9	7.3	100	100
	$+25\%$ Fe_2O_3	6.4	4.5	5.0	57	69
Ferrali–Haplic	Original	6.7	7.4	7.0	100	100
Acrisol	$+25\%$ Fe_2O_3	7.2	2.3	3.4	31	49
Ali–Haplic	Original	5.8	4.8	5.2	100	100
Acrisol	$+15\%$ Fe_2O_3	6.0	4.4	4.7	92	90
	$+25\%$ Fe_2O_3	5.6	2.6	3.6	54	69

[a]$Q = 1.1$ mmol cm^{-2}.

Table 9.4 Effect of Quantity of Diffusing Source (Q) on Diffusion Coefficient of Cations in Soils (Wang and Yu, 1989b)

Cations	Q^a	Soil A[b] 4.2[c]	6.5	Soil B 4.1	5.8	Soil C 4.6	6.8	Soil D 6.0
K^+	0.55	–	0.16	–	0.16	–	0.18	–
	1.11	0.91	0.57	0.83	0.46	0.62	0.19	0.59
	5.55	1.46	0.72	2.3	0.65	1.1	0.44	0.98
	11.20	2.66	1.5	4.6	2.0	1.8	0.77	2.7
Ca^{2+}	0.56	0.77	–	–	–	0.55	–	–
	2.78	0.95	0.23	–	0.23	–	0.07	0.25
	5.60	–	0.48	–	0.43	–	0.18	1.4
	11.20	–	1.0	–	1.4	–	0.85	2.7
	22.20	–	2.5	–	3.2	–	1.4	–

Column header for D: D (cm^2 s^{-1} ×10^6)

[a]Q in unit mmol cm^{-2}.
[b]Soil A, Rhodic Ferralsol; Soil B, Ferrali–Haplic Acrisol; Soil C, Ali–Haplic Acrisol; Soil D, Cambisol.
[c]pH.

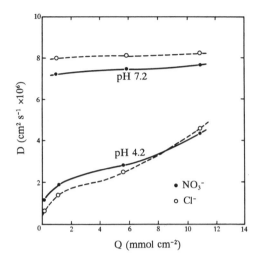

Fig. 9.7. Effect of quantity of diffusing source (Q) on diffusion coefficient of anions in Rhodic Ferralsol (Wang and Yu, 1989b).

Note that for both of the cation species, D increased steadily with the increase in the diffusing source. The change was more remarkable at a high pH. This is because the negative surface charge density of the soil increases and the adsorption for cation becomes stronger when pH is raised.

For anions the relationship between the diffusion coefficient and the quantity of the diffusing source varied with pH and soil type (Figs. 9.7 to 9.9). At pH 4.2–4.4, for both Cl⁻ ions and NO_3^- ions, D increased with the increase in quantity of the diffusing source. This indicates that in this case the adsorption of ions by the soil played a dominant role. At pH 7, D still increased slightly in the Rhodic Ferralsol, but decreased slightly in the Ali–Haplic Acrisol and the Ferrali–Haplic Acrisol. This indicates that for the latter two soils, negative surface charges of clay particles exerted a significant repulsive action on the diffusing ions.

It is worth noting that at about pH 7, D_{Cl} was always larger than D_{NO_3}, regardless of the quantity of the diffusing source. This reflects the difference in mobility between the two ion species in solution. At pH 4.2–4.4, this was still true when the quantity of the diffusing source was large. But, when the quantity of anions was lowered to a certain extent, D_{Cl} became smaller than D_{NO_3}. This suggests again that the adsorption of Cl⁻ ions was stronger than that of NO_3^- ions. When the three soils are compared, it is found that the quantities of the diffusing source, at which the diffusion coefficients of the two anion species equal each other for the Rhodic Ferralsol, Ferrali–Haplic Acrisol and Ali–Haplic Acrisol, were 8.5, 7.5, and 2.0 mmol cm⁻², respectively. This indicates that the power to adsorb anions is of the order Rhodic

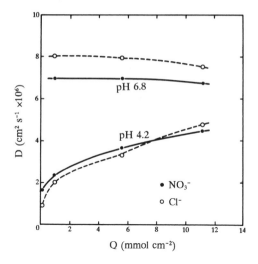

Fig. 9.8. Effect of quantity of diffusing source (Q) on diffusion coefficient of anions in Ferrali–Haplic Acrisol (Wang and Yu, 1989b).

Ferralsol > Ferrali–Haplic Acrisol > Ali–Haplic Acrisol. This conclusion is consistent with that drawn by Wang et al. (1987) when studying anion adsorption with ion–selective electrodes.

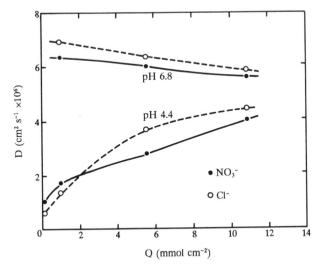

Fig. 9.9. Effect of quantity of diffusing source (Q) on diffusion coefficient of anions in Ali–Haplic Acrisol (Wang and Yu, 1989b).

9.2.5 Effect of Coexisting Ions

In general, there are more than one kind of ions present in soils. These ions would always interact with one another whether they are in diffusion or in a dynamic equilibrium state. Therefore, the diffusion of an ion is always under the influence of its coexisting ions. This can be distinguished in the following three cases.

If coexisting ions carry electric charges opposite to that of the diffusing ions in sign, because of electrostatic attraction, the diffusion of ions may be accelerated by coexisting ions with a larger mobility or retarded by ions with a smaller mobility. If coexisting ions can be adsorbed by the soil, their effect on ion diffusion would also depend on the extent of ion adsorption of these ions. Therefore, when different neutral salts consisting of the same kind of diffusing ions and different kinds of accompanying ions are added to a soil separately, the diffusion rate of that ion species will vary with the kind of accompanying ions.

The effects of accompanying ions on the diffusion coefficient of K^+ ions and NO_3^- ions in variable charge soils are shown in Tables 9.5 and 9.6, respectively. Table 9.5 shows that when accompanying ions were sulfate, phosphate, and biphthalate the diffusion coefficients of potassium ions were much smaller than when the accompanying ion was chloride. This is because the less mobile ions retarded the diffusion of K^+ ions. The data in Table 9.6 show that for the diffusion of nitrate ions the diffusion coefficients in the three soils were of the order $D_{NaNO_3} > D_{KNO_3} > D_{Ca(NO_3)_2}$. This is of the same order in diffusion rate for the three kinds of cations in soils (Nye,

Table 9.5   Effect of Accompanying Ions on Diffusion Coefficient of
Potassium Ions in Soils[a] (Wang and Yu, 1989b)

Accompanying Ions	D (cm^2 s^{-1} $\times 10^6$)		
	Rhodic Ferralsol	Ferrali–Haplic Acrisol	Ali–Haplic Acrisol
Cl^-	0.67	0.55	1.10
SO_4^{2-}	0.33	0.47	0.25
$H_2PO_4^-$	0.23	0.34	0.16
$C_6H_4(COO)_2^{2-}$	–	0.40	0.27

[a]$Q = 5.5$ mmol cm^{-2}. Soil pH: Rhodic Ferralsol, 6.5;
Ferrali–Haplic Acrisol, 6.2; Ali–Haplic Acrisol, 4.6.

Table 9.6 Effect of Accompanying Ions on Diffusion Coefficient of Nitrate Ions in Soils[a] (Wang and Yu, 1989b)

Accompanying Ions	D (cm^2 s^{-1} ×10^6)		
	Rhodic Ferralsol	Ferrali–Haplic Acrisol	Ali–Haplic Acrisol
Na$^+$	8.3	7.6	6.9
K$^+$	7.0	6.5	6.0
Ca^{2+}	5.9	2.9	2.7

[a]Q = 5.5 mmol cm^{-2}.

1979).

During the process of ion diffusion in soils, ions will constantly interact with the charge sites (exchange sites) on the surface of soil particles and will also interact with ions originally adsorbed on the exchange site. Therefore, it may be expected that the preexisting exchangeable cations in the soil will affect ion diffusion. An example is shown in Table 9.7. It should be noted that for a given soil a higher pH also implies a higher proportion of exchangeable sodium or calcium ions and a correspondingly lower proportion of exchangeable hydrogen (aluminum) ions. Bearing this in mind, it can be noted that the diffusion coefficients of potassium ions were of the order H(Al)–saturated soil > Ca–saturated soil > Na–saturated soil. A reversed order was found for anions. Obviously, the difference in the diffusion coefficient of K$^+$ ions among these soils was caused by the difficulty in competing with H(Al) ions and the ease in competing with Na$^+$ ions for exchange sites. In H(Al)–saturated soils the electric double layer of soil colloids is thinner than that in Na–saturated soils. This would be favorable to the diffusion of K$^+$ ions. The situation in Ca–saturated soils is intermediate between H(Al)–saturated soil and Na–saturated soil. The diffusion coefficients of anions in H(Al)–saturated soils were larger than those in Na–saturated soils. This is related to the effect of accompanying ions mentioned in the previous section.

If, in addition to diffusing ions, other neutral electrolytes are present in the soil, both the cations and the anions of the electrolyte will affect the diffusion of these ions in a complicated way. As shown in Table 9.8, the diffusion coefficients of both nitrate and potassium ions were larger in the presence of CaCl$_2$ than those in water. This is primarily caused by the weakening of adsorption of diffusing ions by the soil in the presence of ions of the electrolyte.

Table 9.7 Effect of Exchangeable Cations on Diffusion Coefficient of Ions in Soils[a] (Wang and Yu, 1989b)

Soil	pH	Exchangeable Cations	D (cm^2 s^{-1} ×10^6)		
			Cl$^-$	NO$_3^-$	K$^+$
Ali–Haplic Acrisol	5.0	Na–H(Al)	3.0	3.6	0.58
		Ca–H(Al)	2.9	3.2	0.68
	5.5	Na–H(Al)	4.0	4.5	0.52
		Ca–H(Al)	3.5	3.5	0.57
	6.5	Na–H(Al)	6.5	6.0	0.40
		Ca–H(Al)	5.4	5.8	0.47
Ferrali–Haplic Acrisol	5.2	Na–H(Al)	4.6	4.5	0.78
		Ca–H(Al)	2.8	3.4	1.00
	6.0	Na–H(Al)	5.0	5.6	0.60
		Ca–H(Al)	4.4	5.0	0.75
	6.8	Na–H(Al)	7.4	6.8	0.50
		Ca–H(Al)	6.3	6.5	0.52

[a]Q = 1.1 mmol cm^{-2}.

Table 9.8 Effect of Electrolyte on Diffusion Coefficient of Ions in Soils (Wang and Yu, 1989b)

Soil	pH	Medium	D (cm^2 s^{-1} ×10^6)	
			NO$_3^-$	K$^+$
Ali–Haplic Acrisol	4.4	H$_2$O	2.8	0.72
		CaCl$_2$[a]	3.3	0.77
	6.8	H$_2$O	6.0	0.44
		CaCl$_2$	6.2	0.58
Cambisol	5.3	H$_2$O	5.5	1.0
		CaCl$_2$	6.0	2.1
	6.0	H$_2$O	5.3	0.64
		CaCl$_2$	6.5	1.0

[a]0.04 M CaCl$_2$ solution.

In the above, the effect of coexisting ions on ion diffusion is briefly discussed from three aspects. Actually, these coexisting ions may also affect other surface electrochemical properties of the soil. A soil always contains a variety of cations and anions. Therefore, under field conditions, factors that can affect ion diffusion are rather complicated. The interrelationship between two ion species when diffusing simultaneously will be discussed in a latter section.

9.2.6 Active Fraction of Diffusing Ions

Equation (9-6) may be modified. A term $\theta dC_1/dC$ is defined here as the apparent relative active fraction, or active fraction, of the diffusing ions and is denoted as a. If the D_a term is neglected, equation (9-6) may be rewritten as

$$D = D_l fa \qquad\qquad (9\text{-}14)$$

The a value for the ions at zero adsorption should be equal to 1. Then,

$$D = D_l f \qquad\qquad (9\text{-}15)$$

According to the data of Wang et al. (1987), when soils were treated with a 10^{-3} M solution, the points of zero adsorption of nitrate ions for the Rhodic Ferralsol, Ferrali–Haplic Acrisol and Ali–Haplic Acrisol occurred at pH 7.2, 6.9, and 6.0, respectively. The D_l value of NO_3^- ions in aqueous solution is 1.89×10^{-5} cm^2 s^{-1}. If the D value of NO_3^- ions at zero adsorption is substituted into equation (9-15), the impedance factor f can be calculated. Then, the active fractions of various ions at different pH can be obtained. The relevant data for four soils are listed in Table 9.9. These data show that for anions the dependence of active fractions on soil pH varies with the soil type. It can be calculated that, for the Cambisol carrying only a small proportion of variable charge, the a value increased by a factor of 0 and 0.19 for NO_3^- ions and Cl^- ions, when the pH increased by 1 unit. By contrast, for the three variable charge soils Ali–Haplic Acrisol, Ferrali–Haplic Acrisol, and Rhodic Ferralsol, the corresponding a values for NO_3^- were 0.79, 1.17, and 1.26, respectively, and for Cl^-, 1.04, 1.50, and 1.68, respectively. This is in qualitative agreement with the change in positive surface charge density for these three soils (Yu and Zhang, 1986). Thus, it is apparent that the surface charge of the soil is the principal factor in governing the active fraction of anions in the diffusion.

An a value larger than unity implies a negative adsorption. It can be seen from the table that this is possible for Cl^- ions and NO_3^- ions at high pH, especially for the Ali–Haplic Acrisol. A similar phenomenon has been

Table 9.9 Apparent Active Fraction (a) of Diffusing Ions at Different
pH[a] (Wang and Yu, 1989b)

Soil	f	pH	Cl^-	NO_3^-	K^+	Ca^{2+}
				a		
Rhodic Ferralsol	0.42	4.2	0.19	0.23	0.11	–
		4.6	0.35	0.39	0.099	0.069
		5.2	0.46	0.55	0.090	–
		6.3	0.82	0.78	0.074	–
		6.7	0.90	0.90	0.064	0.024
Ferrali–Haplic	0.37	4.1	0.26	0.32	0.11	0.068
Acrisol		4.7	0.40	0.43	0.096	0.050
		5.9	0.70	0.74	0.060	0.035
		6.2	0.79	0.81	0.043	–
		6.8	0.99	1.00	0.031	–
Ali–Haplic	0.29	4.4	0.24	0.29	0.11	0.089
Acrisol		4.6	0.39	0.46	0.10	0.079
		5.5	0.65	0.82	0.097	0.051
		5.8	0.73	0.89	0.087	–
		6.3	1.03	1.03	0.071	–
		6.7	1.17	1.13	0.033	–
Cambisol	0.40	4.3	0.65	0.85	0.11	–
		5.2	0.85	0.80	0.095	–
		5.7	0.89	0.81	0.082	0.046
		6.4	0.91	0.84	0.067	0.036

[a] $Q_{Cl} = Q_{NO_3} = Q_K = 2Q_{Ca} = 1.1$ mmol cm^{-2}

found by Wang et al. (1987) when studying anion adsorption by these soils
with the direct use of chloride ion–selective and nitrate ion–selective
electrodes.

The difference in active fraction between Cl^- ions and NO_3^- ions is
interesting. The hydrated radii of Cl^- ions and NO_3^- ions are 0.332 and 0.335
nm, respectively. If the active fraction is governed solely by electrostatic
force between ions and soil surface charge, the a values for the two ion
species should be similar. Actually, except when the pH was high, the a
value of nitrate ions was larger than that of chloride ions for the three
variable charge soils, and there was a general trend that the lower the pH

the greater the difference. This means that chloride ions were more retarded than nitrate ions during diffusion. This situation is consistent with the difference in adsorption ratio NO_3^-/Cl^- discussed in Chapter 4. Thus, in the present chapter, the discussions made from a dynamic viewpoint provide a further support to a conclusion reached in Chapter 5, that is, in the interactions between chloride ions and the surface of soil particles, in addition to coulombic force, some kinds of specific forces may be involved.

In the Cambisol, the active fraction of potassium ions increased with the decrease in pH. This may be caused by the competitive adsorption of aluminum ions with the diffusing K^+ ions (cf. Chapter 11). The more pronounced dependence of active fraction of potassium ions on pH for the three variable charge soils is likely to be caused by the combined effect of the competition and the variation in negative surface charge density of the soils.

9.3 SIMULTANEOUS DIFFUSION OF TWO ION SPECIES IN MIXED–SALT SYSTEMS

All the data cited in Section 9.2 reflect the situation in which only a single salt such as KCl or KNO_3 is present in the soil, although the influences of other ions are occasionally considered in the discussions. In the present section, the simultaneous diffusion of two ion species in mixed–salt systems will be examined. Because in such a system the examined ions diffuse under identical environmental conditions, any difference in the diffusion rate between the two ion species could only be ascribed to the difference in properties of the diffusing ions. In the following discussions, emphasis will be placed on diffusion of anions, because this is of great significance in understanding the characteristics of variable charge soils.

9.3.1 Relationship Between Diffusion Coefficient and Ion Species

The diffusion coefficients of several ion species in mixed–salt systems (KCl + KNO_3, or KCl + $CaCl_2$) are given in Table 9.10. The difference in the diffusion coefficient among these ions is similar to that given in Table 9.2, that is, $D_{NO_3} > D_{Cl}$ and $D_K > D_{Ca}$. The diffusion coefficient of potassium ions is about twice as large as D_{Ca}. When a comparison is made between the two tables, it can be seen that the presence of NO_3^- ions facilitates the diffusion of Cl^- ions, and vice versa. Apparently, this is caused by the competitive adsorption of another ion species carrying the same kind of charge. The much larger D values of potassium ions and calcium ions given in Table 9.10 as compared to Table 9.2 are also related to the larger amount of diffusing source in the experiment.

Table 9.10 Diffusion Coefficients of Different Ions in Soils of
Mixed–Salt Systems[a] (Wang and Yu, 1989c)

| Soil | D (cm^2 s^{-1} ×10^6) | | | |
	Cl$^-$	NO$_3^-$	K$^+$	Ca^{2+}
Rhodic Ferralsol (Guangdong)	3.2	4.4	1.5	0.9
Ferrali–Haplic Acrisol (Guangdong)	3.0	4.0	1.5	0.9
Ali–Haplic Acrisol (Jiangxi)	2.6	3.1	1.4	0.8
Rhodic Ferralsol (Hawaii)	2.8	3.4	1.8	1.1
Xanthic Ferralsol (Manaus)	4.5	5.0	–	–
Cambisol (Nanjing)	6.9	6.6	–	–
Mean	3.8	4.4	1.6	0.9

[a]Electrodialyzed soil.
$Q_{Cl} = Q_{NO_3} = 1.1$ mmol cm^{-2}; $Q_K = 2Q_{Ca} = 5.5$ mmol cm^{-2}.

9.3.2 Effect of Surface Charge of Soil Particles

In a mixed-salt system, similar to the case of single–salt systems, when the

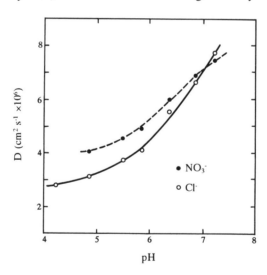

Fig. 9.10. Effect of pH on diffusion coefficient of chloride ions and nitrate
ions in Rhodic Ferralsol ($Q_{Cl} = Q_{NO_3} = 1.1$ mmol cm^{-2}) (Wang and
Yu,1989c).

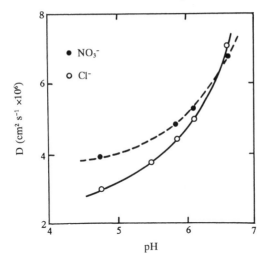

Fig. 9.11. Effect of pH on diffusion coefficient of chloride ions and nitrate ions in Ferrali–Haplic Acrisol ($Q_{Cl} = Q_{NO_3}$ = 1.1 mmol cm^{-2}) (Wang and Yu, 1989c).

surface charge properties of the soil were changed by the difference in pH or by the addition of iron oxide coatings, the diffusion coefficients of anions decreased (Figs. 9.10 to 9.12; Table 9.11) and those of cations increased (Fig. 9.13) with the increase in positive surface charge and the decrease in

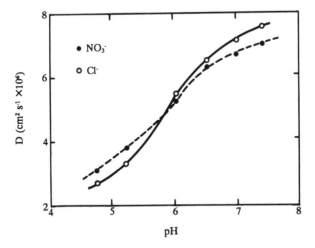

Fig. 9.12. Effect of pH on diffusion coefficient of chloride ions and nitrate ions in Ali–Haplic Acrisol ($Q_{Cl} = Q_{NO_3}$ = 1.1 mmol cm^{-2}) (Wang and Yu, 1989c).

Table 9.11 Effect of Iron Oxides on Diffusion Coefficient of Anions in Soils of Mixed–Salt Systems (Wang and Yu, 1989c)

Soil	Treatment	pH	D (cm^2 s^{-1} $\times 10^6$) Cl^-	NO_3^-	$\dfrac{a_{Cl}}{a_{NO_3}}$
Rhodic Ferralsol	Original	6.8	7.0	6.8	0.97
	+25% Fe_2O_3	6.7	3.3	4.3	0.70
Ferrali–Haplic Acrisol	Original	6.4	6.7	6.9	0.91
	+25% Fe_2O_3	7.2	4.3	5.5	0.74
Ali–Haplic Acrisol	Original	5.8	4.3	4.7	0.84
	+15% Fe_2O_3	6.0	4.0	4.6	0.80
	+25% Fe_2O_3	5.6	2.8	4.3	0.62

negative surface charge of the soil.

The variation of the difference in diffusion between two anion species with the change in surface properties of the soil is interesting. The addition of iron oxides to the soil led to a decrease in the ratio of the active fraction of chloride ions in diffusion to that of nitrate ions. The larger the amount of iron oxides added, the more the decrease of this ratio occurred (Table 9.11). For a given soil, the lower the pH, the smaller the diffusion coefficient of chloride ions as compared to that of nitrate ions (Figs 9.10 to 9.12). These two aspects imply that the more positive the charge properties of a soil, the stronger the adsorptive power for chloride ions over nitrate ions. At a certain pH, D_{Cl} was equal to D_{NO_3}. Beyond that pH, D_{Cl} may become greater than D_{NO_3}, and the higher the pH, the greater this difference. This is a reflection on the difference in diffusion rate between the two ion species in solution plus the possible effect of negative adsorption of anions on diffusion when the negative surface charge properties of the soil is strong. For the Rhodic Ferralsol, Ferrali-Haplic Acrisol, and Ali-Haplic Acrisol, the pH at which the diffusion coefficients of the two ion species were equal to each other was 7.1, 6.4, and 5.8, reflecting the difference in surface charge properties among the three soils.

The materials presented in this section indicate once again that the surface charge properties of a soil are extremely important factors in affecting the diffusion of ions in the soil. On the other hand, they also indicate that, in addition to electrostatic force, a specific adsorption force may be involved in the interactions between variable charge soils and chloride ions. It can be imagined that the Pauling radius and chemical

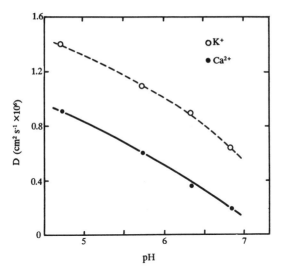

Fig. 9.13. Effect of pH on diffusion coefficients of potassium ions and calcium ions in Ali–Haplic Acrisol ($Q_K = 2Q_{Ca} = 5.5$ mmol cm^{-2}) (Wang and Yu, 1989c).

structure of chloride ions and nitrate ions are different, although they have an identical hydrated radius. In particular, single-atom ion Cl$^-$ is more polarizable and has a higher activity and a larger charge density than NO$_3^-$ ion. At a low pH when the activity of iron (and aluminum) in the soil is high, it is possible to form small amounts of Fe–Cl and Al–Cl complexes. This activity of iron and aluminum, however, will become weaker when the pH is increased.

9.3.3 Effect of Quantity of Diffusing Source

The effect of the quantity of diffusing source on diffusion coefficients of Cl$^-$ ions and NO$_3^-$ ions is shown in Table 9.12. At a low pH, when the soil carried a large quantity of positive surface charge, the diffusion coefficient of anions at a low quantity of the diffusing source was much smaller than that at a high quantity, especially for Cl$^-$ ions, because in this case a larger proportion of diffusing ions might be retarded by adsorption–desorption processes during their diffusion. At a high pH, when the soil did not adsorb anions appreciably, the effect of the quantity of the diffusing source became smaller. For a similar reason, the difference in the diffusion coefficient between the two ion species was larger at a low pH and low quantities of the diffusing source, and it was negligible at a high pH.

Table 9.12 Effect of Quantity of Diffusing Source on Diffusion Coefficient in Soils of Mixed–Salt Systems at Different pH (Wang and Yu, 1989c)

		D (cm^2 s^{-1} ×10^6)			
		Rhodic Ferralsol		Ali–Haplic Acrisol	
Ions	Q (mmol cm^{-2})	pH 4.8	pH 6.8	pH 4.6	pH 6.9
Cl$^-$	0.11	1.4	5.1	1.6	–
	1.11	3.2	6.6	2.5	7.8
	11.1	3.9	6.9	3.7	7.6
	22.2	4.5	7.5	4.5	6.5
NO$_3^-$	0.11	2.8	5.4	2.3	–
	1.11	4.1	6.8	3.0	7.4
	11.1	4.4	6.9	4.0	7.2
	22.2	4.7	7.7	4.3	6.6

9.3.4 Effect of Electrolyte

When a soil contains other electrolytes besides the diffusing ions, the presence of the electrolyte will facilitate ion diffusion, owing to the competitive adsorption between ions of the electrolyte and the diffusing ions. For the diffusion of anions, this effect should be more pronounced at a low pH. This situation can be seen from the data given in Table 9.13.

9.3.5 Comparison Between Mixed–Salt System and Single–Salt System

When comparing the materials cited in this chapter it can be found that, qualitatively speaking, factors that affect the diffusion of cations and anions in both single–salt and mixed–salt systems are essentially the same. However, the diffusion coefficients of anions at a low pH and of cations were larger in mixed–salt systems than in single–salt systems. For two ion species with the same kind of charge, the difference in the diffusion coefficients between anion species became larger and those between cation species became smaller in mixed–salt systems.

The increase in the diffusion coefficient of anions in mixed–salt systems at a low pH might be due to several causes. Because these anions can be adsorbed by the surface of soil particles carrying positive charges when the pH is low, the presence of one anion species will lead to an increase in the fraction of another anion species in soil solution, caused by the competition

Table 9.13 Effect of Electrolyte on Diffusion Coefficient of Anions in Soils of Mixed-Salt Systems (Wang and Yu, 1989c)

| | | | D (cm^2 s^{-1} $\times 10^6$) | |
| | | | | |
Soil	pH	Medium	Cl$^-$	NO$_3^-$
Ferrali-Haplic Acrisol	6.2	Water	5.3	5.8
		Salt solution[a]	6.3	6.8
	6.8	Water	8.2	6.9
		Salt solution	8.4	7.4
Ali-Haplic Acrisol	4.4	Water	2.4	3.9
		Salt solution	3.4	5.3
	6.8	Water	6.2	6.0
		Salt solution	6.4	6.2

[a]0.04 M NaNO$_3$ solution in chloride diffusion;
0.04 M CaCl$_2$ solution in nitrate diffusion.

for exchange sites between these ions. Since the cation species, in this case K$^+$, associated with the anions doubled its quantity in mixed-salt systems, the increased diffusion rate of the cations would also cause the associated anions to move more rapidly. Besides, the total ionic strength and the thickness of the electrical double layer around positive surface charge sites would also be different from those in single-salt systems.

In a mixed-salt system, such as KCl + KNO$_3$ in soil, the increases in the diffusion coefficient of the three ion species as compared to single-salt systems were not of the same magnitude. This can be explained as follows. Because the diffusion coefficient of NO$_3^-$ ions is the largest among the three ion species, NO$_3^-$ ions will diffuse rapidly. At the diffusing front, an electrical potential should be created (Ellis et al., 1970; Rhue, 1992). This potential would accelerate the diffusion of K$^+$ ions and retard the diffusion of Cl$^-$ ions and remaining NO$_3^-$ ions. The appearance of K$^+$ ions near the diffusing front would cause soil pH to decrease, owing to the cation-exchange reaction with adsorbed hydrogen and aluminum ions. This would cause a subsequent increase in positive surface charge. As a result, the Cl$^-$ ions would be retarded more than the NO$_3^-$ ions. The overall result of these reactions would be a continuous segregation between NO$_3^-$ ions and Cl$^-$ ions in the diffusing path. Thus, it can be concluded that at a low pH the increase in positive surface charge of the soil was the principal reason for the larger difference between D_{Cl} and D_{NO_3} in a mixed system than that when they diffused along. This difference tended to vanish when the soil pH became high.

When K^+ ions and Ca^{2+} ions diffused simultaneously in a $KCl + CaCl_2$ system in soil, because the diffusion coefficient of chloride ions is much larger than that of the two cation species, the increased diffusion rate of Cl^- ions due to their increased concentration will cause K^+ ions and Ca^{2+} ions to move more rapidly than in a single-salt system. In this case, because divalent calcium ions associate with Cl^- ions more closely and move slower than monovalent potassium ions, they should be affected to a greater extent by Cl^- ions than the potassium ions. The fast-moving K^+ ions must replace strongly adsorbed H^+ and Al^{3+} ions from the exchange sites during their diffusion, whereas the slow-moving Ca^{2+} ions need only to replace weakly adsorbed K^+ ions. This means that K^+ ions would encounter a stronger retarding effect than Ca^{2+} ions during diffusion (Phillips and Brown, 1964). These effects combined together would cause the difference between D_K and D_{Ca} to become smaller than when they diffuse alone.

It can be seen here that the migration of ions under a concentration gradient is different from that under an externally applied electric field. In a conductivity measurement, the presence of an applied electrical field can weaken co-ion effect and counterion effect among moving ions. In this case only accompanying ions of the ion atmosphere can move with the associated ions. By contrast, in ion diffusion the principle of electroneutrality plays a very important role in affecting the interactions among various ion species and the interactions between these ions and soil surface charges.

In the literature, little attention has been paid to interactions among ions during their diffusion in mixed systems (Rhue, 1992), although this has been considered theoretically (Frere, 1969; Olsen and Kemper, 1968). From the materials presented in this chapter it is apparent that the situation in mixed systems is quite different from that in single-salt systems. In variable charge soil the presence of both positive and negative surface charges, as well as their variability, makes the situation in ion diffusion complicated. Under field conditions, there are always a variety of ion species present. Therefore, the situation in the simultaneous diffusion of these ions is of more practical significance.

BIBLIOGRAPHY

Barraclough , P. B. and Tinker, P. B. (1981) *J. Soil Sci.*, 32:225-236.

Barraclough , P. B. and Tinker, P. B. (1982) *J. Soil Sci.*, 33:13-24.

Barrer, R. M. (1951) *Diffusion in and Through Solids.* Cambridge University Press, London.

Bar-Yosef, B., Posner, A. M., and Quirk, J. P. (1975) *J. Soil Sci.*, 26:1-21.

Bhat, K. K. S. and Nye, P. H. (1973) *Plant and Soil*, 38:161-175.

Brown, D. A., Fulton, B. E., and Phillips, R. E. (1964) *Soil Sci. Soc. Am. Proc.*, 28:628–632.

Conkling, B. L. and Blanchar, R. W. (1986) *Soil Sci. Soc. Am. J.*, 50:14-55–1499.

Conkling, B. L. and Blanchar, R. W. (1989) *Soil Sci. Soc. Am. J.*, 53:16-85–1690.

Conway, B. E. (1981)   *Ionic Hydration in Chemistry and Biophysics.* Elsevier, Amsterdam.

Cottenie, A. and Verloo, M. (1984) *Anal. Chemie*, 317:389–393.

Cussler, E. L. (1984) *Diffusion, Mass Transfer in Fluid Systems.* Cambridge University Press, Cambridge, pp. 146–171.

Ellis, J. H., Barnhisel, R. L., and Phillips, R. E. (1970) *Soil Sci. Soc. Am. Proc.*, 34:866–870.

Frere, M. H. (1969) *Soil Sci. Soc. Am. Proc.*, 33:883–886.

Haan, F. A. M., Bolt, G. H., and Pieters, B. G. M. (1965) *Soil Sci. Soc. Am. Proc.*, 29:529–530.

Habib, R. and Guennelon, R. (1983) *Agronomie*, 3:113–121.

Hira, G. S. and Singh, N. T. (1977) *Soil Sci. Soc. Am. J.*, 41:537–540.

Hira, G. S. and Singh, N. T. (1978) *Soil Sci. Soc. Am. J.*, 42:561–565.

Husted, R. F. and Low, P. F. (1954) *Soil Sci.*, 77:343–353.

Jurinak, J. J., Sandhu, S. S., and Dudley, L. M. (1987) *Soil Sci. Soc. Am. J.*, 51:625–630.

Kemper, W. D. (1960) *Soil Sci. Soc. Am. Proc.*, 24:10–16.

Low, P. F. (1962) *Clays Clay Miner.*, 9:219–228.

Low, P. F. (1981)   in *Chemistry in the Soil Environment* (M. Stelly, ed.). American Society of Agronomy, Madison, pp. 31–45.

Melton, J. R., Mahtab, S. K., and Swoboda, A. R. (1973) *Soil Sci. Soc. Am. Proc.*, 37:379–381.

Nye, P. H. (1979) *Adv. Agron.*, 31:225–271.

Nye, P. H. and Ameloko, A. (1986) *J. Soil Sci.*, 37:191–196.

Nye, P. H. and Tinker, P. B. (1977) *Solute Movement in the Root-Soil System.* Blackwell, Oxford.

Olsen, S. R. and Kemper, W. D. (1968) *Adv. Agron.*, 20:91–151.

Palmer, C. J. and Blanchar, R. W. (1980) *Soil Sci. Soc. Am. J.*, 44:925–929.

Phillips, R. E. and Brown, D. A. (1964) *Soil Sci. Soc. Am. Proc.*, 28:758–763.

Phillips, R. E., Barnhisel, R. I. and Ellis, J. H. (1972) *Soil Sci. Soc. Am. Proc.*, 36:35–39.

Place, G. A., Phillips, R. E., and Brown, D. A. (1968) *Soil Sci. Soc. Am. Proc.*, 32:657–660.

Rhadoria, P. B. S., Kaselowsky, J., Clasen, N., and Jungk, A. (1991) *Z. Pflanzenernähr. Bodenk.*, 154:69–72.

Rhue, D. (1992) *Soil Sci. Soc. Am. J.*, 56:683–689.

Sinha, B. K. and Singh, N. T. (1977) *J. Indian Soil Sci. Soc.*, 25:74-76.
Skogley, E. O. and Schaff, B. E. (1985) *Soil Sci. Soc. Am. J.*, 49:847-850.
Song, K. K. and Ishiguro, M. (1992) *Soil Sci. Plant Nutr.*, 38:477-484.
Staunton, S. (1986) *J. Soil Sci.*, 37:373-377.
Staunton, S. (1990) *J. Soil Sci.*, 41:643-653.
Staunton, S. and Nye, P. H. (1987) *J. Soil Sci.*, 38:651-658.
Stigter, D. (1980) *Soil Sci.*, 130:1-6.
Trefry, J. H. and Metz, S. (1984) *Anal. Chem.*, 56:745-749.
van Schaik, J. C., Kemper, W. D., and Olsen, S. R. (1966) *Soil Sci. Soc. Am. Proc.*, 30:17-22.
Wang, E. J. and Yu, T. R. (1989a) *Soil Sci.*, 147:34-39.
Wang, E. J. and Yu, T. R. (1989b) *Soil Sci.*, 147:91-96.
Wang, E. J. and Yu, T. R. (1989c) *Soil Sci.*, 147:174-178.
Wang, P. G., Ji, G. L. and Yu, T. R. (1987) *Z. Pflanzenernähr. Bodenk.*, 150:17-23.
Wu, J. (1976) in *Electrochemical Properties of Soils and Their Research Methods* (T. R. Yu, ed.). Science Press, Beijing, pp. 302-324.
Yu, T. R. (1987) in *Principles of Soil Chemistry* (T. R. Yu, ed.). Science Press, Beijing, pp. 246-324.
Yu, T. R. and Zhang, X. N. (1986) in *Proceedings of the International Symposium on Red Soils*. Science Press/Elsevier, Beijing/Amsterdam, pp. 409-444.

10

REACTIONS WITH HYDROGEN IONS

F. S. Zhang and T. R. Yu

Hydrogen ion is one kind of cation which possesses many properties common to all cations. Hydrogen ion also has its own characteristic features which are of particular significance for variable charge soils. The interactions between hydrogen ions and the surface of soil particles is the basic cause of the variability of both positive and negative surface charges of variable charge soils. The quantity of hydrogen ions in soils determines the acidity of the soil while the acidity of variable charge soils is among the strongest in all the soils. This strong acidity of variable charge soils affects many other chemical properties of the soil.

In this chapter, the basic properties of hydrogen ions will be briefly discussed. Then, the products and the kinetics of the interaction between hydrogen ions and variable charge soils will be treated. The dissociation of hydrogen ions from the surface of soil particles has already been mentioned in Chapter 2.

10.1 PROPERTIES OF HYDROGEN IONS

10.1.1 Properties of Protons

After the dissociation of an electron, a hydrogen atom becomes a proton (H^+ ion). The ionization energy of hydrogen atoms is 1310 kj mol^{-1}, whereas those of alkali metals, Li, Na, K, and Cs, are 519, 494, 419 and 377 kj mol^{-1}, respectively. This difference in the ionization energy between hydrogen and alkali metals indicates that protons have a particularly strong affinity for electrons. Therefore, protons are apt to form a covalent bond with other atoms by sharing a pair of electrons, or to form a hydrogen bond.

Because of the absence of an electronic shell, a proton has a diameter of the order of 10^{-13} cm, while other ions with electronic shells generally have a diameter of the order of 10^{-8} cm. Because a proton is so small, it is quite accessible to its neighboring ions and molecules. Therefore, there is very little steric hindrance when protons participate in chemical reactions.

The above-mentioned features of proton are the basis for its particular properties.

10.1.2 Hydration of Protons

Free proton in solution is extremely unstable because it is very active. In an aqueous solution it will react with water molecules to form a hydrated proton, H_3O^+. At room temperature, the concentration of free protons is only of the order of 10^{-150} mol L^{-1}. In other words, almost all protons are hydrated. Structurally, H_3O^+ ion is in the form of a trigonal pyramid, in which three hydrogen atoms occupy three of the apexes around a central oxygen atom. The length of O–H bond is 0.102 nm. The distance between each pair of hydrogen atoms is 0.172 nm, and the H–O–H angle is 104.5°.

When a proton combines with a water molecule to form an H_3O^+ ion, the energy change involved would be −712 kj mol^{-1}. The hydration energy of proton is −1114 kj mol^{-1} which suggests that a proton may combine with more than one water molecule in an aqueous solution. According to calculation, the hydration number of hydrated proton is 3; that is, an H_3O^+ ion combines with three water molecules to form an $(H_9O_4)^+$ associated ion.

10.1.3 Transport of Protons

The transport of protons in an aqueous solution is peculiar. Reflections in this respect include:

1. The mobility of hydrated proton is particularly large, attaining a value of 36×10^{-4} cm^2 V^{-1} s^{-1}, quite different from that of most simple inorganic ions, which have a mobility of about 5×10^{-4} cm^2 V^{-1} s^{-1}.

2. According to Stokes' law, the migration velocity of an ion in an electric field should be inversely proportional to the diameter of that ion. Assuming that protons migrate in the form of H_3O^+, the calculated mobility would be 7.6×10^{-4} cm^2 V^{-1} s^{-1}, much smaller than the experimental value.

3. On the other hand, if it is assumed that a proton migrates in solution in a free state, the calculated mobility from its diameter would be much larger than the experimental value. Besides, it is known that there are very few free protons in solution.

4. For most ions, the activation energy of transport is independent of temperature within a wide temperature range. For proton, the activation energy decreases with the increase in temperature.

From the phenomena mentioned above, it can only be assumed that in the transport of protons in a solution a mechanism different from that of other ions must be involved.

According to the modern point of view (Bockris and Reddy, 1970), a proton migrates chiefly in a fashion of successive jumps from one water molecule to another water molecule in solution. In an aqueous solution hydrogen ions are surrounded by water molecules. When the O–H bond of an H_3O^+ ion points toward the lone electron pair of an adjacent water molecule, the proton will transfer from this O–H bond to the lone pair electrons. The original H_3O^+ ion then becomes an H_2O molecule, while the H_2O molecule that accepts the proton becomes an H_3O^+ ion; that is,

$$H_3O^+ + H_2O \rightarrow H_2O + H_3O^+ \qquad (10\text{-}1)$$

This newly formed H_3O^+ ion will transfer its proton to the next H_2O molecule in the same way. The continuous repeat of this process makes a proton migrate in a fashion of successive jumps, although the H_3O^+ ion remains stationary. Because a proton is very small and very light, it can pass through energy barriers by a quantum mechanical tunnel effect when it migrates in solution.

The migration rate of protons through this tunnel effect is very high. It has been calculated that the mobility would be as high as 7.58×10^{-1} cm^2 V^{-1} s^{-1}, much higher than the experimental value, if this mechanism is the rate–determining step during the transport of protons in an aqueous solution. It is thought that the proton transport process consists of several steps. The first step is the orientation of water molecules, i.e., the O–H bond of an H_3O^+ ion points toward the lone pair of electrons of the acceptor H_2O, or the lone pair of electrons of an acceptor H_2O point to the proton of an H_3O^+ ion. Then, in the next step, the transfer of protons from the donor H_3O^+ to the acceptor H_2O is realized through the tunnel effect. After each proton transfer, the water molecule must rotate to orient again, so that it can accept another proton. Therefore, for the whole transport process, the orientation of water molecules is the rate–determining step.

If the proton transport process is realized wholly through the above–mentioned mechanism, the calculated mobility of protons should be 28×10^{-4} cm^2 V^{-1} s^{-1}, slightly lower than the experimental value. This is because H_3O^+ ion itself can also migrate. It has been estimated that this contribution accounts for about 20% of the total mobility of protons.

10.1.4 Transfer of Protons

Chemically, protons are very active and can transfer from one chemical species to another species. In an aqueous solution, the transfer process is relatively simple and can be distinguished into four types:

1. A proton transfers from an ion to a neutral molecule, forming a new ion and a new neutral molecule. For example,

$$NH_4^+ + H_2O \rightarrow NH_3 + H_3O^+ \qquad (10\text{-}2)$$

2. A proton transfers from an ion to a neutral molecule, forming two new ions. The following reaction can be taken as such an example:

$$HCO_3^- + H_2O \rightarrow CO_3^{2-} + H_3O^+ \qquad (10\text{-}3)$$

3. A proton transfers from one neutral molecule to another neutral molecule, forming two ions of opposite charges:

$$H_2O + NH_3 \rightarrow OH^- + NH_4^+ \qquad (10\text{-}4)$$

4. The transfer reaction proceeds between two amphoteric solvent molecules, with one molecule functioning as the proton donator and another molecule as the proton acceptor:

$$H_2O + H_2O \rightarrow H_3O^+ + OH^- \qquad (10\text{-}5)$$

All the above reactions can occur in soil solution, particularly reactions (10-3) and (10-5). In soils consisting of solid, liquid, and gaseous phases, protons can also transfer at the liquid–gas and liquid–solid interfaces (van Breemen et al., 1983). The proton transfer process at the liquid–solid interphase is of great significance in soil science. This will be discussed in the following sections.

10.2 REACTION MECHANISMS OF HYDROGEN IONS WITH SOILS

10.2.1 Sources and Fates of Hydrogen Ions in Soils

Hydrogen ions in soils may come from a variety of sources (Ritchie, 1989; Thomas and Hargrove, 1984; van Breemen, 1991; Yu, 1987). The dissolution of carbon dioxide of the air dissolved in soil solution produces carbonic acid. Rain water usually contains a small quantity of nitric acid. The rain water in industrial areas may contain a large quantity of sulfuric acid, and this type of rain is called *acid rain*. Acid rain may have a pH value of as low as 2.7. In recent years it has been found that the pH value may be 4 for the acid rain in many areas of variable charge soils in China. The absorption of cations by plant roots is accompanied by the release of protons from plant roots to soil solution. Microbiological and botanical activities can also produce carbonic acid and organic acids. Some fertilizers, such as ammonium sulfate, are acid in reaction or can become acidic. In addition to foreign sources of acids, many chemical processes in soils can also produce hydrogen ions, such as the mineralization and nitrification of organic nitrogen, the mineralization and oxidation of organic sulfur, the mineralization of organic phosphorus, the complexation of metal ions, and the nitrification of ammonium ions. When a soil changes to an oxidized state from a reduced state, some reducing substances such as hydrogen sulfide, ferric sulfide, and ferrous ions will be oxidized and hydrogen ions will be produced as a reaction product (Yu, 1985). In a soil solution, during the autodissociation of water molecules, hydrogen ions will be produced [equation (10-5)]. Since this dissociation is a reversible process, hydrogen ions can be produced continuously, although the amount of hydrogen ions produced due to this autodissociation at equilibrium is very small (10^{-7} mol L^{-1}). Therefore, under natural conditions, there are always the presence of hydrogen ions in soils and sometimes the quantity may be quite large.

The presence of hydrogen ions in soils disturbs the original chemical equilibria, and these ions will undergo chemical reactions with other substances (Bruggenwert et al., 1991; Ritchie, 1989; Scheffer and Schachtschabel, 1992; Schwertmann and Fischer, 1982; Thomas and Hargrove, 1984; Ulrich, 1986, 1991). Among these reactions, the most rapid and important one is the exchange reaction between hydrogen ions and metal ions originally adsorbed by the soil. Because these adsorbed hydrogen ions are unstable, they will further react with the solid phase of the soil, releasing metal ions (mainly aluminum) (Chernov, 1957; Eeckman and Laudelout, 1961; Harward and Coleman, 1954; Thomas, 1988). This will be discussed in more detail in Chapter 11. Hydrogen ions can also directly react with primary and secondary minerals in soils if their concentration is high.

Variable charge soils possess some characteristic features in this respect. Under the conditions of high temperature and high rainfall, primary minerals are practically nonexistent due to their easy weathering. Most adsorbed bases have been leached out and the negative exchangeable sites are essentially occupied by hydrogen and aluminum ions. On the other hand, the soil contains large amounts of iron and aluminum oxides, which are capable of accepting protons directly. It is for these reasons that the interactions between hydrogen ions and variable charge soils are quite different from those in constant charge soils.

According to the present knowledge about the chemical properties of variable charge soils, for soils devoid of exchangeable bases, there are three possible outcomes for added hydrogen ions: transformation into positive surface charge of the soil, consumption in the release of soluble aluminum, and transformation into exchangeable acidity. These will be discussed in the following sections.

10.2.2 Transformation into Positive Surface Charge

It has been known that on the broken bonds on edges of clay minerals and

on the surface of hydrated oxides there are OH-containing groups such as
-Fe-OH and -Al-OH. These groups are capable of accepting protons
from the media and acquiring positive charges under acid conditions
(Bowden et al. 1980; Parfitt, 1980). In variable charge soils, -Fe-OH
groups play a more important role. The reaction between -Fe-OH groups
and protons is given as

$$
\begin{bmatrix}
\text{OH} \\
\text{Fe} \\
\text{O} \quad \text{OH} \\
\text{Fe} \\
\text{OH}
\end{bmatrix}^{0}
+ 3H^+ \rightarrow
\begin{bmatrix}
\text{OH}_2 \\
\text{Fe} \\
\text{O} \quad \text{OH}_2 \\
\text{Fe} \\
\text{OH}_2
\end{bmatrix}^{3+}
\tag{10-6}
$$

The changes in positive surface charge after the addition of different
amounts of hydrogen ions are shown in Fig. 10.1 for three variable charge
soils and a kaolinite. The iron oxide contents of Rhodic Ferralsol, Ferrali–
Haplic Acrisol, and Ali–Haplic Acrisol are 17.22%, 5.65%, and 4.86%,
respectively. When 2.875 cmol kg^{-1} of hydrogen ions was added, the
increases in positive surface charge were 0.77, 0.65, and 0.27 cmol kg^{-1},
respectively. This indicates that the increase in positive surface charge is

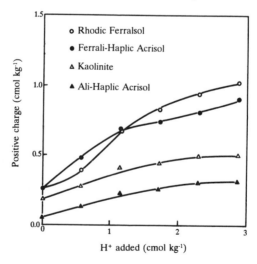

Fig. 10.1. Change in positive surface charge of soils and kaolinite after
addition of hydrogen ions (Zhang et al., 1991a).

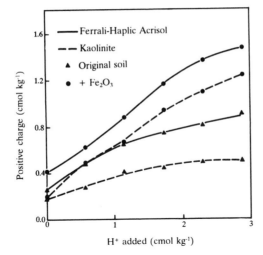

Fig. 10.2. Effect of iron oxides on change in positive surface charge of Ferrali-Haplic Acrisol and kaolinite caused by the addition of hydrogen ions (Zhang et al., 1991a).

closely related to the iron oxide of the soil. The results shown in Fig. 10.2 further verify the function of iron oxides as a proton acceptor. Note that the coating of Ferrali-Haplic Acrisol or kaolinite with iron oxides led to a marked enhancement in the increase of positive surface charge when reacting with hydrogen ions. However, the increase in positive surface charge may not be proportional to the iron oxide content. This is because iron oxides are not mechanically mixed with clay minerals in soils, but are bound through a variety of means (Yu and Zhang, 1986). Besides, the form of iron oxides may vary with the kind of the soil.

In addition to $-Fe-OH$ groups, the role of $-Al-OH$ groups in accepting protons is also important (Boehm, 1971; Schlindler, 1981). Figure 10.1 shows that kaolinite may carry some positive surface charges and may acquire additional positive charges when reacting with H^+ ions, although it almost does not contain an $-Fe-OH$ group. The positive surface charge-H^+ ion concentration curve for the Ferrali-Haplic Acrisol lies markedly above that of the Ali-Haplic Acrisol, although the iron oxide content of the former soil is only slightly higher than that of the latter soil. This may be related to the higher contents of gibbsite and kaolinite in Ferrali-Haplic Acrisol.

In the literature, there are different opinions about the relative importance of iron and aluminum in contributing to variable positive charge in soils (Deshpande et al., 1964; Schofield, 1949; Sumner, 1962) (cf. Chapter 2). From the data presented in this section, it appears that the mechanism through which the surface of soil particles accepts H^+ ions and produces positive charges may depend on the type of the soil and the concentration of H^+ ions. This will be discussed in Section 10.2.5.

10.2.3 Release of Water-Soluble Aluminum

For H,Al-saturated soils, the addition of H^+ ions may cause a pH sufficiently low to dissolve some metal ions from the minerals. Since variable charge soils have been highly weathered, their mineral part consists essentially of clay minerals and quartz. Therefore, the dissolution of alkali and alkaline earth ions can be neglected. If the pH is not too low, the dissolution of iron is also negligible. The content of manganese is very low. Thus, only the dissolution of aluminum needs to be considered (Conyers, 1990; Helyar et al., 1993: Paterson et al., 1991).

The relationship between the amount of soluble Al and the amount of added H^+ for three soils and a kaolinite are shown in Fig. 10.3. For soils and mineral not added with H^+ ions, the amount of soluble Al was nearly zero, although they have been H and Al-saturated through electrodialysis. The amount of soluble Al increased with the increase in added H^+ ions. For the three soils the increment was of the order Ali-Haplic Acrisol > Ferrali-Haplic Acrisol > Rhodic Ferralsol. This order is just opposite to the degree of weathering for these soils, that is, the Al in the Rhodic Ferralsol that has been most highly weathered is the least soluble.

This difference in release of Al can be interpreted in terms of the difference in mineralogical composition.

There are four minerals important for the release of Al. The chemical reactions and the related equilibrium constants for these minerals are as follows (Lindsay, 1979):

Gibbsite:
$$Al(OH) + 3H^+ = Al^{3+} + 3H_2O \qquad pK = 8.04 \qquad (10\text{-}7)$$

Amorphous $Al(OH)_3$:
$$Al(OH)_3 + 3H^+ = Al^{3+} + 3H_2O \qquad pK = 9.66 \qquad (10\text{-}8)$$

Kaolinite:
$$Al_2Si_2O_5(OH)_4 + 6H^+ = 2Al^{3+} + 2H_4SiO_4 + H_2O$$
$$pK = 5.45 \qquad (10\text{-}9)$$

Illite:
$$K_{0.6}Mg_{0.25}Al_{2.3}Si_{3.5}O_{10}(OH)_2 + 8H^+ + 2H_2O$$
$$= 0.6K^+ + 0.25Mg^{2+} + 2.3Al^{3+} + 3.5H_4SiO_4$$
$$pK = 10.35 \qquad (10\text{-}10)$$

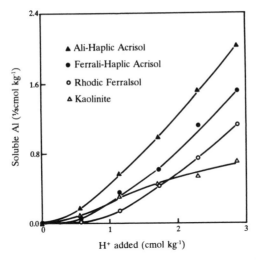

Fig. 10.3. Content of water-soluble aluminum of soils and kaolinite after addition of different amounts of hydrogen ions (Zhang et al., 1991a).

The amount of Al released from these minerals at different pH can be calculated from the related equilibrium constant. The calculated results are graphically shown in Fig. 10.4.

Results in Fig. 10.4 reflect the situation in a pure system where there is

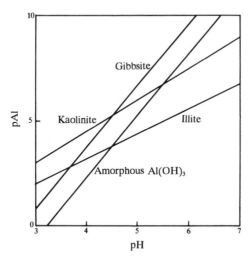

Fig. 10.4. pAl-pH relationship for four aluminum-containing minerals (Zhang et al., 1991a).

only one mineral species, and none of the other cations or anions are added except H^+ ions. The practical situation in soils is much more complex than that in a pure system. Nevertheless, from the solubility curves it is possible to judge the relative ease of releasing soluble Al for the four minerals.

Suppose that the four minerals coexist in a system. At a pH of higher than 4.5 the Al concentration could be controlled by illite, and at a pH below 4.5 it could be controlled by amorphous aluminum hydroxide. Kaolinite can release more Al than does amorphous aluminum hydroxide only when the pH is higher than 5.55. Gibbsite would release more Al than does illite when the pH is lower than 3.7. However, the solubility of gibbsite is always smaller than that of amorphous aluminum hydroxide by about 1.5 orders of magnitude.

Based on the analysis made above, it is not difficult to interpret the results shown in Fig. 10.3. Ali-Haplic Acrisol contains more illite and amorphous aluminum hydroxide than do the other two soils and thus released the highest amount of soluble Al. In contrast, Rhodic Ferralsol released the lowest amount of Al due to the predominance of kaolinite and gibbsite in the mineralogical composition. In Ferrali-Haplic Ferralsol, in addition to kaolinite and gibbsite, there is a small amount of 2:1-type clay minerals. This would be the reason why it released an intermediate amount of Al. At a low pH of the equilibrium solution—that is, in the case where a relatively large amount of H^+ ions was added-kaolinite would release the lowest amount of Al. It can be seen from Fig. 10.3 that the practical situation is consistent with theoretical predictions.

In variable charge soils, Al-containing minerals always coexist with iron oxides. These iron oxides can form coatings on the surface of the aluminum minerals. As a result, the exposed surface area that can release Al is reduced. Besides, the speed of the acceptance of H^+ ions by -Fe-OH groups of iron oxide is much higher than that of the entering of protons in Al-containing minerals. These two factors will cause the decrease in the amount of Al released by Al-containing minerals when reacting with H^+ ions. This prediction is consistent with the data presented in Fig. 10.5. It can be seen from the figure that either Ferrali-Haplic Acrisol or kaolinite, when added with 15% iron oxide to form coatings, released less soluble Al than their original samples during reactions with H^+ ions.

On the other hand, after the removal of iron oxides by chemical means, the soils also released less soluble Al when reacting with H^+ ions (Fig. 10.6), due presumably to the removal of a part of noncrystalline Al (e.g., amorphous aluminum hydroxides) during the chemical treatment.

10.2.4 Transformation into Exchangeable Acidity

Exchangeable acidity in soils is the acid that can be replaced by cations of

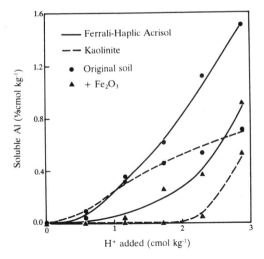

Fig. 10.5. Effect of iron oxides on release of water–soluble aluminum (Zhang et al., 1991a).

neutral salts and then titrated by a base. Generally, exchangeable acidity is dominated by exchangeable Al, accompanied by a small portion of H^+ ions. In principle, the exchangeable acidity of an electrodialyzed soil should not be affected by the addition of foreign H^+ ions, because all the exchange

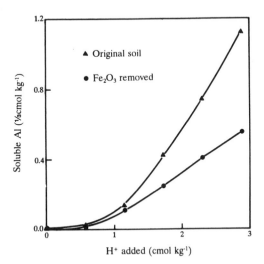

Fig. 10.6. Release of water–soluble aluminum in Rhodic Ferralsol after removal of iron oxides (Zhang et al., 1991a).

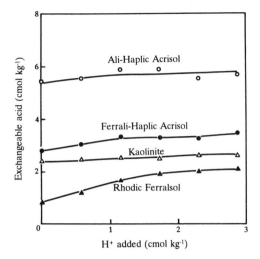

Fig. 10.7. Change in exchangeable acidity after addition of hydrogen ions in H,Al-saturated soils (Zhang et al., 1991a).

sites have already been occupied by H^+ and Al^{3+} ions. In practice, however, the situation is not so simple. As shown in Fig. 10.7, the amount of exchangeable acidity for pure clay mineral kaolinite practically did not change indeed. By contrast, for variable charge soils it increased with the addition of H^+ ions. For example, when 2.875 cmol kg^{-1} of H^+ ions was added, the amount of exchangeable acidity increased by 0.25, 0.68, and 1.24 cmol kg^{-1} for Ali–Haplic Acrisol, Ferrali–Haplic Acrisol, and Rhodic Ferralsol, respectively.

This seems to imply that the increase in exchangeable acidity is related to free iron oxides. This is understandable. As a kind of cementing agent, iron oxide coatings can mask some exchange sites of natural soils. The addition of H^+ ions may result in a sufficiently low initial pH to lead to some changes in the form of such cementing agents, demasking some exchange sites. This assumption is supported by the data shown in Fig. 10.8. Note that the amount of exchangeable acidity increased to 4.62 cmol kg^{-1} from 0.85 cmol kg^{-1} after the removal of free iron oxides from Rhodic Ferralsol. On the other hand, since most of the masked exchange sites have been released during the removal of free iron oxides, the effect of added H^+ ions on the release of exchange sites was not as distinct as on the untreated soil.

According to the modern viewpoint on the forms of free aluminum in soils, the change in forms of aluminum caused by the addition of H^+ ions should be the more important mechanism in the increase in exchangeable acidity. Aluminum, present as either coatings or as nonexchangeable

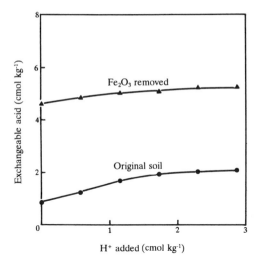

Fig. 10.8. Effect of removal of iron oxides on exchangeable acidity in Rhodic Ferralsol.

hydroxyl Al and polymers on the exchange sites (Coulter, 1969; Huang, 1988; Marcano-Martinez and McBride, 1989; Parker et al., 1979; Thomas, 1988; Thomas and Hargrove, 1984), must be activated by H^+ ions, resulting in an increase in exchangeable Al, that is, exchangeable acidity.

In either of the two cases, the effect of H^+ ions on the increase in exchangeable activity must approach a limit. As can be seen in Figs. 10.7 and 10.8, this tendency occurred for both the Rhodic Ferralsol and the Ferrali-Haplic Acrisol.

10.2.5 Relative Magnitude of Three Consumption Mechanisms

In the previous sections, the three mechanisms of interactions of H^+ ions with variable charge soils have been discussed. In this section, the relative importance of these mechanisms under different conditions will be considered.

As shown in Table 10.1, two days after the addition of H^+ ions into the soil, 90% of added H^+ ions had been consumed. The pH of the equilibrium solution decreased with the increase in added H^+ ions, attaining the lowest pH of 3.8. The sum of the three pathways of consumption accounted for approximately 100% of H^+ ions consumed, irrespective of the amount of H^+ ions added. This indicates that the H^+ ions added to these soils were essentially consumed by being transformed into positive surface charge of the clay, releasing Al ions and transforming them into exchangeable acidity. However, the relative importance of the three mechanisms differed with the

Table 10.1 Contribution of Three Fates to Consumption of Hydrogen Ions Added to Variable Charge Soils (Zhang et al., 1991a)

Soil	H$^+$ (cmol kg^{-1})		Equilibrium pH	H$^+$ Consumption (%)			
	Added	Consumption		A[a]	B	C	Total
Ali–Haplic	0.575	0.49	4.46	16.2	34.4	20.5	71.1
Acrisol	1.150	1.00	4.22	17.3	56.2	43.0	116.5
	1.725	1.53	4.11	13.4	64.1	28.1	105.6
	2.300	2.06	4.02	12.3	73.8	4.4	90.5
	2.875	2.59	3.95	10.3	78.3	9.7	98.3
Ferrali–Haplic	0.575	0.51	4.55	43.6	7.5	45.6	96.7
Acrisol	1.150	0.99	4.20	39.2	35.1	54.4	128.7
	1.725	1.55	4.16	30.9	31.5	32.2	94.6
	2.300	2.02	3.94	27.3	55.3	22.3	104.9
	2.875	2.50	3.83	25.9	60.2	27.2	113.3
Ferrali–Haplic	1.150	1.05	4.40	43.8	3.5	22.9	70.2
Acrisol+Fe$_2$O$_3$	1.725	1.53	4.11	48.8	16.6	28.8	94.2
	2.300	2.07	4.03	46.0	17.7	30.0	93.7
	2.875	2.50	3.82	42.2	36.5	22.8	101.5
Rhodic Ferralsol	0.575	0.53	4.77	24.8	2.4	63.9	91.1
	1.150	1.07	4.47	38.9	12.1	76.1	127.1
	1.725	1.58	4.23	35.8	26.6	67.8	130.2
	2.300	2.09	4.07	32.4	35.4	55.5	123.3
	2.875	2.62	3.99	29.2	42.9	47.4	119.5

[a]A, increase in positive surface charge; B, increase in soluble aluminum; C, increase in exchangeable acidity.

composition of the soil and the amount of H$^+$ ions added. Generally speaking, for a given soil, the percentage of increase in positive surface charge declined while that of soluble Al increased with the increase in addition of H$^+$ ions. This is to be expected, because the transfer of protons to surface OH groups occurs at the clay–solution interface, whereas in the dissolution of Al-containing minerals several steps are involved. Apparently, the former process would proceed much faster than the latter one. Therefore, when H$^+$ ions are added to the soil, initially a large proportion of them would be consumed to increase positive surface charge. It is only when most of the active -Fe-OH and/or -Al-OH groups have been saturated with protons that the free acid can have an opportunity to release Al ions. The process leading to the increase in exchangeable acidity, although also mainly occuring at the clay-solution interface, would proceed more slowly than the protonation of surface -Fe-OH and -Al-OH groups because during the process the transformation of aluminum ions is involved. It can then be concluded that when H$^+$ ions react with the solid phase of variable charge soils the relative importance of the three mechanisms will be determined by relevant reaction rates. This topic shall be discussed in more detail in Section 10.3.

Based on the above discussions, we can explain why the three pathways of the consumption of added H$^+$ ions differed so markedly in the three variable charge soils. For Ali-Haplic Acrisol, the percentage in consumption by the release of soluble Al was the highest and that by the increase of positive surface charge the lowest. This can be related to the high contents of illite and amorphous aluminum hydroxide and the low content of iron oxides. For Rhodic Ferralsol, the consumption by the increase in the exchangeable acidity occupied the largest part and that by the increase in soluble Al was the smallest. This must be caused by the high content of iron oxides and the predominance of kaolinite and gibbsite, because the attack by acid on these Al-containing minerals is unlikely. It was only when the added amount of acid was sufficiently large that the consumption by the increase in soluble Al might exceed that of the positive charge. The properties of Ferrali-Haplic Acrisol is intermediate between Ali-Haplic Acrisol and Rhodic Ferralsol. For this soil the relative importance of the three mechanisms was dependent on the quantity of H$^+$ ions added. At low quantities its behavior resembled that of Rhodic Ferralsol, while at high quantities it resembled that of Ali-Haplic Acrisol.

Since iron oxides are the principal acceptors of protons in variable charge soils, it can be expected that the addition of iron oxides to the soil would result in an increase in the increment of positive surface charge accompanied by a corresponding decrease in the increment of water-soluble aluminum as compared to the untreated soil, as shown in Table 10.1.

In the literature, most reports on the mechanisms of the interactions of hydrogen ions with soils were based on experiments conducted with constant charge soils or pure clay minerals. Some authors emphasized the importance of the transformation of aluminum and others emphasized the importance of iron oxides in accepting protons. In practice, for variable charge soils with a complex composition, the relative importance of each mechanism will depend on the mineralogical composition of the soil. If organic matter, which always exists in all soils, is involved or various types of soils are examined, the mechanisms would be even more complex (Ritchie, 1989).

10.3 REACTION KINETICS OF HYDROGEN IONS WITH SOILS

We have been concerned with the reaction products in the interactions between hydrogen ions and the solid phase of the soil under equilibrium conditions. Then, based on the relative proportions of these products, the reaction mechanisms are deduced. When explaining the relative importance of these mechanisms, it has been pointed out that the protonation of OH groups on the surface of soil particles is a fast reaction, while the dissolution of Al-containing minerals is a slow process with the transformation of hydrogen ions into exchangeable acidity occupying an intermediate position in the reaction rate. In the present section, based on the rate of production of reaction products, the conclusions made in the previous section shall be verified.

Among reaction products, soluble Al can be extracted from the reaction system at intervals during the reaction and then analyzed. The positive surface charge can be calculated from adsorption of Cl⁻ ions by the soil. So, these two species will be examined in the present section. Since in variable charge soils goethite, gibbsite, and kaolinite are the principal minerals that are likely to react with hydrogen ions, the interactions of hydrogen ions with these minerals when present separately will also be considered.

10.3.1 Kinetics of Reaction with Minerals

Because the chemical compositions of goethite, gibbsite, and kaolinite differ markedly, their reaction rates with H^+ ions would be different. If the reaction mechanism is the protonation of OH groups on the surface of minerals, the reaction rate must be fast, because in this case the rate-determining step is the diffusion of protons from the bulk solution to the mineral surface, which proceeds rapidly through a quantum mechanical tunneling effect. In contrast, if the reaction mechanism is the dissolution of Al- or Fe-containing minerals, the reaction rate will be slow. This is because in the process the breakdown of the mineral and the diffusion of dissolved ions from the mineral surface to the bulk solution are involved.

For goethite, it can be calculated from its solubility product that the concentration of Fe^{3+} ions in the equilibrium solution should be as low as 10^{-9} mol L⁻¹ when the initial concentration of H^+ ions is 1.189 mmol L⁻¹, namely when the pH is about 3. Thus, the dissolution of the mineral would be negligible and the protonation of −Fe−OH groups would play the dominant role in consuming H^+ ions (Astumian et al., 1981). Because this latter reaction is very rapid, when H^+ ions react with goethite, the H^+ ion concentration would drop sharply at first and then remain almost unchanged, as shown in Fig. 10.9. Note that the H^+ ion concentration decreased sharply from an initial value of 0.2375 mmol L⁻¹ or 1.189 mmol

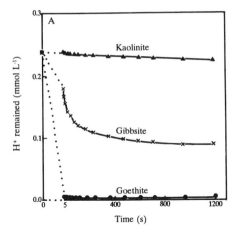

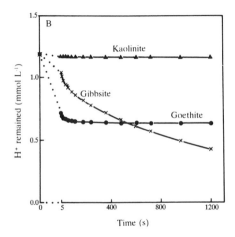

Fig. 10.9. Change in hydrogen ion concentration during reaction with three minerals (Zhang et al., 1991b). (A) Initial H^+ ion concentration = 0.2378 mmol L^{-1}. (B) Initial H^+ ion concentration = 1.189 mmol L^{-1}.

L^{-1} to 6.6×10^{-3} or 0.713 mmol L^{-1} within the first 5 s, and then showed only a slight decrease with time.

For gibbsite, it can be shown that both the protonation of surface $-Al-OH$ groups and the dissolution of Al^{3+} ions will be involved in its reaction with H^+ ions. When a small amount of H^+ ions is added, the majority of these ions will be consumed through protonation of surface $-Al-OH$ groups, leaving only a small portion consumed by the slow dissolution of the mineral. Thus, the H^+ ion concentration-time curve would be similar to that for goethite. This is the situation shown in Fig.

10.9A. Conversely, if a large amount of H^+ ion is added, assuming that the protonation of surface $-Al-OH$ groups is secondary in importance, at the first stage of reaction only a small portion of H^+ ions will be consumed, leaving the majority to be consumed by the slow dissolution of the mineral at the later stage of reaction until an equilibrium is reached. This is the situation shown in Fig.10.9B. If the log $K°$ value in the dissolution of gibbsite is taken as -8.04, it can be calculated that, when the initial concentration of H^+ ions is 1.189 mmol L^{-1} and the solid/liquid ratio is 1/250, the equilibrium concentration should be 0.148 mmol L^{-1}. Actually, however, as can be seen from Fig.10.9B, at 5 s the concentration of H^+ ions was only reduced to 1.03 mmol L^{-1}. This means that during the reaction only 0.16 mmol L^{-1} of H^+ ions were consumed. After this stage of rapid reaction, H^+ ions were consumed continually through the decomposition of the mineral and their concentration decreased slowly. Thus, the kinetic study verified the supposition made in Section 10.2, that is, when the quantity of added H^+ ions is small, most of these ions will transform into positive surface charge, whereas when the quantity is large, these H^+ ions will be consumed mainly by the dissolution of aluminum.

For kaolinite, the quantity of surface hydroxyl groups is much smaller than that of goethite and gibbsite. The solubility is lower than that of gibbsite. Therefore, as shown in Fig. 10.9, when H^+ ions were added to it, the H^+ ion concentration only decreased slightly during a long period.

10.3.2 Kinetics of Reaction with Soils

The reactions of H^+ ions with soils are more complex than those with single minerals. Nevertheless, since in variable charge soils the mineral part consists essentially of the three minerals discussed above and the content of organic matter is very low, it would not be difficult to understand the behavior of these soils from the reaction mechanisms of the minerals.

Figure 10.10 shows the concentration of H^+ ions remaining in solution as a function of reaction time for the three variable charge soils. It can be seen that, as in the case of minerals, there were also a rapid process and a slow process.

As mentioned previously, the reality of the rapid reaction is the protonation of surface hydroxyl groups. The protonation at the surface must result in an increase in the positive surface charge, which in turn would be accompanied by the adsorption of Cl^- ions. Figure 10.11 shows this clearly.

When comparing the three soils it can be seen that the magnitude in the decrease in Cl^- concentration was in the order Rhodic Ferralsol > Ferrali-Haplic Acrisol > Ali-Haplic Acrisol. This order agrees with the order of the content of iron oxides for the three soils. It has been shown in Section 10.2 that surface $-Fe-OH$ groups are the main groups in accepting

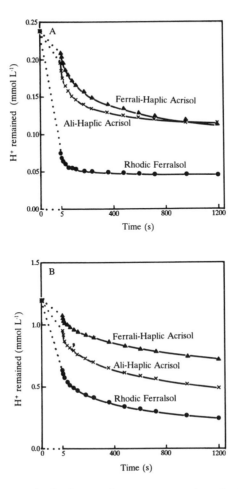

Fig. 10.10. Change in hydrogen ion concentration during reaction with variable charge soils (Zhang et al., 1991b). (A) Initial H^+ ion concentration = 0.2378 mmol L^{-1}. (B) Initial H^+ ion concentration = 1.189 mmol L^{-1}.

protons and acquiring positive surface charges. However, when considering the quantity of H^+ ions consumed at the stage of the fast reaction, it can be seen that Ali–Haplic Acrisol consumed more H^+ ions than did Ferrali–Haplic Acrisol. This reversed order between the two soils may be caused by the presence of a large amount of permanent negative surface charge in Ali–Haplic Acrisol. The aluminum ions adsorbed by negative surface charges must also consume H^+ ions through ion exchange.

The slow stage of reaction between H^+ ions and soils may last more than one hour. Because at this stage the release of soluble Al is the principal process in consuming H^+ ions, Ali–Haplic Acrisol released the largest

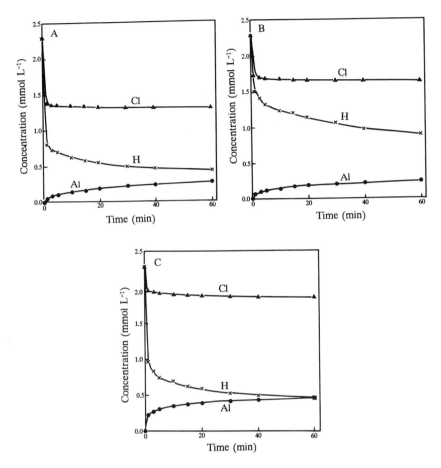

Fig. 10.11. Change in concentrations of H^+, Cl^-, and water–soluble aluminum ions during reaction of hydrogen ions with variable charge soils (Initial H^+ ion concentration = 2.38 mmol L^{-1}) (Zhang et al., 1991b). (A) Rhodic ferralsol; (B) Ferrali–Haplic Acrisol; (C) Ali–Haplic acrisol.

amount of Al among the three soils.

From the discussions made in this section it is clear that when variable charge soils react with H^+ ions the kinetics of reaction are closely related to their mineralogical composition. Generally speaking, both rapid protonation of surface hydroxyl groups and slow dissolution of Al-containing minerals are involved. For Rhodic Ferralsol and Ferrali–Haplic Acrisol with relatively larger proportions of variable charges, the fast reaction is predominantly the protonation of surface hydroxyl groups. For Ali–Haplic Acrisol carrying a large proportion of constant charge, the exchange of aluminum ions by H^+ ions also has a significant contribution to

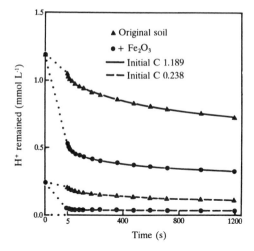

Fig. 10.12. Effect of iron oxides on consumption of hydrogen ions in Ferrali–Haplic Acrisol (Zhang et al., 1991b). (Initial H^+ ion concentration is expressed in millimoles per liter).

the fast reaction. The role of the slow reaction in the consumption of H^+ ions depends upon the content and activity of aluminum in the soil.

In the present section, the consumption of H^+ ions through the increase in exchangeable acidity caused by the transformation of some aluminum and/or iron compounds has not been examined. It can be assumed that the rate of this process is intermediate between those of surface protonation and the dissolution of Al–containing minerals.

10.3.3 Effect of Iron Oxides

Because $-Fe-OH$ groups are the principal proton acceptors during the rapid reaction between the soil and H^+ ions, the addition of iron oxides to a soil should result in an enhancement in the consumption of H^+ ions at this stage. This assumption is supported by the results shown in Fig. 10.12. Correspondingly, the quantity of the positive surface charge, as reflected by the decrease in Cl^- concentration in solution, would increase in a more pronounced manner when compared with the original soil (Fig. 10.13). From the figure it can be seen that the addition of iron oxides to Ferrali–Haplic Acrisol led to a reduced release of soluble Al at the slow stage of reaction. This may be caused by two factors. The enhanced consumption of H^+ ions at the first stage resulted in a reduced acidity of the medium. The surface of Al–containing minerals may be masked partly by iron oxide coatings. As a result, H^+ ions have less opportunity to react with the minerals.

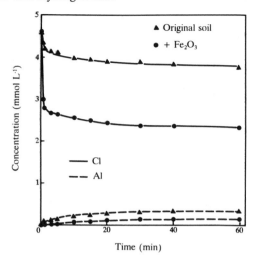

Fig. 10.13. Effect of iron oxides on change in concentrations of hydrogen ions and water–soluble aluminum during reaction of hydrogen ions with Ferrali–Haplic Acrisol (Zhang et al., 1991b).

Conversely, if free iron oxides are removed from the soil, because the protonation of surface –Fe–OH groups is reduced, the decrease in H^+ ion concentration during the rapid stage of reaction would become slower as compared with the untreated soil (Fig. 10.14). Meanwhile, during the slow

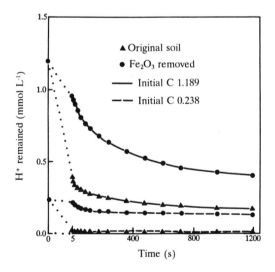

Fig. 10.14. Effect of removal of iron oxides on consumption of hydrogen ions in Hyper–Rhodic Ferralsol (Zhang et al., 1991b). (Initial H^+ ion concentration is expressed in millimoles per liter).

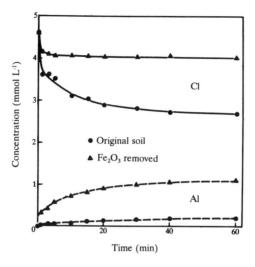

Fig. 10.15. Effect of removal of iron oxides on change in concentrations of chloride ions and water-soluble aluminum during reaction of hydrogen ions with Hyper-Rhodic Ferralsol (Zhang et al., 1991b).

stage of reaction, the release of aluminum would become faster as compared with the untreated soil (Fig. 10.15).

In light of the above explanations, when a variable charge soil reacts with H^+ ions, iron oxide may affect the kinetics of the reaction by affecting the relative contributions of protonation of surface -Fe-OH groups, masking exchange sites and releasing soluble Al from Al-containing minerals to the consumption of H^+ ions. Thus, the kinetic considerations presented in this section confirm the conclusions drawn in Section 10.2.

10.3.4 Empirical Equation of Reaction Kinetics

In soil science, many authors have proposed a variety of kinetic equations to describe chemical reactions in soils (Sparks, 1989), including reactions with acids (Aringhieri and Pardini, 1989; Bloom and Erich, 1987; Susser and Schwertmann, 1991). The most commonly used equations are as follows:

1. First-order kinetic equation:

$$\ln C = -kt + \ln C_0 \tag{10-11}$$

2. Parabolic diffusion equation:

$$C = a + b(t)^{\frac{1}{2}} \tag{10-12}$$

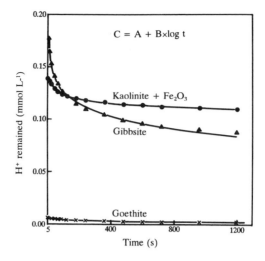

Fig. 10.16. Conformity between measured (dots, triangles, and crosses) and simulated (solid line) H^+ concentration-time curves using the Elovich equation at low initial H^+ ion concentration (Zhang et al., 1991b).

3. Modified Elovich equation:

$$C = a + b \ln t \tag{10-13}$$

4. Two-constant rate equation:

$$\ln C = a + b \ln t \tag{10-14}$$

where C is concentration, t is reaction time, and a, b, and k are constants.

These equations may also be used to describe the kinetics of the reaction of H^+ ions with minerals and soils. It was found that for the goethite, gibbsite, and kaolinite added with iron oxide coatings, the data at low initial H^+ ion concentrations fitted both the Elovich equation (Fig. 10.16) and the two-constant rate equation (Fig. 10.17) fairly well, whereas at high initial H^+ ion concentrations the best equation was the parabolic diffusion equation for kaolinite and gibbsite, although the former two equations still fitted better for goethite (Fig. 10.18).

When H^+ ions react with soils or minerals, the whole process may consist of three steps: diffusion of H^+ ions to the reaction sites, chemical reactions at the sites, and diffusion of reaction products from the sites to the bulk solution. If the initial H^+ ion concentration is high, the overall reaction can still proceed with a fairly high rate after the first several seconds. In this case the diffusion of reaction products would be the rate-determining step.

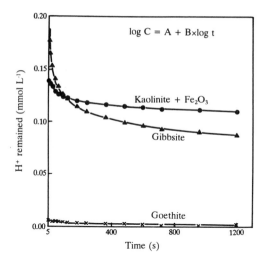

Fig. 10.17. Conformity between measured (dots, triangles, and crosses) and simulated (solid line) H^+ concentration–time curves using two–constant rate equation at low initial H^+ ion concentration (Zhang et al., 1991b).

This is the reason why the parabolic diffusion equation fitted the data well for kaolinite and gibbsite. If the initial H^+ ion concentration is low, the reaction after the first several seconds can proceed only at a low rate

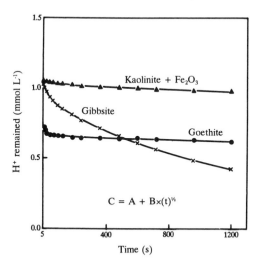

Fig. 10.18. Conformity between measured (dots, triangles, and crosses) and simulated (solid line) H^+ concentration-time curves using parabolic diffusion equation at high initial H^+ ion concentration (Zhang et al., 1991b).

because the majority of H^+ ions have been consumed within this period. Under such circumstances the slow breakdown of the minerals (Berner, 1981; Bloom, 1983; Bloom and Erich, 1987; Stumm, 1986) and the consequent release of soluble Al would become the rate-determining step, and therefore the Elovich equation fitted the data well. For goethite, because it can provide abundant surface hydroxyl groups to consume most of the protons rapidly, the subsequent reaction would proceed with a very slow rate, even when the initial H^+ ion concentration is quite high. Thus, when H^+ ions react with the three minerals, it is the reaction mechanism that determines both the rate of consumption of H^+ ions and which kinetic equation would be more suitable to fit the experimental data.

The reaction course of soils with H^+ ions should be closely related to the mineralogical composition of the soil. Although the mineralogical composi-

Table 10.2 Calculated Parameters in Reaction Kinetics of Hydrogen Ions with Variable Charge Soils Expressed by Two Equations (Zhang et al., 1991b)

Soil	Initial H^+ Concentration[a]	$C_t = a + b(t)^{1/2}$			$C_t = a + b\log(t)$		
		$a \times 10^3$	$b \times 10^3$	r	$a \times 10^3$	$b \times 10^3$	r
Rhodic	0.119	43	−0.24	−0.868	49	−1.8	−0.976
Ferralsol	0.238	61	−0.58	−0.880	74	−44.4	−0.982
	0.476	147	−2.64	−0.930	204	−19.4	−0.997
	1.189	599	−11.06	−0.985	810	−76.1	−0.989
	2.378	1500	−18.58	−0.992	1844	−125.9	−0.982
Ferrali−Haplic	0.119	121	−1.14	−0.943	145	−8.3	−0.999
	0.238	202	−2.92	−0.976	259	−20.3	−0.993
Acrisol	0.476	401	−5.77	−0.993	507	−38.9	−0.980
	1.189	1050	−9.66	−0.998	1223	−64.3	−0.971
	2.378	2088	−12.64	−0.995	2317	−84.7	−0.974
	0.119	122	−0.81	−0.918	140	−6.0	−0.995
Ali−Haplic	0.238	176	−2.14	−0.937	222	−15.6	−0.998
Acrisol	0.476	332	− 5.63	−0.968	444	−39.7	−0.997
	1.189	932	−13.69	−0.990	1188	−93.3	−0.986
	2.378	1997	−20.62	−0.992	2379	−139.8	−0.983

[a]$(mmol L^{-1})$.

tion of soils is rather complex, it is still possible to choose a proper kinetic equation to describe the consumption of H^+ ions. The relevant data in fitting the reaction kinetics of three soils with the parabolic diffusion equation and the Elovich equation are given in Table 10.2. It can be seen that the parabolic diffusion equation fits the data better than the Elovich equation when the initial H^+ ion concentration is high. This is the case for gibbsite. The reverse is true when the initial H^+ ion concentration is low. This is similar to the case for goethite. This means that in the former case the ion diffusion is the dominant rate-determining step, whereas in the latter case the chemical reactions at the solid surface play the principal role in determining the overall reaction rate. It also means that in the second stage of the reaction between H^+ ions and variable charge soils the dominant course is the release of Al ions.

BIBLIOGRAPHY

Aringhieri, R. and Pardini, G. (1989) *Soil Sci.*, 147:85-90.

Astumian, R. D., Sasaki, M., Yasunaga, T., and Schelly, Z. A. (1981) *J. Phys. Chem.*, 85:3832-3835.

Atkinson, R. J., Posner, A. M., and Quirk, J. P. (1967) *J. Phys. Chem.*, 71:550-558.

Berner, R. A. (1981) in *Kinetics of Geochemical Processes* (A. C. Lasaka and R. J. Kirkpatch, eds.). Mineral Society of America, Washington D.C., pp. 111-134.

Bloom, P. R. (1983) *Soil Sci. Soc. Am. J.*, 47:164-168.

Bloom, P. R. and Erich, M. S. (1987) *Soil Sci. Soc. Am. J.*, 51:1131-1136.

Bockris, J. O'.M. and Reddy, A. (1970) *Modern Electrochemistry*. MacDonald, London, Chapter 5.

Boehm, H. P. (1971) *Disc. Faraday Soc.*, 52:264-275.

Bowden, J. W., Posner, A. M., and Quirk, J. P. (1980) in *Soils with Variable Charge* (B. K. G. Theng, ed.). New Zealand Society of Soil Science, pp. 147-166.

Bruggenwert, M. G. M., Hiemstra, T., and Bolt, G. H. (1991) in *Soil Acidity* (B. Ulrich and M. E. Sumner, eds.). Springer-Verlag, Berlin, pp. 8-27.

Chernov, V. A. (1957) *Nature of Soil Acidity* (translated into Chinese by T. R. Yu). Science Press, Beijing.

Coleman, N. T. and Craig, D. (1961) *Soil Sci.*, 91:14-18.

Conyers, M. K. (1990) *J. Soil Sci.*, 41:147-156.

Coulter, B. S. (1969) *Soils Fert.*, 32:215-223.

Deshpande, T. L., Greenland, D. J., and Quirk, J. P. (1964) *Trans. 8th Intern. Congr. Soil Sci.*, III:1213-1225.

Duquette, M. and Hendershot, W. (1993) *Soil Sci. Soc. Am. J.*, 57:1222-

1226.

Eeckman, J. P. and Laudelout, H. (1961) *Koll. Z.*, 178:99-107.

Harward, M. E. and Coleman, N. T. (1954) *Soil Sci.*, 79:181-188.

Hiemstra, T., van Reimsdijk, W. H., and Bolt, G. H. (1989) *J. Colloid Interface Sci.*, 133:91-104.

Hiemstra, T, deWit, J. C. M., and Reimsdijk, W. H. (1989) *J. Colloid Interface Sci.*, 133:105-117.

Huang, P. M. (1988) *Adv. Soil Sci.*, 8:1-78.

Helyar, K. R., Conyers, M. K., and Munns, D. N. (1993) *J. Soil Sci.*, 44:317-333.

Lindsay, W. L. (1979) *Chemical Equilibria in Soils*. John Wiley & Sons, New York.

Marcano-Martinez, E. and McBride, M. B. (1989) *Soil Sci. Soc. Am. J.*, 53:1041-1045.

Miller, R. J. (1965) *Soil Sci. Soc. Am. Proc.*, 29:36-39.

Parfitt, R. L. (1980) in Soils with Variable Charge" (B. K. G. Theng, ed.). New Zealand Society of Soil Science, pp. 167-194.

Parker, J. C. Zelazny, L. W., Sampath, S., and Harris, W. G. (1979) *Soil Sci. Soc. Am. J.*, 43:668-674.

Paterson, E., Goodman, B. A., and Farmer, V. C. (1991) in *Soil Acidity* (B. Ulrich and M. E. Sumner, eds.). Springer Verlag, Berlin, pp. 97-124.

Ritchie, G. S. P. (1989) in *Soil Acidity and Plant Growth* (A. D. Robson, ed.). Academic Press, Sydney, pp. 1-60.

Schaffer, G. and Fisher, W. R. (1985) *Z. Pflanzenernähr. Bodenk.*, 148: 471-480.

Scheffer, F. und Schachtschabel, P. (1992) *Lehrbuch der Bodenkunde*. Enke, Stuttgart, pp. 113-126.

Schlindler, P. W. (1981) in *Adsorption of Inorganic at Solid-liquid Interfaces* (M. A. Anderson and A. J. Rubins, eds.) Ann Arbor Science Publishers.

Schofield, R. K. (1949) *J. Soil Sci.*, 1:1-8.

Schulthess, C. P. and Sparks, D. L. (1986) *Soil Sci. Soc. Am. J.*, 50:1406-1411.

Schulthess, C. P. and Sparks, D. L. (1987) *Soil Sci. Soc. Am. J.*, 51:1136-1144.

Schwertmann, U. and Fischer, W. R. (1982) *Z. Pflanzenernähr. Bodenk.*, 145:221-223.

Sparks, D. L. (1985) *Adv. Agron.*, 38:231-266.

Sparks, D. L. (1989) *Kinetics of Soil Chemical Processes*. Academic Press, San Diego.

Stumm, W. (1986) *Geoderma*, 38:19-30.

Sumner, M. E. (1962) *Agrochimica*, 6:183-189.

Susser, P. and Schwertmann, U. (1991) *Geoderma*, 49:63-76.

Thomas, G. W. (1988) *Comm. Soil Sci. Plant Anal.*, 19:833-856.

Thomas, G. W. and Hargrove, W. L. (1984) in *Soil Acidity and Liming* (F. Adams, ed.). American Society of Agronomy, Madison, WI, pp. 3-56.

Ulrich, B. (1986) *Z. Pflanzenernähr. Bodenk.*, 149:702-717.

Ulrich, B. (1991) in *Soil Acidity* (B. Ulrich and M. E. Sumner, eds.). Springer Verlag, Berlin, pp. 28-79.

van Breemen, N. (1991) in *Soil Acidity* (B. Ulrich and M. E. Sumner, eds.). Springer Verlag, Berlin, pp. 1-7.

van Breemen, N., Mulder, J., and Driscoll, C. T. (1983) *Plant and Soil*, 75:283-308.

Wang, J. H. and Yu, T. R. (1976) in *Electrochemical Properties of Soils and Their Research Methods* (T. R. Yu, ed.). Science Press, Beijing, pp. 325-353.

Yu, T. R. (1985) *Physical Chemistry of Paddy Soils*. Science Press/Springer Verlag, Beijing/Berlin.

Yu, T. R. (1987) in *Principles of Soil Chemistry* (T. R. Yu, ed.). Science Press, Beijing, pp. 325-364.

Yu, T. R. (1990) in *Chemical Processes in Soil Genesis* (T. R. Yu and Z. C. Chen, eds.). Science Press, Beijing, pp. 96-132.

Yu, T. R. and Zhang, X. N. (1986) in *Proceedings of the International Symposium on Red Soils*. Science Press/Elsevier, Beijing/Amsterdam. pp. 409-441.

Zhang, F. S., Zhang, X. N., and Yu, T. R. (1991a) *Soil Sci.*, 151:436-443.

Zhang, F. S., Ji, G. L., and Yu, T. R. (1991b) *Soil Sci.*, 152:25-32.

11

ACIDITY

X. L. Kong, X. N. Zhang, J. H. Wang, and T. R. Yu

For variable charge soils, acidity is a property that is of equal importance as the surface charge. These two properties may affect each other, with the effect of the former on the latter more remarkable than the reverse. In the previous chapters it was shown that pH affects many other properties of the soil by affecting the surface charge. Therefore, soil acidity is more significant than surface charge in some aspects. Owing to a similar reason, the importance of acidity for variable charge soils may exceed that for constant charge soils.

Soil acidity generally manifests itself in the form of hydrogen ions. Actually, these hydrogen ions are chiefly the product of the hydrolysis of aluminum ions. Therefore, when examining soil acidity it is necessary to examine the properties of aluminum ions. In the previous chapter the transformation of hydrogen ions into aluminum ions has already been mentioned. In this chapter the relationship between aluminum ions and hydrogen ions will be discussed in greater detail.

Another difference between variable charge soils and constant charge soils with respect to acidity is that, not only hydrogen ions, but also hydroxyl ions can participate in chemical reactions between the solid phase and the liquid phase. In constant charge soils the quantity of hydroxyl ions is an induced variable and is determined by the quantity of hydrogen ions in the solution and the ionic product of water. In variable charge soils, on the other hand, the quantity is also determined by the chemical equilibrium of that ion species itself at the solid–solution interface. Thus, hydroxyl ions can, in turn, affect the quantity of hydrogen ions in solution.

In this chapter the nature of acidity of variable charge soils will be discussed mainly from these characteristics.

11.1 ALUMINUM IONS

11.1.1 Exchangeable Hydrogen and Exchangeable Aluminum

In the field of soil chemistry, there has been an interesting history with regard to the nature of soil acidity. Soon after the recognition of the relationship between acid reaction and hydrogen ions in chemistry, this concept of the nature of acidity was introduced into soil science, and the significance of hydrogen ions was invariably associated with it whenever soil acidity was considered. In particular, after the introduction of the concept of pH by Sörensen in 1909, beginning in 1913, the concentration of hydrogen ions in soils was determined by the application of potentiometric method (Jenny, 1961). Bradfield (1923) advanced the hydrogen ion theory in soil science further by regarding soil colloid as one kind of weak colloidal acid, and calculated the apparent dissociation constant of soil colloidal acid from the potentiometric titration curve. With a similar idea, Wiegner and Pallman (1930) explained the "suspension effect" by assuming that the concentration of hydrogen ions in the diffuse double layer of negatively charged soil colloids is higher than that of the free solution. Thus, within a considerably long period of time the generally accepted theory in soil chemistry was that the acid reaction of soils was caused by the dissociation of exchangeable hydrogen ions.

In 1904, Veitch discovered that there were aluminum ions in the extract of NaCl solution from acid soils and assumed that these aluminum ions were replaced by sodium ions. In 1914, Daikuhara, based on the phenomenon that the quantity of titratable acid in the extract from soils was close to the amount of aluminum in the solution, considered that soil acidity was caused by adsorbed aluminum ions. In 1916, Kappen further assumed that aluminum ions participate in cation–exchange reaction in the form of trivalent ions and that these ions were adsorbed by the soil more tightly than did hydrogen ions. However, despite these suggestions, the aluminum theory has not attracted wide attention in soil chemistry for a long time.

This situation remained until 1947 when Chernov, based on his systematic research, advanced the theory that soil acidity was caused chiefly by adsorbed aluminum ions (Chernov, 1957). Chernov showed that adsorbed hydrogen ions accounted for only a very small portion of exchangeable acidity and that hydrogen ions added to a soil would transform into aluminum ions spontaneously. After the introduction of Chernov's work to the United State in the early 1950s, Coleman and Harward (1953) obtained similar conclusions. After further research by other authors (Thomas and Hargrove, 1984), the exchangeable aluminum theory of soil acidity was generally accepted in soil science by the late 1950s.

When reviewing this history of soil acidity, Jenny (1961) called it a "merry-go-round." Of course, our present knowledge about the nature of soil acidity is much greater than that of Veitch or Chernov (Thomas, 1988).

The research cited above was conducted primarily with constant charge soils of temperate regions or clay minerals. One may raise a question: How about the situation with variable charge soils of tropical and subtropical regions? This is the topic that is of concern in this chapter.

Early studies (Ling and Yu, 1957) showed that the exchangeable acidity of various types of variable charge soils in south China was mainly composed of aluminum ions. Hydrogen ions generally accounted for less than 4% of the total exchangeable acidity. Only in the surface layer where the content of organic matter was high there may be the presence of a considerable amount of exchangeable hydrogen. Similar conclusions have been obtained in recent years by the use of more accurate methods (Zhang et al., 1990). The relationship between exchangeable acidity and exchangeable aluminum for some variable charge soils is graphically shown in Fig. 11.1. It is seen that the correlation coefficient is 0.989. These data confirm again that the acid reaction of the mineral part of variable charge soils is chiefly caused by aluminum ions, even though the type of these soils differs greatly.

The slope of the straight line in the figure is 1.100. Instead of zero, the intercept point is 0.189. This implies that the quantity of exchangeable acidity is higher than that of exchangeable aluminum by 10% or more on average. Since the quantity of other metal ions that can undergo hydrolysis in the extract is very limited, this surplus part of acidity can only be

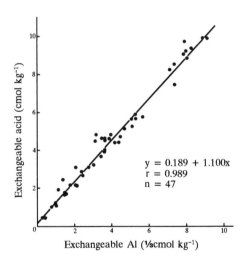

$$y = 0.189 + 1.100x$$
$$r = 0.989$$
$$n = 47$$

Fig. 11.1. Relationship between exchangeable acidity and exchangeable aluminum in variable charge soils.

attributed to hydrogen ions that have been replaced from the soil. Despite the hypothesis that the H$^+$ ions present in salt extracts of acid soils are largely the product of hydrolysis of hydroxyl Al (Kissel et al., 1971; Veith, 1977), the occurrence of a certain amount of exchangeable hydrogen ions should be real. Generally speaking, the ratio of exchangeable hydrogen ions to aluminum ions in various soils is related to the content of organic matter, with the tendency that the higher the content, the larger the ratio (Ling and Yu, 1957; Yuan, 1963). This is understandable because there is a very strong affinity between protons and weak acid groups of organic matter.

Because exchangeable aluminum is the principal source of soil acidity, its quantity is closely related to pH of the soil. As can be seen in Fig. 11.2, when the pH is higher than about 5.5, the soil almost does not contain exchangeable aluminum. However, this does not necessarily mean that at this time the soil does not contain aluminum ions that are adsorbed by the soil so tightly that they cannot be replaced by potassium ions. Here, a question about the forms of aluminum ions is involved.

11.1.2 Properties of Aluminum Ions

It is generally assumed that in aqueous solutions one aluminum ion is associated with six hydrated water molecules. These aluminum ions may undergo hydrolysis, producing hydrogen ions.

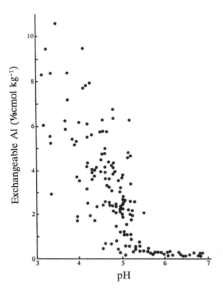

Fig. 11.2. Relationship between exchangeable aluminum and pH in soils (Ling and Yu, 1957).

The hydrolysis of aluminum ions proceeds in three steps:

$$Al^{3+} + H_2O \rightleftharpoons AlOH^{2+} + H^+ \tag{11-1}$$

$$Al^{3+} + 2H_2O \rightleftharpoons Al(OH)_2^+ + 2H^+ \tag{11-2}$$

$$Al^{3+} + 3H_2O \rightleftharpoons Al(OH)_3 + 3H^+ \tag{11-3}$$

For the first-step hydrolysis, the hydrolytic constant K_{h1} is

$$K_{h1} = [H^+][AlOH^{2+}]/[(Al^{3+}) - (H^+)] = [H^+]^2/[(Al^{3+}) - (H^+)] \tag{11-4}$$

$$pH = 0.5pK_{h1} - 0.5 \log[(Al^{3+}) - (H^+)] \tag{11-5}$$

The numerical value of pK_{h1} is about 5.0.

For solutions of aluminum salts such as $AlCl_3$ and $Al_2(SO_4)_3$, the hydrolysis reaction stops at the first step. Under certain conditions the hydrolysis can also proceed according to equation (11-2) or equation (11-3).

For the second- and the third-step hydrolysis, the relevant equations should be

$$pH = 0.33 \log 2 + 0.33pK_{h2} - 0.33 \log[(Al^{3+}) - 0.5(H^+)] \tag{11-6}$$

$$pH = 0.25 \log 3 + 0.25pK_{h3} - 0.25 \log[(Al^{3+}) - 0.33(H^+)] \tag{11-7}$$

It is thus seen that, in addition to Al^{3+} ions, there is the possibility of the existence of hydroxyl aluminum ions such as $AlOH^{2+}$ and $Al(OH)_2^+$ in the solution. These hydroxyl aluminum ions can polymerize, forming various forms of polynuclear substances. At present it is still not exactly known about the composition and structure of these substances. A variety of formulas, such as $Al_2(OH)_2^{4+}$, $Al_6(OH)_{15}^{3+}$, $Al_7(OH)_{17}^{4+}$, $Al_8(OH)_{20}^{4+}$, $Al_{13}(OH)_{32}^{7+}$, and $[Al_{13}O_4(OH)_{24}(H_2O)_{12}]^{7+}$, have been suggested (Coutler, 1969; Frink, 1972; Huang, 1988; Jackson, 1963; Singh, 1982; Thomas and Hargrove, 1984). Generally speaking, the composition of these substances is chiefly determined by the concentration and the basicity of the aluminum salts, with a high concentration and a high basicity favoring the formation of polynuclear hydroxyl aluminum.

If there are anions such as SO_4^{2-} or F^- present in the solution, aluminum ions can form complexes with these anions. Many organic anions can also form complexes with aluminum ions (Lundström, 1993; Huang, 1988; Ritchie, 1989; Thomas and Hargrove, 1984).

An aluminum ion is one kind of weak acid because it can produce hydrogen ions on hydrolysis [equations (11–1) to (11–3)]. When aluminum ions are titrated with an alkali, a titration curve, typical of weak acids, can be obtained. A typical curve for the titration of $Al_2(SO_4)_3$ solution with NaOH is shown in Fig. 11.3. The whole curve can be distinguished into three regions. The pH of the original solution is 3.53. When a small amount of alkali is added, the pH rises markedly. Then, in the second region until the amount of added alkali corresponds to 80–85% of the aluminum present in the system, the change in pH is small, starting from 4.0–4.1 (at which a precipitate begins to form) and reaching 4.7–5.0 when the precipitation ends. Within the third region of 80–100% neutralization, the pH of the solution increases sharply to 10. The whole curve reflects the characteristic features of weak acids titrated with a strong alkali. Note that within the second region the increment of pH with the increase in alkali is very small, especially at about 50% neutralization. This type of buffering action of aluminum ions is significant for acid soils.

When the pH is higher than 10, aluminum hydroxide can transform into aluminate and is dissolved.

11.1.3 Forms of Aluminum in Soils

The content of aluminum in variable charge soils is higher than that in constant charge soils. Because, in addition to aluminosilicates, there is a large amount of free aluminum in the soil. Many attempts have been made to distinguish different forms of these nonsilicate aluminum ions (Helyar et al., 1993; Huang, 1988; Moore and Ritchie, 1988; Patterson et al., 1991;

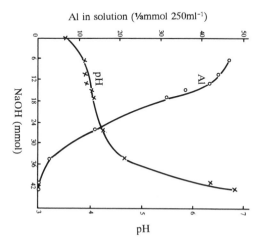

Fig. 11.3. Titration curve of $Al_2(SO_4)_3$ solution with alkali (Chernov, 1957).

Ritchie, 1989; Soon, 1993; Thomas and Hargrove, 1984), including those in soil solution (Hodges, 1987; Munns et al., 1992). Generally speaking, this free aluminum can be classified into four forms: exchangeable form, organo-complexed form, amorphous form, and crystalline form. Exchangeable aluminum can be replaced by cations of an unbuffered neutral salt solution such as KCl solution. Organo-complexed aluminum is complexed by organic groups of the solid phase of the soil. Amorphous aluminum is hydroxyl aluminum with different basicities, including some hydroxyl aluminum ions. Gibbsite may be taken as a typical example of crystalline aluminum. In various types of variable charge soils, the absolute quantity of free aluminum and the relative proportions among different forms of aluminum may vary widely. The approximate distribution of different forms of free aluminum in four representative soils is shown in Fig. 11.4. Exchangeable aluminum generally accounts for less than 10% of the total free aluminum, with the amount depending chiefly on the quantity of the negative surface charge and pH. The amount of organo-complexed aluminum is closely related to the organic matter content of the soil. The particularly high proportion of organo-complexed aluminum in the Xanthic-Haplic Acrisol is caused by a high content (more than 4%) of organic matter in the whole profile. By contrast, for the Ali-Haplic Acrisol with an organic matter content of less than 1%, the amount of organo-complexed aluminum is quite low. The reason for the higher amount of organo-complexed aluminum in the surface layer of Rhodic Ferralsol and

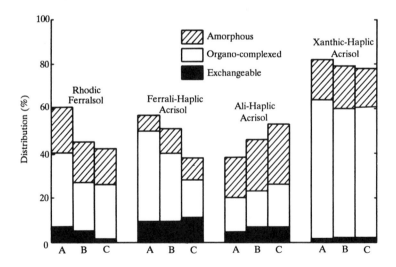

Fig. 11.4. Distribution of various forms of free aluminum in soils (A, B, and C represent surface layer, subsurface layer, and substratum, respectively).

Ferrali–Haplic Acrisol as compared to the subsoil is that the organic matter content of this layer is relatively high. In the figure, the remaining proportion of 100% distribution not represented in the above three forms is the crystalline aluminum. However, Fig. 11.4 can only show an approximate distribution pattern of aluminum because at present no satisfactory method for the characterization of different forms of aluminum in soils is available.

All four forms of aluminum are related to the acidity of soils to a certain extent. Exchangeable aluminum is the direct producer of hydrogen ions. Organo–complexed aluminum can dissociate to a certain degree. Amorphous aluminum and even crystalline aluminum can be transformed into aluminum ions under certain conditions. In particular, mononuclear and polynuclear hydroxyl aluminum ions can exert important influence on soil acidity and other surface properties of the soil.

Early experimental results obtained by Chernov showed that the adsorption tightness of various forms of aluminum and hydrogen ions was of the order $Al(OH)_2^+ > AlOH^{2+} > Al^{3+} > H^+$ (Chernov, 1957). This order is also generally true for clay minerals (Bloom et al., 1977; Thomas and Hargrove, 1984). Apparently, hydroxyl aluminum ions can be adsorbed specifically because the order of adsorption tightness of the three forms of aluminum is opposite to that of the number of the positive charge carried by the ions. Later research has proven the existence of polynuclear hydroxyl aluminum ions and showed that some hydroxyl aluminum ions, once adsorbed by soils, are very difficult to replace by cations of a neutral salt (Huang, 1988; Thomas and Hargrove, 1984). Some authors (Schwertmann and Jackson, 1964; Volk and Jackson, 1964) have also interpreted the presence of more than two buffering ranges in acid soils in terms of the functioning of hydroxyl aluminum ions. In the majority of the variable charge soils, the abundant existence of interlayer hydroxyl aluminum, as found in montmorillonite and vermiculite, would be unlikely. On the other hand, owing to the characteristic features of the solid phase, it is probable that the amount of hydroxyl aluminum existing in the form of specific adsorption is quite large. It was shown in Chapter 10 that the quantity of exchangeable acidity in three electrodialyzed and H,Al–saturated variable charge soils increased after the addition of hydrogen ions to the soil, and that for the Rhodic Ferralsol in particular the increase accounted for more than 50% of the hydrogen ions added. This can only be interpreted as the result of the transformation of some mononuclear or polynuclear hydroxyl aluminum ions into aluminum ions, which are replaced by cations of a neutral salt. The increase in apparent cation–exchange capacity of Haplic Acrisols after treatment with acids (Ling and Yu, 1957) may be partly caused by this reason. Thus, amorphous aluminum, particularly hydroxyl aluminum ions, may play a very important role in variable charge soils.

11.1.4 Hydrogen–Aluminum Transformation

The basic cause of the very small proportion of exchangeable hydrogen ions in the total acidity of soils under natural conditions is that hydrogen–saturated soils are unstable, and hydrogen ions can react spontaneously with aluminum–containing compounds of the solid phase, including crystalline aluminum in clay minerals, releasing equivalent amount of aluminum ions. The speed of this transformation is quite high. Chernov (1957) has made a detailed study in this respect.

Studies with clay minerals show that the hydrogen–aluminum transformation rate is actually controlled by the interdiffusion rate between aluminum ions at the edges of the mineral crystals and the hydrogen ions on the crystal surface, independent of clay concentration of the system. The diffusion coefficient at room temperature is about 10^{-20} cm^2 s^{-1} (Eeckman and Laudelout, 1961). The rate is affected by temperature. Generally, an increase in 10^{0}C induces a doubling in rate (Coleman and Craig, 1961).

The hydrogen–aluminum transformation rate in most variable charge soils is higher than that in kaolinite and in constant charge soils (Yu, 1976). This is probably related to the form of aluminum. Experiments showed that when a part of aluminum with a high activity was removed from the soil by chemical means, the transformation rate dropped sharply, although the amount removed only accounted for 0.2–1% of the total aluminum. Two examples in this respect are shown in Fig. 11.5. In the figure, the transformation rate in the Ali–Haplic Acrisol is higher than that in the Rhodic

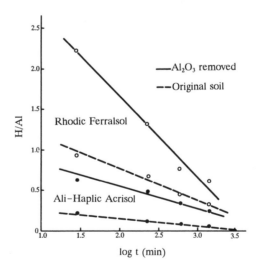

Fig. 11.5. Effect of amorphous aluminum oxides on hydrogen–aluminum transformation rate in soils (Ling and Yu, 1957).

Ferralsol initially. This could be caused by the higher activity of aluminum in the former case. On the other hand, the reaction of the hydrogen–aluminum transformation can occur even after the removal of most of the free aluminum from the soil and the destruction of a part of crystalline minerals by boiling with concentrated HCl (Ling and Yu, 1957). Thus, it may be concluded that the transformation of hydrogen–saturated soils into aluminum–saturated soils is a universal phenomenon and that this is also the basic reason why aluminum ions always constitute the major part of exchangeable acidity of mineral soils under natural conditions.

In addition to the form of aluminum, organic matter can also have some effect on hydrogen–aluminum transformation. It has been mentioned that some weak acid groups in organic matter possess a strong affinity for hydrogen ions. This should retard the process of the hydrogen–aluminum transformation. This is actually the case. As can be seen in Fig. 11.6, the rate of the hydrogen–aluminum transformation for different horizons of the profile of an Ali–Haplic Acrisol is the lowest in the surface layer with an organic matter content of 15.6% and the highest in the substratum containing only 0.5% of organic matter. This difference in the transformation rate could be caused mainly by the difference in the organic matter content.

Under field conditions there is always the continuous production of hydrogen ions in soil (cf. Chapter 10). Therefore, it is not possible to have

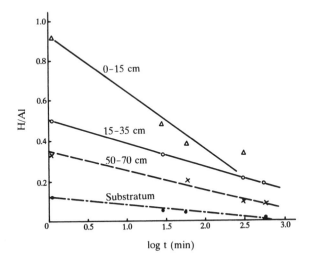

Fig. 11.6. Change in H/Al ratio in soil samples of different horizons of Ali–Haplic Acrisol after washing with acid (organic matter contents: 0–15 cm, 15.6%; 15–35 cm, 4.1%; 50–70 cm, 1.3%; substratum, 0.5%) (Ling and Yu, 1957).

a necessary condition for the establishment of a reaction equilibrium in the hydrogen–aluminum transformation. Besides, some hydrogen ions combined tightly with organic matter are difficult to participate in the hydrogen–aluminum transformation. Hence, in soils there is always the presence of a certain amount of hydrogen ions. This causes a widely observed phenomenon: the higher the organic matter content of the soil, the larger the proportion of hydrogen ions in the total acidity.

11.1.5 Adsorption and Desorption of Aluminum Ions

It has been mentioned several times in the previous chapters that aluminum ions possess a very high affinity energy with variable charge soils. This is because the charge/size function (z^2/r) of Al^{3+} ions is 43.6 $(\times 10^{-28}$ C^2 $m^{-1})$, much larger than 2.2 for Na^+, 10.3 for Ca^{2+}, 12.5 for Mn^{2+}, 13.3 for Fe^{2+}, and 14.3 for Mg^{2+} ions (Thomas and Hargrove, 1984). When aluminum ions polymerize, the difference in this function would be even larger. Therefore, as has been well established, in ion–exchange for clay minerals and constant charge soils, aluminum is a strong competitor for monovalent and divalent cations Na, K, Mg, and Ca (Bloom et al., 1977; Clark and Turner, 1965; Chernov, 1957; Coulter and Talibudeen, 1968; Foscolos, 1968; Nye et al., 1961).

The quantities of adsorbed aluminum ions for the Na–saturated Ali–Haplic Acrisol and the Na–saturated Hyper-Rhodic Ferralsol as related to the quantity of aluminum ions added are shown in Figs. 11.7 and 11.8, respectively. When the quantity of added aluminum ions was smaller than one symmetry value of the apparent negative surface charge sites, these ions were adsorbed almost completely by the soil. This means that aluminum ions are extremely strong in competing with sodium ions for the negative charge sites in the soil. In this case, aluminum ions are probably adsorbed chiefly in the form of Al^{3+}. When added aluminum ions exceeded one symmetry value in quantity, they could be adsorbed by the soil continuously, although the increment in adsorption decreased with the increase in the amount of added aluminum ions. Since the pH of the equilibrium solution even decreased slightly with the increase in amount of added aluminum ions, the increase in the aluminum adsorption would not be caused by the increase in the number of adsorption sites in the soil. It can only be suggested that, following the increase in aluminum ions, there was more aluminum adsorbed in the form of hydroxyl aluminum ions. The desorption curve of adsorbed aluminum shown in the figure can support this suggestion. When the quantity of the adsorbed aluminum exceeded one symmetry value, there were more and more portions of aluminum ions that could not be replaced by potassium ions progressively.

The desorption of adsorbed aluminum ions is also related to the

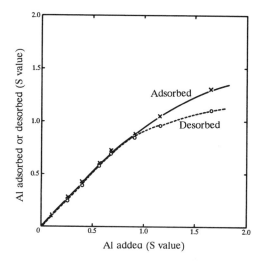

Fig. 11.7. Adsorption and desorption of aluminum in Na-saturated Ali-Haplic Acrisol in relation to amount of aluminum ions added.

concentration of the replacing agent (KCl). If the properties of the adsorbed aluminum ions can fulfill the requirement of the general principles in ion-exchange, when extracted with either a concentrated KCl solution or a dilute KCl solution, the final extracted amounts should equal each other, although in the latter case the required number of extractions would be

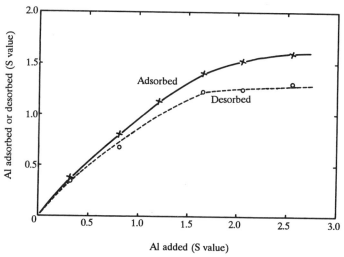

Fig. 11.8. Adsorption and desorption of aluminum in a Na-saturated Hyper-Rhodic Ferralsol in relation to the amount of added aluminum ions.

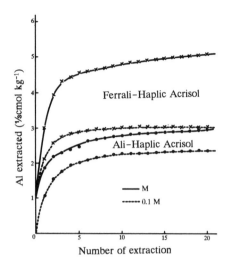

Fig. 11.9. Accumulated amount of aluminum extracted consecutively with different concentrations of KCl solutions from Ali–Haplic Acrisol and Ferrali–Haplic Acrisol (Soil:solution ratio for each extraction, 1:10).

larger. Actually, this is not the case. As is seen in Fig. 11.9, if the variable charge soil was treated with two KCl solutions of different concentrations consecutively, the quantity of aluminum extracted with dilute KCl solution did not increase after 10 extractions. The total quantity in 20 extractions was smaller than that extracted with a concentrated KCl solution extracted 2–3 times. This indicates that the aluminum extracted by the 0.1 M KCl solution and by the M KCl solution is different not only in quantity but also in form. The reason for this phenomenon is not known at present. Several possibilities exist. The hydrolysis of nonexchangeable hydroxyl aluminum (Kissel et al., 1971; Veith, 1977) may be enhanced by high concentrations of the electrolyte. The competitive ability of potassium ions for adsorbing sites with the aluminum ions may be related to the ratio between the two ion species. Some hydroxyl aluminum ions that are nonexchangeable may transform into chloro–aluminum complexes in the presence of a high concentration of chloride ions, and thus they can be replaced by potassium ions from the adsorption sites of variable charge soils.

The data shown in Figs. 11.7 and 11.8 cannot directly prove that when the quantity of aluminum ions adsorbed by the soil is smaller than one symmetry value of the adsorption sites, these ions are adsorbed chiefly in the form of Al^{3+}, although this point has been well established for some clay minerals (Clark and Turner, 1965; Nye et al., 1961; Thomas and Hargrove, 1984; Veith, 1977) and constant charge soils (Chernov, 1957). In order to verify this point, it is necessary to know the quantitative relationship

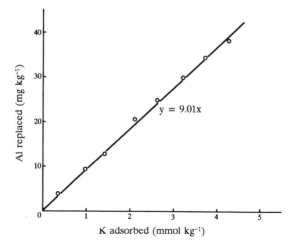

Fig. 11.10. Quantitative relationship in potassium–aluminum exchange for Ali–Haplic Acrisol.

between the replaced aluminum ions and the adsorbed replacing cations participating in the ion–exchange reaction. Two comparisons in this respect are shown in Figs. 11.10 and 11.11, respectively. It can be seen that for the Ali–Haplic Acrisol and the Rhodic Ferralsol, one mole of potassium ions replaced 9.01 and 9.29 g of aluminum ions, respectively, approximately

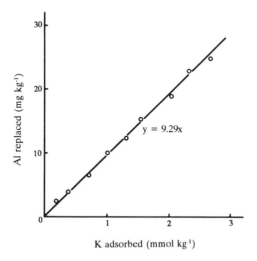

Fig. 11.11. Quantitative relationship in potassium–aluminum exchange for Rhodic Ferralsol.

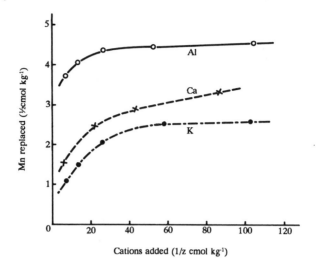

Fig. 11.12. Comparison of replacing power among aluminum, calcium and potassium ions in replacing adsorbed manganese ions from Rhodic Ferralsol (Yu, 1987).

corresponding to one-third of the atomic weight of aluminum, 9.0. This can directly prove that in this case adsorbed aluminum ions are essentially in the form of Al^{3+}.

A comparison of the power for replacing manganese ions from a Mn-saturated Rhodic Ferralsol among aluminum, calcium, and potassium ions is shown in Fig. 11.12. When the quantity of added cations is equivalent to one symmetry value of the adsorption sites of the soil, the replaced amount of the manganese ions by the aluminum ions is about 2.4 times of that replaced by the calcium ions and 3.7 times of that replaced by the potassium ions. When the amount of replacing cations is low, the strong replacing power of aluminum ions is even more pronounced. One can also compare the difference in the replacing power between the aluminum ions and the calcium ions by preparing an Al-saturated soil and a Ca-saturated soil separately and then replacing the adsorbed cations with calcium ions and aluminum ions. As is seen in Fig. 11.13, the replacing power of the aluminum ions is much stronger than that of the calcium ions for both the Ali-Haplic Acrisol and the Rhodic Ferralsol.

It can be concluded that the replacing power of the trivalent Al^{3+} ions in replacing other exchangeable cations is stronger than that of the alkali metal ions and alkaline earth metal ions, even though aluminum ions are not adsorbed specifically by variable charge soils.

When there are two or more ion species present together in the soil system, aluminum ions can also affect the adsorption ratio among these

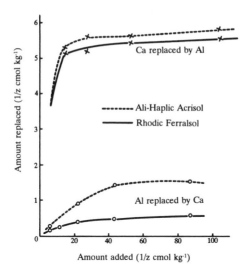

Fig. 11.13. Comparison of replacing power between aluminum ions and calcium ions (Yu and Zhang, 1985).

ions. Table 11.1 shows that, on one hand, the amounts of the potassium ions and the calcium ions adsorbed by the H,Al-saturated soils are much smaller than those adsorbed by the Na-saturated soils. On the other hand, the presence of the aluminum ions causes the difference in adsorption between potassium ions and calcium ions smaller than that in the presence of sodium ions. This implies that aluminum ions can retard the adsorption of both potassium and calcium ions, whereas sodium ions exert a much weaker retarding effect on the adsorption of calcium ions than on the adsorption of potassium ions. In view of the principle of Donnan theory, this is easy to understand. The practical significance of this point is that, as the principal base cation species in variable charge soils, calcium ions are easy to leach out in the presence of a large amount of aluminum ions under field conditions. Thus, the presence of a large amount of aluminum ions in variable charge soils, as a result of the intensive leaching loss of base cations, is also an important cause for the easy loss of base cations in these soils.

The strong bonding energy between aluminum ions and the surface of soil particles may also be the cause for the slow neutralization rate of acid soils. It has long been known that when acid soils are neutralized with alkalies it frequently requires hours or even days to attain a reaction equilibrium. In the past, no satisfactory explanation could account for this phenomenon. Some authors (Wiklander, 1955) supposed that this slow reaction may be a reaction between the alkali and the silicates. However,

Table 11.1 Effects of Sodium Ions and Hydrogen and Aluminum Ions on Adsorption of Potassium and Calcium Ions by Variable Charge Soils (Yu et al., 1989)

Soil	K+Ca Added[a]	Na-Saturated Soil			H,Al-Saturated Soil		
		pK	pCa	$a_K/2a_{Ca}$	pK	pCa	$a_K/2a_{Ca}$
Ali–Haplic	0.1	4.76	6.14	12.0	3.47	3.62	0.71
Acrisol	0.2	4.26	5.50	8.7	3.04	3.20	0.71
	0.3	3.86	5.00	6.9	2.80	2.96	0.73
	0.4	3.59	4.60	5.1	2.63	2.80	0.72
	0.5	3.36	4.22	3.6	2.50	2.68	0.76
Rhodic Ferralsol	0.1	4.19	5.12	4.3	3.56	3.82	0.91
	0.2	3.72	4.64	4.2	3.12	3.40	0.95
	0.3	3.41	4.26	3.5	2.87	3.16	0.98
	0.4	2.98	3.54	1.8	2.68	2.94	0.91
	0.5	2.95	3.60	2.2	2.56	2.82	0.91

[a]Symmetry value; K/0.5Ca ratio = 1.

since this slow reaction can proceed even in a nearly neutral medium, it appears that this supposition is not plausible. The reduced hydrolysis rate of the aluminum ions caused by their strong adsorption, as was supposed earlier (Thomas and Hargrove, 1984), also appears groundless. According to the modern viewpoint about the kinetics of the ion–exchange reactions, it should be the slow diffusion rate of the aluminum ions, caused by the strong bonding energy between these ions and the negative surface charge sites of the soil, that plays the dominant role in inducing this slow rate of overall reaction. This is because when an acid soil is neutralized with alkali, such as NaOH, the outward diffusion of the replaced aluminum ions, particularly when polymerized, either in a fashion of particle diffusion or in a fashion of film diffusion, would be strongly retarded by the negative charge sites at the surface of soil particles, although sodium ions can attain the adsorption sites easily through ion diffusion and the rate of ion–exchange reaction between sodium ions and aluminum ions at the adsorption sites is fast. It can then be suggested that it is the slow diffusion rate of aluminum ions that determines the slow reaction rate when acid soils are neutralized by alkalies. A comparison shown in Fig. 11.14 can support this suggestion, in which the neutralization rate of the Al-saturated soil colloid is slower than that of the H-saturated soil colloid. Since in this case the so-called "H-saturated" soil colloid may contain a considerable amount of

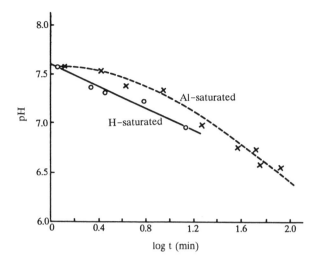

Fig. 11.14. Comparison of neutralization rate between H-saturated and Al-saturated Ali-Haplic Acrisols (clay fraction) (Yu, 1976).

aluminum ions, the actual difference in reaction rate would be more remarkable than that shown in the figure, if the soil colloid contains only hydrogen ions. A direct comparison in ion-exchange rate between the two ion species is shown in Fig. 11.15. The ion-exchange reaction between calcium ions and H-saturated strongly acidic (sulfonic)-type ion-exchange

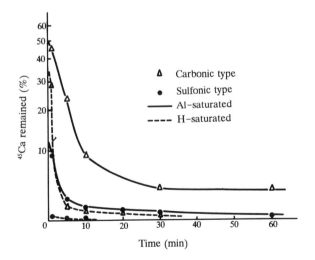

Fig. 11.15. Exchange rate of calcium ions with H-saturated and Al-saturated ion-exchange resins (Yu, 1976).

resin proceeds instantaneously. The reaction for H-saturated weakly acidic (carbonic)-type resin proceeds a little slower, lasting five minutes. By contrast, the reaction rates of Al-saturated resins are much lower, especially for the weakly acidic-type resin. These results can qualitatively explain the reason of the low neutralization rate of acid soils containing a large amount of exchangeable aluminum ions.

It can be concluded that aluminum ions are the principal cation species in acid variable charge soils and that the characteristic properties of these ions may deeply affect a series of interactions between other ions and soil colloids. Since the interactions at the solid-liquid interface are the core of chemical properties of a soil, it can be suggested that the significance of aluminum ions in variable charge soils is not confined to their being the principal producer of hydrogen ions.

11.2 SOIL pH

Although aluminum ions are the basic cause of acid reaction in soils, this acid reaction invariably manifests itself in the form of hydrogen ions. Besides, because aluminum ions together with the solid part of the soil behaves like a weak acid in many aspects, such a system may be regarded as one kind of weak acid. This is also the case for variable charge soils. The most commonly used index for expressing the strength of acid is the pH. Therefore, in soil science, pH has been a routine determination. In the following, several important factors in affecting pH of soils will be discussed.

11.2.1 In Relation to Composition of Ions

Just as the pH of a mixture of a weak acid and its salt is determined by the ratio of the two components and the nature of the base ions and the acid group, the pH of a soil is determined chiefly by the ratio of hydrogen and aluminum ions to base ions. This can be illustrated by the change in pH when a Rhodic Ferralsol is titrated with NaOH shown in Fig. 11.16. Before titration, the pH of the electrodialyzed soil colloid is 5.2. Following the addition of NaOH, the aluminum ions are neutralized and the pH is increased. After the complete neutralization of all of the exchangeable aluminum ions, the pH rises sharply. Afterward, because of the presence of free OH^- ions, the pH may rise to a quite high value. The whole titration curve is similar to that when a weak acid is titrated with NaOH and is also similar to the titration curve of $Al_2(SO_4)_3$ solution with NaOH shown in Fig. 11.3, with the exception that the change in the pH within the middle range is not as remarkable as the case for pure aluminum salt. The cause for this difference is that except for Al^{3+} ions, hydroxyl aluminum ions of different basicities in the soil can also be neutralized. Besides, the surface of the solid

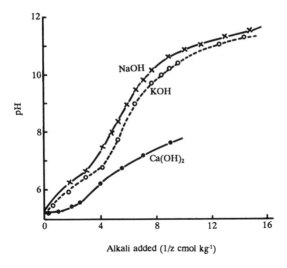

Fig. 11.16. Potentiometric titration curves of the clay fraction of Rhodic Ferralsol.

part of the soil can undergo de-protonation reaction. Generally, when variable charge soils are neutralized with an alkali the change in pH is not as sharp as in the case of clay minerals or constant charge soils. The larger the variability in surface charge of the soil, the less distinct the change in the pH. For some variable charge soils the inflection point of the pH change may even be invisible (Wang and Yu, 1986). This is one of the characteristics of variable charge soils.

The nature of the base cations can also have some effect. It can be seen from the figure that when the colloid of a Rhodic Ferralsol is titrated with KOH the pH is always lower than the corresponding pH when titrated with NaOH. The pH is even lower when the soil is titrated with $Ca(OH)_2$. For example, when the quantity of $Ca(OH)_2$ is equivalent to one symmetry value of the cation-adsorbing sites, the pH is only about 7.0 (Fig. 11.16). This is the reason why under field conditions the pH of the soil seldom exceeds 7.0 when various variable charge soils are saturated with base cations, chiefly calcium ions, through fertilization.

11.2.2 In Relation to Soil Type

When a weak acid is neutralized, the pH is related to the degree of neutralization:

$$pH = pK_a + \log([salt]/[acid]) \tag{11-8}$$

where pK_a is the dissociation constant of the acid. Likewise, if soils are regarded as one kind of weak acid, it is possible to compare the characteristics in acidity properties among various types of soils based on the titration curve.

The neutralization curves of the colloid fraction of two types of variable charge soils and a Cambisol when titrated with NaOH are shown in Fig. 11.17. On the curve it is possible to distinguish three threshold pH: ultimate pH, half-neutralization pH, and neutralization pH.

According to the concept of Mattson (Mattson, 1931, 1932), the pH of a soil colloid completely devoid of adsorbed base ions after electrodialysis may be called "ultimate pH." This is the lowest pH that a soil can have because in this case the soil is saturated by hydrogen and aluminum ions and therefore is equivalent to the situation for a weak acid not mixed with its salt. Strictly speaking, this "ultimate pH" of an electrodialyzed soil is generally higher than that of the same soil freshly treated with a H-saturated ion-exchange resin and will increase slightly upon aging. Nevertheless, this pH may be used as a valuable parameter to characterize the acidity properties of soils. Figure 11.17 shows that the ultimate pH of the Rhodic Ferralsol, Ali-Haplic Acrisol, and Cambisol is 5.0, 4.5 and 3.9, respectively. This difference is related to differences in clay minerals and other chemical compositions of these soils. The ultimate pH of montmorillonite is about 3.5, whereas that of kaolinite is 4.5-5.0. The presence of iron oxides causes the pH to be higher, while the presence of organic matter causes the pH to be lower. The highest ultimate pH of the Rhodic Ferralsol shown in the

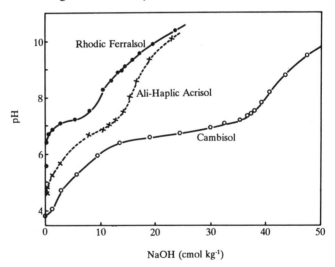

Fig. 11.17. Neutralization curve of the clay fraction of Rhodic Ferralsol, Ali-Haplic Acrisol, and Cambisol (Yu, 1987).

figure is caused by the predominance of kaolinite in the clay mineralogical composition and the high content of iron oxides.

The pH at which half of the acid is neutralized may be called the "half-neutralization pH." This pH corresponds to the pH when the ratio of the acid to its salt is equal to 1. It can be seen from equation (11-8) that this pH is also equivalent to the negative logarithm of the dissociation constant of the acid pK_a. This is the theoretical basis when Bradfield (1923) treated soil colloids as one kind of weak acid. In Fig. 11.17 the numerical values of the half-neutralization pH of the Rhodic Ferralsol, Ali-Haplic Acrisol, and Cambisol are 7.2, 6.7, and 6.5, respectively, reflecting the difference in acid strength among the three soil colloids. The Half-neutralization pH is affected by the nature of the base cations. Figure 11.16 shows that the half-neutralization pH is the highest when the base cations are sodium and the lowest when they are calcium. For field soils, the half-neutralization pH of various soils is related to the type of the soil to a certain extent, although the composition of exchangeable base cations differs considerably. According to statistics, the half-neutralization pH of paddy soils derived from Rhodic Ferralsols is 5.49 on an average, and those for paddy soils derived from Ali-Haplic Acrisols and for paddy soils derived from alluvial materials are 5.06 and 4.51. Therefore, the half-neutralization pH of soils shows a tendency toward decreasing from the south to the north of China, reflecting a zonal pattern in geographical distribution (Yu, 1985a). This difference is consistent with the difference in the clay mineralogical composition of the soils.

The pH at which all acid groups of the soil are neutralized may be called the "neutralization pH". At this point the effect of the addition of an acid or an alkali on the pH is most remarkable. Figure 11.17 shows that the neutralization pH of the three soil colloids is 8.2 ± 0.1, nearly independent of the type of the soil. This is because the difference in dissociation constant among the three acid colloids is not sufficiently large and therefore it is difficult to distinguish the difference in the neutralization pH. The neutralization pH of the same soil differs with the kind of the alkali used.

According to the principle of acid-base equilibrium, the quantity of alkali consumed at the neutralization pH is equivalent to the quantity of acid groups of the soil. It can be seen from Fig. 11.17 that for the colloid of the Rhodic Ferralsol, Ali-Haplic Acrisol and Cambisol, these quantities are 9.5, 13.5, and 41.0 cmol kg^{-1}, respectively. For clay minerals and constant charge soils, this quantity is often called the "cation-exchange capacity". For variable charge soils, this quantity can also reflect the approximate quantity of negative surface charges carried by the soil at a certain pH, although the terminology is not very exact.

11.2.3 Effect of Submergence

As shall be seen in Chapter 13, reduction processes will occur when a soil is submerged. The most remarkable influence of reduction processes on soil acidity is the rise in the pH.

Let us examine the most important reduction processes in soils:

$$\frac{1}{4}O_2 + H^+ + e \rightleftharpoons \frac{1}{2}H_2O \qquad (11\text{-}9)$$

$$\frac{1}{2}MnO_2 + 2H^+ + e \rightleftharpoons \frac{1}{2}Mn^{2+} + H_2O \qquad (11\text{-}10)$$

$$Fe(OH)_3 + 3H^+ + e \rightleftharpoons Fe^{2+} + 3H_2O \qquad (11\text{-}11)$$

$$\frac{1}{2}NO_3^- + H^+ + e \rightleftharpoons \frac{1}{2}NO_2^- + \frac{1}{2}H_2O \qquad (11\text{-}12)$$

$$\frac{1}{8}SO_4^{2-} + 5/4H^+ + e \rightleftharpoons \frac{1}{8}H_2S + \frac{1}{2}H_2O \qquad (11\text{-}13)$$

In all the above reduction processes there is invariably the participation of protons. This would consume hydrogen ions of the medium, resulting in a rise in the pH.

However, the relationship between the change in the pH and the reduction processes is not so simple. This is because when a soil is submerged, owing to the activities of microorganisms, there may be the production of acid substances such as CO_2 and organic acids. Therefore, the changing pattern of the pH when a soil is submerged is determined by the relative intensity of the pH-increasing and pH-decreasing factors (Yu, 1985a).

The change in the pH for an Ali-Haplic Acrisol during submergence is shown in Fig. 11.18. The pH increased remarkably during the 2-3 days of submergence. Within the following several days, owing to the production of large amounts of acid substances during the vigorous decomposition of organic matter, the pH fell slightly. The pH rose drastically again after about 10 days until a steady value was reached. If green manures were added to the soil, because of the production of large amounts of organic reducing substances and the subsequent processes expressed in equations (11-9) to (11-13), particularly those in equations (11-11) and (11-13), the rise in the pH might be even more remarkable within the two weeks of submergence. The changing pattern of the pH and the effect of organic matter when a soil is submerged (shown in Fig. 11.18) are quite representative in variable charge soils. For these strongly acid soils with a pH of lower than 5.5, the pH may rise to the neutral range 6.5-7.0 after submergence.

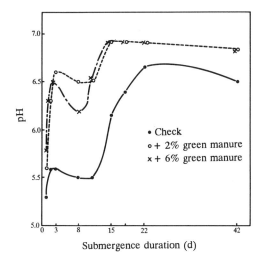

pH

• Check
o + 2% green manure
x + 6% green manure

Submergence duration (d)

Fig. 11.18. Change in pH of Ali–Haplic Acrisol after submergence (Yu, 1985a).

11.2.4 Suspension Effect

The concept of pH in chemistry is quite clear. In soil systems, however, the situation is not so simple. In addition to the theoretical consideration as to how the activities of hydrogen ions as sensed by an indicator electrode in the determination of pH represent the hydrogen ions in different parts of the electric double layer, there are some practical problems in the determination of soil pH. In this respect, variable charge soils behave differently from constant charge soils in some respects.

It has been known for a long time that when the pH of soils is determined with the potentiometric method, the pH of the suspension or paste is frequently different from that of the equilibrium solution. For soils in temperate regions, the pH of the suspension is generally lower. Wiegner and Pallman (1930) conducted a detailed study on this subject, and called the phenomenon the "suspension effect." According to their explanation, the electrode senses hydrogen ions not only in the solution but also in the diffuse layer of the colloid. Because the hydrogen ion concentration in the diffuse layer for colloids carrying negative charge is higher than that in the free solution, the pH would be relatively low. If the colloid is positively charged, the pH of the suspension may be higher than that of the equilibrium solution.

Jenny et al. (1950) proposed an alternative interpretation. They suggested that the suspension effect is caused by a liquid–junction potential between the salt bridge of the reference electrode and the soil suspension. When the

KCl of the salt bridge is in contact with the soil suspension, the mobilities of K^+ ions and Cl^- ions are changed by charged colloidal particles, resulting in an inequality in the distribution of electric charge between the two sides of the interface. This is the origin of the liquid–junction potential. At present, based on extensive research conducted by subsequent authors, it is generally considered that the interpretation of Jenny et al. is the more reasonable explanation (Yu, 1985b).

Variable charge soils possess two features. The sign of surface charge carried by these soils may be positive or negative. Second, the presence of large amounts of exchangeable aluminum ions may cause the sign of liquid–junction potential opposite to what would be expected only based on the surface charge of the soil colloid.

The relationship between the suspension effect and the pH for two Rhodic Ferralsols is shown in Fig. 11.19. When the pH is high, the pH of the soil paste is lower than that of the equilibrium solution, and the reverse is true when the pH is low. This can be qualitatively explained as follows. If the net surface charge of soil particles is negative, the diffusion rate of K^+ ions will be faster and that of Cl^- ions slower, resulting in surplus positive charges at the suspension (paste) side of the interface and surplus negative charges at the salt bridge side. Therefore, the interface potential is negative in sign. Hence, the superficial phenomenon would be that the activity of hydrogen ions in the suspension is increased. If the net surface charge of the soil is positive in sign, the situation would be reversed.

If the surface charge of the soil is the only factor in determining the liquid–junction potential, the potential would be zero at the zero point of

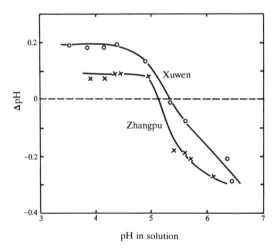

Fig. 11.19. Suspension effect of Rhodic Ferralsol in relation to pH (ΔpH = pH$_{suspension}$ − pH$_{equilibrium\ solution}$). (Xuan and Yu, 1964).

charge of the soil. It can be seen in Fig. 11.19 that, for the Rhodic Ferralsol of Guangdong Province and that of Fujian Province, the pH values of the equilibrium solution are 5.3 and 5.1, respectively, when the ΔpH is zero, consistent with the respective zero points of charge as determined by the ion adsorption method. However, the relationship between the magnitude including the sign of liquid–junction potential and the surface charge of the soil is not so simple, because it has been found that the presence of polyvalent aluminum ions may cause the ΔpH of negatively charged particles (montmorillonite, kaolinite, cation–exchange resin) to become positive in sign (Xuan and Yu, 1964). Lanthanum ions can cause a similar effect (Panicolau, 1970).

In recent years, it has been found that soil particles can exert their effect on the liquid–junction potential even if the salt bridge is kept at a certain distance from the soil without direct contact (Ji and Yu, 1985). This effect becomes less noticeable with the increase in distance. It can be seen from Fig. 11.20 that for the Rhodic Ferralsol and the Cambisol the signs of the liquid–junction potential were positive and negative, respectively. When the volume charge density of the system was decreased by the addition of silicon dioxide which had only a very small effect on the liquid–junction potential, the numeral value of the liquid–junction potential decreased. This once again reflects the relationship between the liquid–junction potential and the surface charge properties of the soil.

The materials presented above indicate that the suspension effect in the determination of soil pH is of both theoretical and practical significance. It

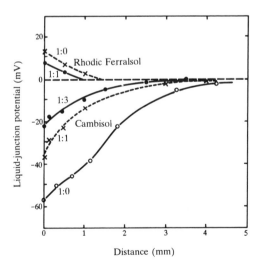

Fig. 11.20. Effect of soil on liquid–junction potential in relation to distance (Numbers in the figure are soil clay:silicon dioxide ratio) (Ji and Yu, 1985).

makes the pH value determined by usual method conditional to a certain extent, although the concept of pH itself is unambiguous. At present, no method can overcome this problem. Therefore, caution is needed when interpreting the results of pH determinations, despite the fact that the concept of pH has been widely applied in soil science. Since in general the lower the concentration of electrolyte in the system, the more remarkable the suspension effect, the above statement is particularly true for variable charge soils with a very low concentration of electrolyte.

11.3 EXCHANGEABLE ACIDITY AND EXCHANGEABLE ALKALINITY

11.3.1 Significance

There is an important difference in acidity properties between variable charge soils and constant charge soils. In addition to exchangeable hydrogen ions and aluminum ions, there may also be the presence of a large amount of exchangeable hydroxyl ions. These two kinds of ions can participate in the chemical equilibria between the solid phase and the liquid phase, with the result that both exchangeable acid and exchangeable alkali can appear when a neutral salt is added to a soil. This feature is caused by the characteristics in the composition of the mineral part of variable charge soils. The mechanism in the release of hydroxyl ions during coordination adsorption has been explained in Chapter 6. In this section focus will be paid to the interrelations between hydrogen and hydroxyl ions when released simultaneously.

As early as in the early 1930s, Mattson (1931, 1932) regarded soil colloids as being composed of acidoids and basoids, and for a given soil colloid the acidity property is determined by the relative acidity strengths and the relative quantities of the two types of colloids. Later, Mattson and Wiklander (1940) made a detailed study on "exchange acidity" and "exchange alkalinity." Unfortunately, this concept was practically ignored during the subsequent decades. It is apparent that this concept needs to be reconsidered when examining the acidity properties of variable charge soils.

11.3.2 Effect of Nature of Replacing Ions

The ability of various cations in replacing hydrogen and aluminum ions is different. This is also true for various anions in replacing hydroxyl ions. Hence, the relative ability of the two constituent ion species of different neutral salts in replacing either hydrogen and aluminum ions or hydroxyl ions would also be different. Therefore, the acidity status of a soil in neutral salt solutions varies with the nature of the two ion species of the salt and

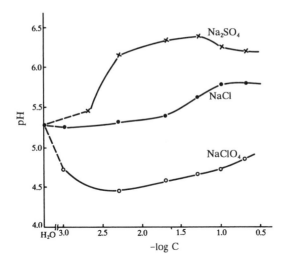

Fig. 11.21. The pH of electrodialyzed Hyper–Rhodic Ferralsol in sodium salt solutions of different concentrations (C is concentration in $1/z$ mol L^{-1}) (Zhang et al., 1989).

is also affected by its concentration.

Let us examine the effect of the nature of anions first. The pH of a Hyper–Rhodic Ferralsol in three sodium salt solutions with different concentrations is shown in Fig. 11.21. The decrease of soil pH in the NaClO$_4$ solution is essentially a reflection of the appearance of exchangeable acid caused by the addition of Na$^+$ ions. In this case ClO$_4^-$ ions can only release a very small amount of hydroxyl ions even when the concentration is considerably high. For this type of Ferralsol, the role of chloride ions in releasing hydroxyl ions may equal or even exceed the role of the sodium ions in the release of hydrogen and aluminum ions. Therefore, the pH of the suspension is close to or slightly higher than the pH in water. It has already been shown in Chapter 6 that sulfate ions possess a strong ability to replace hydroxyl ions from the surface of soil particles. Therefore, the pH of the soil in the Na$_2$SO$_4$ solution may be higher than that in water by more than 1 pH unit, although at this time sodium ions could still possess the ability to lower the pH.

If the cations of the neutral salt are Ba^{2+}, the ability to replace hydrogen and aluminum ions will be much stronger than that of the Na$^+$ ions. This point can be clearly seen from the comparison of the pH in the BaCl$_2$ solution and in the NaCl solution for two soils shown in Fig. 11.22.

From the above discussions it is clear that whether the pH of a soil in a neutral salt solution is higher or lower than that in water is determined principally by the relative ability of the two constituent ion species of the

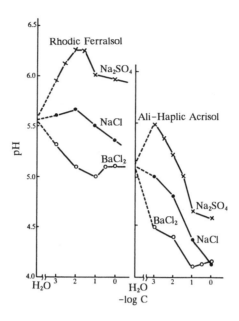

Fig. 11.22. The pH of electrodialyzed Rhodic Ferralsol and Ali–Haplic Acrisol in different salt solutions (C is concentration in $1/z$ mol L^{-1}) (Yu and Zhang, 1985).

salt to replace either hydrogen and aluminum ions or hydroxyl ions from the surface of soil particles.

11.3.3 In Relation to Surface Properties of Soil

It can also be seen in Fig. 11.22 that the effect of the same neutral salt on the pH of a Rhodic Ferralsol and an Ali–Haplic Acrisol is quite different. For the Rhodic Ferralsol, Na_2SO_4 always causes the pH to increase. For the Ali–Haplic Acrisol, on the other hand, the pH may increase or decrease, depending on the concentration of the Na_2SO_4 solution. NaCl always causes the pH of the Ali–Haplic Acrisol to decrease, whereas the effect for the Rhodic Ferralsol is not distinct. The difference in the effect of $BaCl_2$ on the pH in different types of soils is even more remarkable. Apparently, the difference in the variation of the pH in different neutral salt solutions for the three types of soils is caused by the difference in surface properties of the soils. This also indicates that the basoids of the Hyper–Rhodic Ferralsol are the largest in quantity or the strongest in strength, and that the acidoids of the Ali–Haplic Acrisol are the largest in quantity or the strongest in strength, with the Rhodic Ferralsol intermediate in position. This difference is consistent with the differences in the ultimate pH and the half-

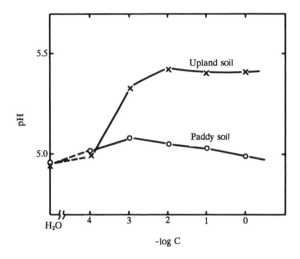

Fig. 11.23. The pH of upland and paddy Rhodic Ferralsols in Na_2SO_4 solutions (C is concentration of Na_2SO_4 in $\frac{1}{2}$mol L^{-1}) (Yu, 1985a).

neutralization pH among these soils cited in Section 11.2.2.

The carriers of basoids in variable charge soils are chiefly iron oxides. This is the reason why the pH of a paddy soil derived from Rhodic Ferralsol in Na_2SO_4 may be close to that in water (Fig. 11.23), due to the drastic decrease in basoids as a result of the reductive eluviation of iron oxides.

In the determination of soil pH, it is a generally adopted procedure to disperse the soil in a salt solution, such as M KCl or 0.01 M $CaCl_2$ instead of water (Aitken and Moody, 1991; Conyers and Davey, 1988; Little, 1992). From the discussions made in this section and in section 11.3.2, it is apparent that the addition of any cations or anions would invariably disturb the original chemical equilibria of the soil, including those involving hydrogen and hydroxyl ions. The effects would be more complex for variable charge soils than for constant charge soils.

11.3.4 Quantitative Estimation

Figures 11.21 to 11.23 show the variation of exchangeable acidity and exchangeable alkalinity of soils when pH is used as an index of intensity factor. Since the pH is related to the concentration of hydrogen ions in a logarithmic manner, the differences in the hydrogen ion concentration within different pH ranges as expressed by the difference in the same pH unit would not equal one another. It would be more reasonable if comparisons are made on the basis of the actual quantity of hydrogen or

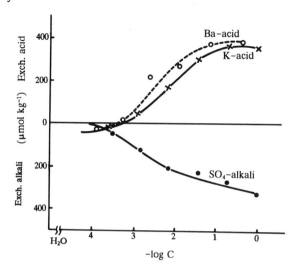

Fig. 11.24. Exchangeable acidity and exchangeable alkalinity of electrodialyzed Xanthic–Haplic Acrisol (*C* is concentration of the neutral salt in $1/z$ mol L⁻¹) (Zhang et al., 1990).

hydroxyl ions. Such comparisons are shown in Figs. 11.24 and 11.25. In the figures, the quantities of both exchangeable acid and exchangeable alkali increase with the increase in concentration of neutral salt, until a nearly constant value is reached when all the exchangeable acid or alkali are

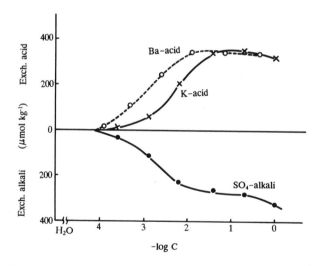

Fig. 11.25. Exchangeable acidity and exchangeable alkalinity of electrodialyzed Ferrali–Haplic Acrisol (Zhang et al., 1990)

replaced. When the two cation species are compared, it can be found that within the low concentration range the replacing ability of barium ions is stronger than that of potassium ions, whereas the difference in the replacing ability disappears when the concentration is increased, because in this latter case all the exchangeable hydrogen and aluminum ions can be replaced, irrespective of the kind of the cation species used for replacement. It is interesting that the exchangeable acidity of a Xanthic-Haplic Acrisol derived from Quaternary clay in the south part of Guizhou Province of China is negative in sign when the concentration of the neutral salt solution is very low (with a pC value of larger than 3.5), implying that in this case the quantity of the exchangeable alkali released from the soil is larger than that of the exchangeable acid.

11.4 SOIL ACIDIFICATION

The causes for the production of hydrogen ions in soils have been mentioned in Chapter 10. Owing to these causes, under field conditions the process of soil acidification would proceed constantly, even if most of the variable charge soils are already strongly acid in reaction. In recent years there appeared another complicated factor, namely, the effect of acid precipitation coming from the atmosphere. In some regions of south China where the soils are dominated by variable charge soils, the pH of rainwater may be as low as 4.

Variable charge soils generally do not contain primary minerals, except quartz. Therefore, there are only two kinds of components that can have a buffering action against hydrogen ions produced within the soil or entered into the soil from external sources: exchangeable base cations and secondary minerals, including free oxides. For most variable charge soils exchangeable base cations only account for less than 20% of the total adsorbed cations. A representative example shown in Fig. 11.26 indicates that the quantity of exchangeable acid of the original sample of a Ferrali--Haplic Acrisol derived from sandstone-shale in Xishuangbanna region accounts for approximately 85% of that of the soil after electrodialysis. This means that the degree of base-saturation is 15%. Besides, the quantity of the negative surface charge carried by variable charge soils is generally low. These two factors combined together make the buffering effect of exchangeable base ions against hydrogen ions in variable charge soils much weaker than that in constant charge soils. Thus, the principal outcome of hydrogen ions would be to react with the solid phase of the soil, that is, to transform into positive surface charge, to transform into exchangeable acid (mainly aluminum) and to release water-soluble aluminum, as discussed in Chapter 10.

The buffering capacity of various variable charge soils against acid is

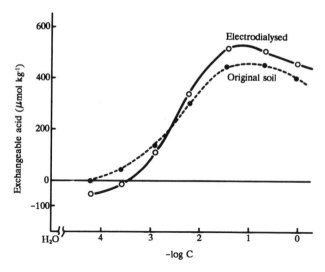

Fig. 11.26. Comparison of exchangeable acidity between electrodialyzed and original Ferrali–Haplic Acrisol (Xishuangbanna) (Zhang et al., 1990).

determined chiefly by clay content and the mineralogical composition of the clay fraction. One of the important causes for the difference in the buffering capacity against hydrogen ions among the three representative variable charge soils shown in Fig. 11.27 is the difference in the mechanical composition. For the Ferrali–Haplic Acrisol derived from granite and containing about 30% of clay, the buffering capacity is the smallest. The Rhodic Ferralsol and the Ali–Haplic Acrisol contain about 65% and 40% of clay, respectively, and the buffering capacity is relatively large, with the former soil larger than the latter soil.

Among the mineral components of the soil, iron oxides play a very important role in the buffering action against hydrogen ions. This is because iron oxides are the chief materials in accepting protons. As shown in Fig. 11.28, the addition of iron oxide coatings to a Ferrali–Haplic Acrisol leads to an increase in the buffering capacity against the hydrogen ions. Conversely, for the glei horizon of a paddy soil derived from Quaternary red clay in which most of the original iron oxides of the red clay have been leached out, the buffering capacity against hydrogen ions is much smaller than that of the illuvial horizon enriched with iron oxides (Fig. 11.29).

In order to characterize the buffering ability of various soils against acid, the concepts of "acid–neutralizing capacity" and "acid–buffering capacity" have been suggested (van Breemen et al., 1983). For the calculation of an acid–neutralizing capacity, it is required to conduct a total analysis for all of the elements. Then, based on a specified "reference pH", the sum of the equivalent of all of the acid–forming elements is subtracted from that of all

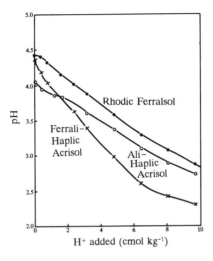

Fig. 11.27. Buffering capacity of three representative variable charge soils against hydrogen ions.

of the alkali–forming elements. If pH 5 is selected as the reference pH, it is only necessary to consider calcium, magnesium, potassium, sodium, manganese, sulfur, and phosphorus. If a reference pH of 3 is selected, aluminum must also be considered. Apparently, in doing so the operation would be very tedious. For practical use, one can refer to the quantity of

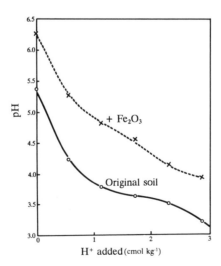

Fig. 11.28. Effect of iron oxides on buffering property of Ferrali–Haplic Acrisol.

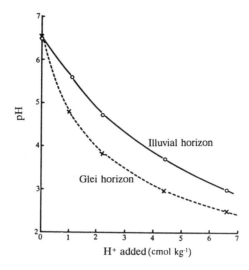

Fig. 11.29. Comparison of buffering property against hydrogen ions between glei horizon and illuviation horizon of a paddy soil derived from Ali-Haplic Acrisol (Yu, 1985a).

acid required to reduce the soil pH to a given level as the acid-buffering capacity of the soil. In this case, one can use the required quantity of acid to reduce the pH by one pH unit as the index, or use the required quantity of acid to reduce the soil pH to a given reference value such as 4 as the index. The former index has the disadvantage that the difference by one pH unit within different pH ranges does not correspond to the same quantity of acid.

The acid-buffering capacity of variable charge soils differs greatly. Based on some available data in China, the capacity may be grouped into four classes. The acid-buffering capacities of grades I, II, III, and IV correspond to < 1, 1-3, 3-5, and > 5 cmol kg^{-1}, respectively. Among various variable charge soils, those belonging to grade I are mostly Ferrali-Haplic Acrisols derived from sandstone. Those belonging to grade II include Ferrali-Haplic Acrisols derived from granite and schist and Ali-Haplic Acrisols derived from Quaternary red clay. Those belonging to grade III include Rhodic Ferralsols derived from basalt and Ferrali-Haplic Acrisols derived from shallow-sea deposits. The acid-buffering capacity of Terra Rossas derived from limestone and some Ferrali-Haplic Acrisols derived from clayey sea deposit may attain the capacity range of grade IV (Wang and Pan, 1992). Generally speaking, the acid-buffering capacity of variable charge soils is small compared to the constant charge soils of temperate regions. The reason for this is that the number of adsorbing sites for cations is small and the degree of base-saturation is low, although the surface of soil particles

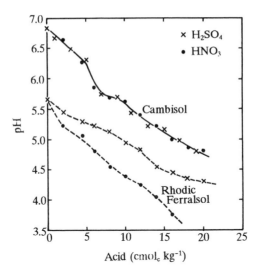

Fig. 11.30. Change in pH after reaction with HNO_3 or H_2SO_4 for a constant charge soil (Cambisol) and a variable charge soil (Rhodic Ferralsol) (Wang and Yu, 1996).

possesses the ability of accepting protons. These factors, together with the fact that most of the variable charge soils are already strongly acid in reaction, make soil acidification a realistic problem in ecology.

In addition to the quantity of acid precipitation, the acidification pattern of a soil is also affected by the anion composition of the rainwater. As has been discussed in Chapter 6, the coordination adsorption of sulfate can lead to the release of hydroxyl ions to the solution. These hydroxyl ions would neutralize a part of the hydrogen ions. Such a reaction occurs only for variable charge soils, but not for constant charge soils such as the Cambisol, as is shown in Fig. 11.30.

The anion composition of rainwater differs greatly. In south China, the SO_4/NO_3 ratio is as large as 5.4-19.6 with an average of 10.9 on an equivalent basis, much larger than the ratio for most acid rain in north Europe and north America where the ratio ranges from 0.8 to 4.9 with an average of 2.8 (Wang and Yu, 1996).

The concurrent occurrence of the predominance of sulfate in rainwater and the wide-spread distribution of variable charge soils in south China induces a practical problem: the release of a considerable amount of hydroxyl ions is of significance in retarding the rate of soil acidification. It was found that this effect varied with the soil type. It was estimated that when acid rain chiefly containing H_2SO_4 was deposited on variable charge soils the acidification rate might be slower by 20-40% than that when the acid rain chiefly contained HNO_3 for soils with a high organic matter content, and that the rate might be half of that caused by HNO_3 with a low organic matter content, especially for Rhodic Ferralsols (Wang and Yu, 1996).

BIBLIOGRAPHY

Aitken, R. L. and Moody, P. W. (1991) *Aust. J. Soil Res.*, 29:483-491.

Baver, L. D. and Scarseth, G. D. (1931) *Soil Sci.*, 31:159-173.

Bloom, P. R., McBride, M. B., and Chadbourne, B. (1977) *Soil Sci. Soc. Am. J.*, 41:1068-1072.

Bradfield, R. J. (1923) *J. Am. Chem. Soc.*, 45:2669-2678.

Brinkman, R. (1970) *Geoderma*, 3:199-206.

Chernov, B. A. (T. R. Yu translator) (1957) *Nature of Soil Acidity*. Science Press, Beijing.

Clark, J. S. and Turner, R. C. (1965) *Soil Sci. Soc. Am. Proc.*, 29:271-274.

Coleman, N. T. and Craig, D. (1961) *Soil Sci.*, 91:14-18.

Coleman, N. T. and Harward, M. E. (1953) *J. Am. Chem. Soc.*, 75:6045-6046.

Coleman, N. T. and Jenny, H. (1950) *Soil Sci. Soc. Am. Proc.*, 15:106-114.

Conyers, M. K. and Davey, B. G. (1988) *Soil Sci.*, 145:29-36.

Coutler, B. S. (1969) *Soils Fert.*, 32:215-223.

Coutler, B. S. and Talibudeen, O. (1968) *J. Soil Sci.*, 19:237-250.

Deshpande, T. L., Greenland, D. J., and Quirk, J. P. (1968) *J. Soil Sci.*, 19:108-122.

Eeckman, J. P. and Laudelout, H. (1961) *Koll. Z.*, 178:99-107.

Foscolos, A. E. (1968) *Soil Sci. Soc. Am. Proc.*, 32:350-354.

Frink, C. R. (1972) in *Acid Sulfate Soils* (H. Dost, ed.). International Institute of Land Reclamation and Improvement, Wageningen, pp. 131-168.

Harward, M. E. and Coleman, N. T. (1954) *Soil Sci.*, 78:181-188.

Helyar, K. R., Conyers, M. K., and Munns, D. N. (1993) *J. Soil Sci.*, 44:317-333.

Hodges, S. C. (1987) *Soil Sci. Soc. Am. J.*, 51:57-64.

Hodges, S. C. and Zelazny, L. W. (1983) *Soil Sci. Soc. Am. J.*, 47:221-225.

Huang, P. M. (1988) *Adv. Soil Sci.*, 8:1-78.

Jackson, M. L. (1963) *Soil Sci. Soc. Am. Proc.*, 27:1-10.

Jenny, H., Nielsen, T. R., Coleman, N. T., and Williams, D. E. (1950) *Science*, 112:164-167.

Jenny, H. (1961) *Soil Sci. Soc. Am. Proc.*, 25:428-432.

Ji, G. L. and Yu, T. R. (1985) *Soil Sci.*, 139:166-171.

Kissel, D. E., Gentzsch, E. P., and Thomas, G. W. (1971) *Soil Sci.*, 111:293-297.

Lindsay, W. L. (1979) *Chemical Equilibria in Soils.* John Wiley & Sons, New York.

Ling, Y. X. and Yu, T. R. (1957) *Acta Pedol. Sinica*, 5:234-244.

Little, I. P. (1992) *Aust. J. Soil Res.*, 30:587-592.

Lundström, U. S. (1993) *J. Soil Sci.*, 44:121-133.

Mattson, S. (1931) *Soil Sci.*, 32:343-365.

Mattson, S. (1932) *Soil Sci.*, 34:209-240.

Mattson, S. and Wiklander, L. (1940) *Soil Sci.*, 49:109-154.

McBride, M. B. and Bloom, P. R. (1977) *Soil Sci. Soc. Am. J.*, 41:1073-1077.

Menzies, N. W., Bell, L. C., and Edwards, D. G. (1994) *Aust. J. Soil Res.*, 32:269-283.

Miller, R. J. (1965) *Soil Sci. Soc. Am. Proc.*, 29:36-39.

Moore, C. S. and Ritchie, G. S. P. (1988) *J. Soil Sci.*, 39:1-8.

Munns, D. N., Helyar, K. R., and Conyers, M. K. (1992) *J. Soil Sci.*, 43:441-446.

Nye, P., Craig, D., Coleman, N. T., and Ragland, J. L. (1961) *Soil Sci. Soc. Am. Proc.*, 25:14-17.

Overbeek, J. Th. G. (1953) *J. Colloid Sci.*, 8:593-605.

Pallman, H. (1930) *Koll. Chem. Beih.*, 30:334-405.

Panicolau, E. P. (1970) *Z. Pflanzenernähr., Dung. Bodenk.*, 126:33-42.

Patterson, E., Goodman, B. A., and Farmer, V. C. (1991) in *Soil Acidity* (B. Ulrich and M. E. Sumner, eds.). Springer-Verlag, Berlin, pp. 97-124.

Paver, H. and Marshall, C. E. (1934) *J. Soc. Chem. Ind.*, 12:750-760.

Reuss, J. O. and Johnson, D. W. (1986) *Acid Deposition and the Acidification of Soils and Waters.* Springer-Verlag, N. Y.

Ritchie, G. S. P. (1989) in *Soil Acidity and Plant Growth* (A. D. Robson, ed.). Academic Press, Sydney, pp. 1-60.

Schwertmann, U. and Jackson, M. L. (1964) *Soil Sci. Soc. Am. Proc.*, 28:179-182.

Singh, S. S. (1982) *Can. J. Soil Sci.*, 62:559-569.

Soon, Y. K. (1993) *Commun. Soil Sci. Plant Anal.*, 24:1683-1708.

Thomas, G. W. (1988) *Commun. Soil Sci. Plant Anal.*, 19:833-856.

Thomas, G. W. and Hargrove, W. L. (1984) in *Soil Acidity and Liming* (F. Adams, ed.). American Society of Agronomy, Madison, WI, pp. 3-56.

Ulrich, B. (1986) *Z. Pflanzenernähr. Bodenk.*, 149:702-717.

van Breemen, N., Mulder, J., and Driscoll, C. T. (1983) *Plant and Soil*, 75:283-308.

Veitch, T. B. (1904) *J. Am. Chem. Soc.*, 26:637-662.

Veith, J. A. (1977) *Soil Sci. Soc. Am. J.*, 41:865-870.

Veith, J. A. (1978) *Clays Clay Miner.*, 26:45-50.

Volk, V. V. and Jackson, M. L. (1964) *Clays Clay Miner.*, 12:281-285.

Wang, J. H. and Yu, T. R. (1976) in *Electrochemical Properties of Soils and*

Their Research Methods (T. R. Yu, ed.), Science Press, Beijing, pp. 325-353.

Wang, J. H. and Yu, T. R. (1986) *Z. Pflanzenernähr. Bodenk.*, 149:598-607.

Wang, J. H. and Pan, X. Z. (1992) in *Proceedings of the International Symposium on Managagement and Development of Red Soils in the Asia and Pacific Regions*, Nanjing.

Wang, J. H. and Yu, T. R. (1996) *Pedosphere*, 6:11-23.

Wiegner, G. and Pallman, H. (1930) *Z. Pflanzenernähr. Dung. Bodenk.*, A16:1-57.

Wiklander, L. (1955) in *Chemistry of the Soil* (F. E. Bear, ed.). Reinhold, New York, pp. 107-148.

Xuan, J. X. and Yu, T. R. (1964) *Acta Pedol. Sinica*, 12:307-319.

Yu, T. R. (1976) *Electrochemical Properties of Soils and Their Research Methods*, Science Press, Beijing.

Yu, T. R. (1985a) *Physical Chemistry of Paddy Soils*. Science Press/Springer Verlag, Beijing/Berlin, pp. 131-156.

Yu, T. R. (1985b) *Ion-Selective Electrode Rev.*, 7:165-202.

Yu, T. R. (1987) in *Principles of Soil Chemistry* (T. R. Yu, ed.). Science Press, Beijing, pp. 325-364.

Yu, T. R. (1988) in *Soil Analytical Chemistry* (T. R. Yu and Z. Q., Wang, eds.). Science Press, Beijing, pp. 224-262.

Yu, T. R. (1990) in *Chemical Processes in Soil Genesis* (T. R. Yu and Z. C. Chen, eds.). Science Press, Beijing, pp. 96-132.

Yu, T. R. (1991) in *Plant-Soil Interactions at Low pH"* (R. J. Wright et al., eds.). Kluwer Academic, Neitherlands, pp. 107-112.

Yu, T. R. and Zhang, X. N. (1985) in *Proceedings of the International Symposium on Red Soils* (Institute of Soil Science, ed.). Science Press /Elsevier, Beijing/Amsterdam, pp. 409-441.

Yu, T. R., Beyme, B., and Richter, J. (1989) *Z. Pflanzenernähr. Bodenk.*, 152:359-365.

Yuan, T. L. (1963) *Soil Sci.*, 95:155-163.

Zhang, H., Kong, X. L. and Ji, G. L. (1990) *Soils*, no. 1, 43-47.

Zhang, X. N., Zhang, G. Y., Zhao, A. Z., and Yu, T. R. (1989) *Geoderma*, 44:275-286.

Zhang, X. N., Zhao, A. Z., Zhang, G. Y., and Zhang, H. (1990) *Acta Pedol. Sinica*, 27:270-279.

Zhao, A. Z. and Zhang, X. N. (1989) *Soils*, 21:5-9.

12

LIME POTENTIAL

J. H. Wang

The properties of hydrogen and aluminum ions have been examined in Chapters 10 and 11. These two ion species are ions that directly induce the acid reaction in soils. In soils devoid of soluble salts, the content of cations is constant and the negative surface charges are saturated by, besides hydrogen and aluminum ions, alkali metal and alkaline earth metal ions. These ions are called *base ions*. The acidity of a soil is determined chiefly by the ratio of the quantity of hydrogen and aluminum ions to that of base ions. Among these base ions, calcium ions occupy the most important position, because they generally account for 65–80% of the total amount of base ions in variable charge soils. Therefore, calcium is an ion species closely related to the acidity of soils. In addition to the parameter pH that directly reflects the concentration of hydrogen ions, one other desirable way is to find a parameter that can reflect the ratio of the hydrogen ions to the calcium ions. This parameter is the lime potential.

Since the introduction of the concept of lime potential 40 years ago, little practical application has been made in soil science, although some further theoretical considerations were advanced in the 1950s and the 1960s. Actually, as shall be seen in this chapter, for strongly acid soils, such as variable charge soils, because the quantity of hydrogen ions is too high and at the same time the quantity of calcium ions is too low, lime potential that can reflect the relative ratio of these two ion species is of significance not only in theory but also in practice.

12.1 SIGNIFICANCE OF LIME POTENTIAL

12.1.1 Definition

The mathematical expression of lime potential is pH–0.5pCa. Lime potential is a simple function of the chemical potential of calcium hydroxide, lime. Hence it may be called *lime potential*. The physical meaning of pH–0.5pCa can be derived as follows.

When both the numerator and the denominator of the activity ratio of the calcium ions to the hydrogen ions $(a_{Ca})^{1/2}/a_H$ are multiplied by a_{OH}, we can get

$$\frac{a_{Ca}^{1/2}}{a_H} = \frac{a_{Ca}^{1/2} \cdot a_{OH}}{a_H \cdot a_{OH}} \qquad (12\text{-}1)$$

Since the ionic product of water $a_H \cdot a_{OH}$ at 20°C is $10^{-14.2}$, when equation (12-1) is expressed in the logarithmic form, the following equation can be obtained:

$$pH - 0.5pCa = 14.2 - 0.5pCa - pOH$$
$$= 0.5 \log a_{Ca(OH)_2} + 14.2 \qquad (12\text{-}2)$$

According to the definition of chemical potential in physical chemistry, the chemical potential of $Ca(OH)_2$, $\mu_{Ca(OH)_2}$ is related to the activity as

$$\mu_{Ca(OH)_2} = \mu^0_{Ca(OH)_2} + 2.303\,RT \log a_{Ca(OH)_2} \qquad (12\text{-}3)$$

where $\mu^0_{Ca(OH)_2}$ is the chemical potential of $Ca(OH)_2$ at standard state.

Therefore, pH–0.5pCa is a simple function of the chemical potential of $Ca(OH)_2$.

The above derivation applies to aqueous solution. In soils, because when in equilibrium with the solution the chemical potential of one component of the solution equals that at the surface of soil particles, pH–0.5pCa is also a function of the chemical potential of calcium hydroxide at the surface of the solid phase.

The concept of the lime potential was suggested by Aslyng (1954) and by Schofield and Taylor (1955a). Later, Turner and colleagues conducted an extensive study on factors affecting the lime potential, using clay minerals and constant charge soils (Nichol and Turner, 1957; Turner, 1965a,b; Turner and Clark, 1965; Turner and Nichol, 1958, 1962a,b; Turner et al, 1963). Ulrich (1961a) regarded the lime potential as one kind of "nutrient potential." The introduction of the nutrient potential concept, including phosphate potential, was based upon the following considerations (Schofield, 1955; Schofield and Taylor, 1955a,b): (A) The availability of a nutrient in a soil is determined chiefly by the chemical potential and its declining rate of that nutrient. A soil may behave like a water pool, and the work that must be done when a nutrient is taken by plants is determined by the level of the chemical potential of that nutrient. Hence, the nutrient potential is an intensity parameter for expressing the availability of nutrients. (B) In theory, the activity of a single ion species in solution cannot be determined accurately, whereas the determination of the mean activity of a molecule entity such as $Ca(OH)_2$ is unambiguous in thermodynamics. (C) Under conditions that the "ratio law" can hold, the ratio of two ion species in soil solution is not affected by the content of water or the addition of an extraneous salt. Therefore, the nutrient potential is a desirable concept when considered from both theoretical and practical viewpoints.

12.1.2 In Relation to Soil Acidity

In the last chapter it was mentioned that for a long time it has been customary to take pH as the intensity parameter of soil acidity. What is the relationship between the lime potential, another intensity parameter, and soil acidity?

As in the potentiometric titration with NaOH, one can titrate an H,Al-saturated soil with $Ca(OH)_2$ and observe the changes in pH, pCa, and pH–0.5pCa. An example of such a titration is shown in Fig. 12.1. In the figure, the pH curve is similar to the pH curve when titrated with NaOH shown in Chapter 11, with the exception that the inflection point lies at about pH 6.6, lower than the NaOH titration curve. The trend in change of

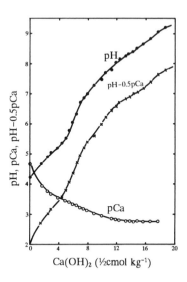

Fig. 12.1. Changes in pH, pCa, and pH–0.5pCa for Ali-Haplic Acrisol titrated with $Ca(OH)_2$ (Wang and Yu, 1981).

pCa is opposite to that of pH, that is, the pCa value decreases continuously with the increase in amount of $Ca(OH)_2$ added. There also appears an inflection point on the curve, although the inflection is not as distinct as that on the pH curve.

pH-0.5pCa is a comprehensive reflection of the pH and pCa. Because the latter two parameters change in opposite directions, the increase in pH-0.5pCa with the increase in amount of $Ca(OH)_2$ is more pronounced than that in the pH. On the curve in Fig. 12.1 three interesting points can be distinguished. Before the addition of $Ca(OH)_2$, the pH, pCa and pH-0.5pCa are 4.3, 4.6, and 2.0. At a pH of 7, the pCa and pH-0.5pCa are 3.2 and 5.4. At a pH-0.5pCa of 7, the pH and pCa are 8.4 and 2.8. These three values for the three parameters cover the variation ranges of acidity for variable charge soils and the majority of other types of soil. Within these ranges the magnitudes of variation of the pH, pCa, and pH-0.5pCa are 4.1, 1.8, and 5.0 units. However, the absolute value of these parameters does not have universal significance because, as shall be seen in the next section, the numerical value of the three threshold points may vary with the type of the soil and other factors.

Let us examine the relationship between pH-0.5pCa and pH within the whole acidity range. If the ratio ΔpCa to ΔpH caused by the increase in a certain amount of $Ca(OH)_2$ remains constant within the whole titration range, a straight line should be obtained when pH-0.5pCa is plotted against the pH. Actually, the situation is not so simple. As shown in Figs. 12.2 and 12.3, within the lower pH-0.5pCa range the curve is slightly convex and

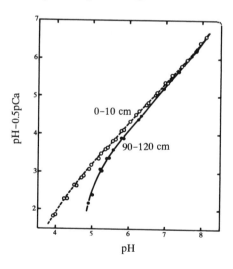

Fig. 12.2. pH-0.5pCa in relation to pH for Rhodic Ferralsol (Australia) (Wang and Yu, 1986).

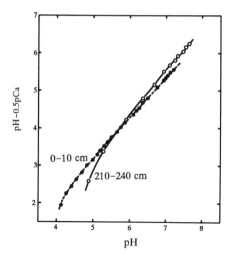

Fig. 12.3. pH–0.5pCa in relation to pH for Ferric Acrisol (Australia) (Wang and Yu, 1986).

within the higher pH–0.5pCa range slightly concave. The line is nearly straight only within the middle range. This tendency is more conspicuous for the subsoil of variable charge soils with little organic matter. The reason for these different trends in pH and pH–0.5pCa curves is that the magnitudes of change in pH and pCa with the increase in amount of alkali added may differ within different pH–0.5pCa ranges. Within the lower range, because the quantity of calcium ions present is small, the variation of pCa value will be large when expressed in a logarithmic form. On the contrary, within the higher range, the variation of pH value will be large.

Despite the complexities in the relationship between pH–0.5pCa and pH, there is a close correlation between these two indexes of soil acidity. Figure 12.4 shows this correlation for 250 pairs of data, although the correlation is not strictly linear.

From the above discussions it can be concluded that lime potential may be used as an intensity index of soil acidity. Lime potential reflects both the hydrogen ion status and the calcium ion status of a soil. The comprehensive reflection of the status of these two ion species essentially represents the base-saturation status of a soil (Turner and Clark, 1965; Turner and Nichol, 1962a,b; Webster and Harward, 1951). When characterizing the acidity of different types of soils, as expected from theoretical discussions, the difference in lime potential would be more remarkable than the difference in pH. Some examples in this respect are given in Table 12.1.

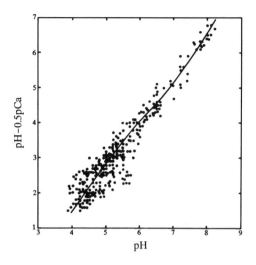

Fig. 12.4. Correlation between pH–0.5pCa and pH for variable charge soils (Wang and Yu, 1986).

12.1.3 Principles of Determination

For the determination of lime potential of soils, there are three methods available for selection.

In one method the pH is determined by the usual procedure, and the concentrations of calcium and all other ions in the solution are determined by chemical methods. After computing the total ionic strength, the activity coefficient and thus the activity of calcium ions are calculated (Schofield and Taylor, 1955a). This has been the generally adopted method in soil

Table 12.1 Comparison of pH and pH–0.5pCa Between Paddy Soils and
Their Parent Soils (Yu, 1985)

Soil type	pH[a]			pH–0.5pCa		
	Paddy	Parent	Δ	Paddy	Parent	Δ
Rhodic Ferralsol	5.23	5.12	0.11	3.40	2.29	1.11
Ali-Haplic Acrisol	6.56	5.15	1.41	4.93	3.02	1.91
Cambisol	6.83	5.71	1.12	5.32	3.91	1.41

[a]Soil:water = 1:1.

science for a long time. Besides tediousness in its operation, there are two important sources of error inherent in this method. It has been shown in Chapter 11 that in the determination of pH a liquid-junction potential between the reference electrode and the soil system is involved. In physical chemistry, the activity coefficient of a single ion species cannot be calculated accurately. This is particularly true for polyvalent calcium ions, and especially when the ionic strength of the solution is high.

Following the advancement in the technology of ion-selective electrodes, some Russian authors, such as Krupsky et al. (1975), determined the pH and pCa separately with the electrochemical method and then calculated the pH–0.5pCa. In this case the liquid-junction potential in the two determinations still cannot be avoided. In principle, if the numeral values of the two liquid-junction potentials are equal to each other, these two potentials can be automatically canceled in the calculation. However, in practice, because liquid-junction potential is a time-dependent variable, particularly in such a complex system as the soil, it would not be possible to entirely eliminate the measurement error caused by liquid-junction potential.

Another feasible method would be to determine the lime potential directly by the use of a measuring cell consisting of two ion-selective electrodes (Wang and Yu, 1981).

In this case, because in a complex soil system the ratio of the activity of hydrogen ions to that of calcium ions is neither an integer nor a constant, it is not possible to use a calibration curve method with a series of standard solutions with known pH or pCa. Another means must be sought.

According to the Nernst equation we have

$$E_{Ca} = E_{Ca}^0 + S_{Ca} \log a_{Ca} + E_j \tag{12-4}$$

$$E_H = E_H^0 + S_H \log a_H + E_j \tag{12-5}$$

where E_{Ca} and E_H are the electrode potentials of the calcium ion-selective electrode and the hydrogen ion-selective electrode, respectively, E_{Ca}^0 and E_H^0 are their standard potentials, and E_j is the liquid-junction potential. S_{Ca} and S_H are constants at a given temperature.

When subtracting equation (12-5) from equation (12-4), we can get

$$\begin{aligned} E_{Ca} - E_H &= (E_{Ca}^0 - E_H^0) + S_{Ca} \log a_{Ca} - S_H \log a_H \\ &= (E_{Ca}^0 - E_H^0) + S_H pH - S_{Ca} pCa \end{aligned} \tag{12-6}$$

or

$$pCa = \frac{S_H}{S_{Ca}} pH - \frac{(E_{Ca} - E_H) - (E_{Ca}^0 - E_H^0)}{S_{Ca}} \qquad (12\text{-}7)$$

$$pH - 0.5pCa = \frac{(E_{Ca} - E_H) - (E_{Ca}^0 - E_H^0)}{2S_{Ca}}$$
$$+ pH \left(1 - \frac{S_H}{2S_{Ca}}\right) \qquad (12\text{-}8a)$$

$$pH - 0.5pCa = \frac{(E_{Ca} - E_H) - (E_{Ca}^0 - E_H^0)}{S_H}$$
$$+ 0.5pCa \left(\frac{2S_{Ca}}{S_H} - 1\right) \qquad (12\text{-}8b)$$

Superficially, equations (12-8a) and (12-8b) are rather complex. Actually, they are quite simple to use. $(E_{Ca} - E_H)$ is the potential difference between the calcium ion-selective electrode and the hydrogen ion-selective electrode, which can be measured directly with an mV-meter. E_{Ca}^0 and E_H^0 are constant for a given electrode and can be obtained by an extrapolating method, in which the electrode potentials in a series of standard solutions with known pH or pCa are measured, and then the respective values are obtained by extrapolating to pH=0 or pCa=0 graphically or by calculation. The ratio of S_H to $2S_{Ca}$ is close to 1 for electrodes with a normal function. Therefore, the second term on the right side of the equation is close to zero. S_H and S_{Ca} can be calculated from the calibration curve. Thus, when lime potential of a soil is determined, only the potential difference between the two electrodes inserted in the soil system needs to be measured. In the case where the function of one of the electrodes is not perfect, a correction for the second term of the right side of the equation can be made.

All the pH-0.5pCa data cited in this chapter were obtained by this method.

12.2 FACTORS AFFECTING LIME POTENTIAL

Since the numerical value of lime potential is determined by the relative magnitude of the two parameters pH and 0.5pCa, all factors that can affect pH and pCa as well as their interrelations will affect the lime potential value. Factors that can affect soil pH have been discussed in the previous chapter. In this chapter, factors that affect pCa value and the relationship between pH and pCa will be examined. These factors include the composition of the solid phase of the soil, the composition of adsorbed ions, and environmental conditions. As shall be seen, variable charge soils frequently behave differently from constant charge soils in this respect.

12.2.1 Concentration of Ions and Soil-to-Solution Ratio

Schofield suggested that under certain conditions the lime potential of a soil is not affected by the soil-to-solution ratio or the concentration of ions in the system. This suggestion is based on the "ratio law," a law derived from the Gouy theory of the electrical double layer (Schofield, 1947). The ratio law states that when the cations in solution are in equilibrium with a large number of exchangeable ions, a change in the concentration of the solution will not disturb the equilibrium if (a) the concentrations (activities) of all monovalent ions are changed in one ratio, (b) those of all divalent ions are changed in the square of that ratio, and (c) those of all trivalent ions are changed in the cube of that ratio. For the relationship between monovalent hydrogen ions and divalent calcium ions, when the soil suspension is diluted with water or when the solution concentration is changed by the addition of a neutral salt, the magnitudes in change of the two ion species will fulfill the above requirement (Schofield, 1947; Yu, 1976), and therefore the pH-0.5pCa will not change.

The establishment of the "ratio law" requires two essential conditions: (A) The quantity of exchangeable cations is much larger than that in the solution; (B) the soil only carries negative charges, or the negative charge sites are much more than the positive charge sites, so that negative surface charges are chiefly compensated for by the accumulation of cations. For clay minerals and constant charge soils, when the concentration of ions in the solution is not high, the above two conditions can essentially be fulfilled. This point has been verified by Schofield and Taylor (1955a), Larsen (1965), and Turner and Nichol (1962a).

Now, let us examine the characteristic features of variable charge soils in this respect. On the surface of variable charge soils there are both negative charge sites and positive charge sites. Therefore, the soil can have both adsorption and repulsion with cations. There is also the adsorption of anions. As a consequence, the distribution of cations between the electric double layer and the free solution could be not as distinct as in the case for constant charge soils, and the redistribution patterns of hydrogen and calcium ions when the concentration of ions in the solution is changed would be not as simple as is predicated by the Donnan theory. The result of these complexities is that the concentration range within which the "ratio law" can apply is narrower than that for constant charge soils. This point has been demonstrated by Schofield (1947) in his original paper. Apparently, the larger the ratio of positive surface charge sites to negative surface charge sites, the lower the upper limit of concentration range that the law could apply. According to the data of Sumner and Marques (1968), for soils having high contents of iron and aluminum oxides, when the quantity of positive surface charge was less than 20% of negative surface charge and the electrolyte concentration of the solution was lower than 10^{-2} mol L^{-1}, the "ratio law" held.

Variable charge soils generally carry less negative surface charges than do constant charge soils. Hence, the upper limit of electrolyte concentration at which pH-0.5pCa is independent of the concentration of electrolyte would be lower than that of constant charge soils.

For the same variable charge soil, the lower the pH, the larger the ratio of positive surface charge sites to negative surface charge sites, and hence the lower the threshold electrolyte concentration at which the "ratio law" can hold.

For the same reason, the soil-to-solution ratio range within which the pH-0.5pCa value remains independent of this ratio for variable charge soils would be narrower than that for constant charge soils.

All of the above-mentioned points can be illustrated by the data given in Table 12.2, in which the effects of soil-to-solution ratio and $CaCl_2$ concentration on pH and pH-0.5pCa for two types of variable charge upland soils and paddy soils are shown. In the table, the relevant data for a paddy soil developed on Cambisol are also given for comparison.

It can be seen from the table that, for the constant charge soil Cambisol, within the soil-to-solution ratio range of 1:0.4 to 1:5 and the $CaCl_2$ concentration range of 10^{-4} M to 10^{-2} M, the patterns of change in the distribution of hydrogen ions and calcium ions between the solid phase and the liquid phase are in agreement with those predicted by the Gouy theory, and therefore the pH-0.5pCa remains constant. The same is basically true for the paddy soil derived from Ali-Haplic Acrisol. For the upland Ali-Haplic Acrisol, the pH-0.5pCa value is markedly higher in 10^{-2} M $CaCl_2$ solution than in dilute solution, and the larger the solution-to-soil ratio, the more remarkable this difference. The difference in behavior between these two types of Ali-Haplic Acrisols is caused by two reasons. The pH of the paddy soil is higher by more than 1 pH unit than that of the upland soil. Besides, owing to the reductive eluviation in the paddy soil, the content of iron oxides is lower than that of the upland soil (Yu, 1985). These two factors make the paddy soil independent of the soil-to-solution ratio and the electrolyte concentration of the solution under certain conditions, whereas for the upland soil the presence of a considerable proportion of positive surface charge makes the "ratio law" invalid under the same conditions. For the Ferrali-Haplic Acrisol with a larger positive surface charge to negative surface charge ratio, lime potential is affected by the electrolyte concentration to a certain extent even for the paddy soil. For the upland soil the effect is even more pronounced.

Table 12.2 Effects of Solution-to-Soil Ratio and $CaCl_2$ Concentration on pH and pH-0.5pCa (Wang and Yu, 1981)

Soil	Utilization	Solution :soil	pH 10^{-4a}	pH 10^{-3}	pH 10^{-2}	pH-0.5pCa 10^{-4}	pH-0.5pCa 10^{-3}	pH-0.5pCa 10^{-2}
Cambisol	Paddy	0.4	6.73	6.78	6.52	5.22	5.24	5.23
		0.5	6.79	6.80	6.53	5.27	5.26	5.21
		0.75	6.81	6.85	6.52	5.30	5.25	5.22
		1	6.85	6.81	6.51	5.24	5.25	5.22
		2.5	6.96	6.94	6.53	5.21	5.28	5.22
		5	7.06	6.93	6.53	5.24	5.24	5.24
Ali-Haplic Acrisol	Upland	0.4	5.03	5.00	4.63	2.98	3.00	3.10
		0.5	5.08	4.97	4.60	2.99	2.98	3.14
		0.75	5.06	4.85	4.52	2.96	3.01	3.14
		1	5.11	4.95	4.56	2.99	3.00	3.19
		2.5	5.22	5.01	4.62	3.02	3.05	3.23
		5	5.26	4.97	4.69	3.00	3.04	3.28
	Paddy	0.4	6.44	6.41	6.21	4.91	4.90	4.94
		0.75	6.51	6.36	6.21	4.95	4.81	4.93
		1	6.58	6.46	6.21	4.92	4.87	4.94
		2.5	6.70	6.46	6.23	5.00	4.93	4.93
		5	6.72	6.51	6.25	4.94	4.94	4.99
Ferrali-Haplic Acrisol	Upland	0.4	5.40	5.14	4.45	2.46	2.68	2.99
		0.5	5.26	5.02	4.42	2.33	2.51	3.00
		0.75	5.00	4.91	4.35	2.43	2.46	2.97
		1	4.91	4.83	4.26	2.42	2.48	3.00
		2.5	4.84	4.73	4.27	2.46	2.65	3.07
		5	4.83	4.69	4.30	2.47	2.69	3.11
	Paddy	0.4	4.97	4.92	4.61	3.27	3.34	3.42
		0.5	5.02	4.90	4.61	3.30	3.30	3.43
		0.75	5.03	4.94	4.58	3.28	3.32	3.43
		1	5.02	4.95	4.55	3.28	3.32	3.42
		2.5	5.11	4.97	4.55	3.26	3.32	3.45
		5	5.22	5.00	4.58	3.30	3.33	3.46

[a]Mol L^{-1} of $CaCl_2$ solution.

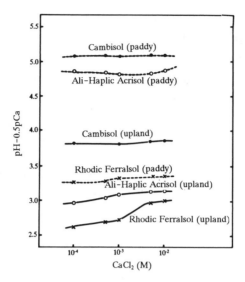

Fig. 12.5. Effect of CaCl$_2$ concentration on pH–0.5pCa of soils (Wang and Yu, 1981).

Apparently, for the same soil, the higher the ion concentration of the solution, the greater its effect on pH–0.5pCa, as is clearly seen in Fig. 12.5.

The lime potential changes readily with the change in solution concentration for variable charge soils. This is related to their characteristic feature that the quantity of the negative surface charge is generally small. When the ion concentration of the system is increased to exceed a certain limit, the amount of free electrolyte in the solution increases markedly. In this case, the increment of calcium ions over that of hydrogen ions (actually aluminum ions) is larger than that predicated by the Donnan principle, with the result that 0.5pCa and pH do not change in a synchronous way. This is the reason for the increase in pH–0.5pCa with the increase in CaCl$_2$ concentration, as shown in Fig. 12.6.

Since most of the surface charges of soils are concentrated in the fine particle fraction, it can be expected that, at the same electrolyte concentration, the finer the soil particles, the less the effect of electrolyte on pH–0.5pCa. In practice, this is just the case (Fig. 12.7). Thus, for the same type of variable charge soil, clay soils would be less affected by solution concentration than sandy soils.

From the above discussions it can be concluded that the lime potential in variable charge soils is more apt to be affected by electrolyte concentration and soil–to–solution ratio than that in constant charge soils, and that the extent of this effect varies with soil type. This point is one of the important differences between variable charge soils and constant charge

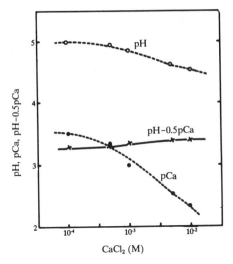

Fig. 12.6. Effect of CaCl$_2$ concentration on pH, pCa and pH-0.5pCa for a paddy soil derived from Ferrali-Haplic Acrisol (Yu, 1985).

soils.

In the determination of lime potential, the generally employed method was to use 0.01 M CaCl$_2$ solution as the extractant (Conyers et al., 1991; Schofield and Taylor, 1955a) and to use a solution-to-soil ratio of 5:1. The adoption of such a condition was based chiefly on the consideration in analytical technology. The use of 0.01 M CaCl$_2$ solution was based on the assumption that such a concentration is approximately close to the

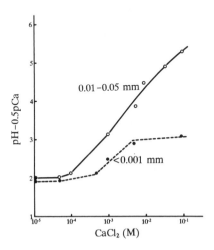

Fig. 12.7. Effect of particle size of soil on pH-0.5pCa (Rhodic Ferralsol).

electrolyte concentration of soil solutions of temperate regions. For constant charge soils, such experimental conditions may be plausible in some respect. For variable charge soils, on the other hand, as has been discussed in this section, such a solution concentration and such a large solution volume would be inappropriate, because this condition differs too much from the natural condition of the soil. Instead, a more reasonable way could be to use water as the extractant with the smallest water–to–soil ratio, so that the experimental condition is close to the natural condition of the soil. If the determination is carried out with two ion-selective electrodes as described in Section 12.1.3, such experimental conditions would be feasible. The data presented in Section 12.3 of this chapter were obtained by the use of such a method.

In spite of the effect of the solution concentration and water content on lime potential of variable charge soils under certain conditions, the effect may not be large, provided that the concentration of electrolytes is not very high, as is seen in Table 12.2 and Figs. 12.5 and 12.6. In particular, the effect is invariably smaller than the effect on pH. This is one of the advantageous points when lime potential is used as an index of soil acidity as compared to pH.

12.2.2 Composition of Soil Solid Phase

For the same composition of cations, the dissociation of both hydrogen (aluminum) ions and calcium ions from soils of different compositions of the solid phase may differ markedly. This would affect the lime potential. The situation with hydrogen ions has been discussed in Chapter 11. In this section the dissociation of calcium ions and its effect on lime potential will be examined.

It has already been discussed in Chapter 3 that the degree of dissociation of adsorbed alkali metal ions from soils is generally of the order Rhodic Ferralsol > Ferrali-Haplic Acrisol > Ali-Haplic Acrisol, and that the degree of dissociation of these ions in variable charge soils is larger than that in constant charge soils, such as Cambisol. For divalent calcium ions, the degree of dissociation is also of the order Rhodic Ferralsol > Ali-Haplic Acrisol > Cambisol (Wang and Yu, 1981). One example in this respect is shown in Fig. 12.8.

When $Ca(OH)_2$ is added to a soil, the magnitude of two parameters will change. The quantity of calcium ions increases and the pH becomes higher. For constant charge soils the effect of the second factor on the equilibrium adsorption-dissociation of calcium ions is comparatively small. In contrast, for variable charge soils–particularly those with a very great variability in surface charge–the increase in negative surface charge caused by the rise in pH may have a marked effect on the dissociation of adsorbed calcium

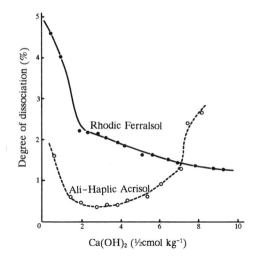

Fig. 12.8. Degree of dissociation of calcium ions in Ali-Haplic Acrisol and Rhodic Ferralsol (Wang and Yu, 1986).

ions. The different trends in the change of dissociation of calcium ions with the change in amount of $Ca(OH)_2$ added for the two types of soils shown in Fig. 12.8 are an overall reflection of these complex factors.

For the Ali-Haplic Acrisol, when the amount of $Ca(OH)_2$ added is small, the strong competition of aluminum ions for negative surface charge sites at low pH increases the dissociation of calcium ions. When the amount of $Ca(OH)_2$ is increased to a certain extent, owing to the presence of a large amount of calcium ions, the degree of dissociation of adsorbed calcium ions increases markedly, caused by a cation-saturation effect. The whole dissociation curve is in a concave form, similar to the curve for the degree of dissociation of monovalent cations.

The dissociation pattern of calcium ions in the Rhodic Ferralsol is different from that in Ali-Haplic Acrisol. As seen in Fig. 12.8, the degree of dissociation in the whole curve decreases steadily with the increase in amount of added $Ca(OH)_2$. The cation-saturation effect does not appear even when the quantity of calcium ions in the system is quite high. Apparently, for this soil with a rather great variability in surface charge, the increase in negative surface charge caused by the rise in pH plays the dominant role in determining the equilibrium adsorption-dissociation of calcium ions. Such a phenomenon for monovalent cations has already been discussed in Chapter 3. Here, for calcium ions, this phenomenon is even more pronounced.

The phenomenon that the degree of dissociation of adsorbed calcium ions decreases steadily with the increase in amount of $Ca(OH)_2$ added is

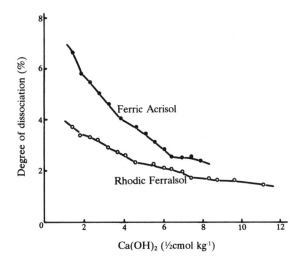

Fig. 12.9. Degree of dissociation of calcium ions in Rhodic Ferralsol (Australia) and Ferric Acrisol (Australia) (Wang, 1986).

quite representative for soils with a great variability in surface charge. As seen in Fig. 12.9, such phenomenon also appears for two Australian highly weathered variable charge soils. Such phenomenon has not been reported in the literature for constant charge soils and clay minerals.

After the addition of $Ca(OH)_2$ to the soil, the change in amount of calcium ions in solution is smaller in magnitude than that of hydrogen ions. Besides, lime potential is related to pH and $0.5pCa$, but not to pH and pCa. Therefore, the changing pattern of $pH-0.5pCa$ is determined chiefly by the change in pH. Thus, at the same cation-saturation status the lime potential for the three types of variable charge soils should be of the order Rhodic Ferralsol > Ferrali-Haplic Acrisol > Ali-Haplic Acrisol, because in this case the pH of the Rhodic Ferralsol is the highest and Ali-Haplic Acrisol the lowest (cf. Chapter 11). Figure 12.10 shows such a comparison. Here again this reflects the role of the composition of soil solid phase in affecting lime potential of the soil.

In addition to clay minerals, iron oxides and organic matter are the other two important components of variable charge soils in affecting the surface charge properties. These should have some effects on lime potential.

Iron oxides may affect lime potential in three ways. It has been shown in the last chapter that iron oxides are the principal basoids of the soil and can cause the ultimate pH and half-neutralization pH high. The presence of iron oxides lowers the net negative surface charge density. This would be favorable to the dissociation of adsorbed calcium ions, resulting in the decrease of pCa value. These two factors make the lime potential high.

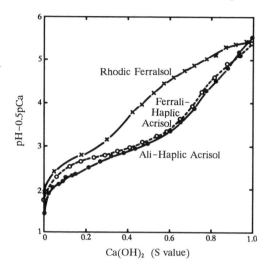

Fig. 12.10. Comparison of lime potential among three variable charge soils (one symmetry value corresponds to one apparent CEC) (Wang and Yu, 1986).

Besides, the surface hydroxyl groups of iron oxides can adsorb hydrogen ions and can also release hydrogen ions, thus possessing a buffering effect under both acid and alkaline conditions.

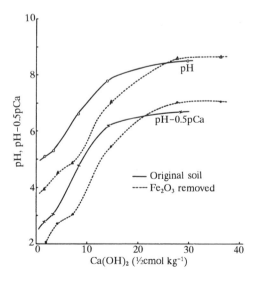

Fig. 12.11. Change in pH and pH–0.5pCa after removal of iron oxides for Xanthic-Haplic Acrisol (Pingba).

The change in pH and lime potential for a Xanthic-Haplic Acrisol after the removal of iron oxides is shown in Fig. 12.11. Within the whole pH–0.5pCa range lower than about 6.5, both pH and lime potential decreased after the removal of iron oxides. This should be proof of the functioning of iron oxides in increasing lime potential. However, when the pH–0.5pCa was higher than 6.5, the reverse became true. This implies that at this time the buffering action of iron oxides against hydroxyl ions gradually plays a more important role. Based upon the same reasoning, it can be expected that after the addition of iron oxide coatings to a soil lime potential should increase within the lower pH–0.5pCa range and decrease within the higher pH–0.5pCa range. Figure 12.12 shows such a situation for a Ferrali-Haplic Acrisol.

With regard to the role of organic matter, it can decrease the pH of a soil and can also decrease the degree of dissociation of adsorbed calcium ions (Wang, 1986). All these two factors lower the pH–0.5pCa of the soil. Since the quantity of negative surface charge sites carried by the mineral part of variable charge soils is generally smaller as compared to constant charge soils, a small difference in content of organic matter may manifest its effect on lime potential appreciably. As can be seen from Figs. 12.13 and 12.14, for the Rhodic Ferralsol containing 0.75% of organic matter, lime potential increased markedly after the removal of organic matter and decreased remarkably when the organic matter content was increased to 1.23% through incubation with green manure.

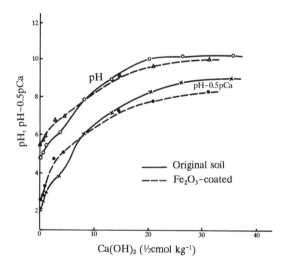

Fig. 12.12. Effect of iron oxide coating on lime potential of Ferrali-Haplic Acrisol.

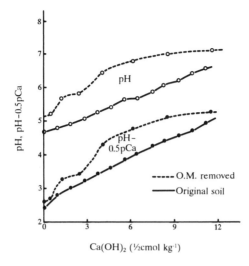

Fig. 12.13. Change in lime potential after removal of organic matter from Rhodic Ferralsol.

The organic matter content of the subsoil of variable charge soils is generally very low. On the other hand, due to the influence of vegetation, the surface layer may have a high content of organic matter. This difference in organic matter content between the two layers may be reflected by lime potential, as is shown in Fig. 12.15 for two soils. The organic matter

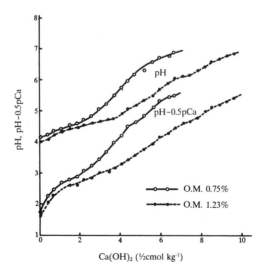

Fig. 12.14. Effect of organic matter on lime potential of Rhodic Ferralsol.

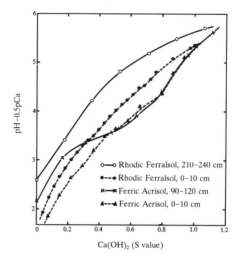

Fig. 12.15. Comparison of lime potential between surface layer and substratum (Wang and Yu, 1986).

contents of the 0- to 10-cm and 210- to 240-cm layers of the Rhodic Ferralsol are 10.9% and 0.4%, respectively, and those of the Ferrali-Haplic Acrisol are 5.0% and 0.5%, respectively. As seen in the figure, the lime potential of the two surface layers are all lower than that of the lower layers.

From the above discussions it can be concluded that the composition of the solid phase of variable charge soils may exert an important effect on lime potential. Among the components of the solid phase, clay minerals, iron oxides, and organic matter play different roles. These roles manifest themselves chiefly through affecting surface charge properties of the soil.

12.2.3 Composition of Cations

For simple systems composed chiefly of hydrogen (aluminum) ions and calcium ions, the lime potential of a soil is determined principally by the ratio of the two ion species. The change in $pH-0.5pCa$ when a soil is titrated with $Ca(OH)_2$ is a reflection of the change in this ratio.

For soils under field conditions, in addition to calcium ions, there is also the presence of some other base cations. These cations may affect the equilibrium adsorption-dissociation of calcium ions. Besides, their effect on pH of the soil may be different from that of calcium ions. Therefore, the presence of these ions may affect the lime potential of a soil. Among these cations, magnesium occupies an important position, because, except for salinized soils, magnesium is the second abundant in the adsorbed base cation species. For variable charge soils this is also the case. For this type of soil the ratio of magnesium ions to alkali metal ions is generally larger than the ratio in constant charge soils, caused by the slower eluviation of magnesium ions as compared to monovalent cations, such as sodium and potassium.

With regard to the effect of magnesium ions on lime potential of variable charge soils, the data in Tables 12.3 and 12.4 can show some tendencies. At the same $CaCl_2$ concentration, the higher the apparent degree of Mg-saturation, the higher the lime potential. This is caused by two reasons. The pH of the soil rises with the increase in degree of Mg-saturation. The competition of magnesium ions for negative surface charge sites makes more calcium ions existing in soil solution, resulting in a low pCa value. On the other hand, if the total quantity of calcium ions and magnesium ions is kept constant, an increase in the amount of magnesium ions would be accompanied by a corresponding decrease in amount of calcium ions. Thus, the lime potential would decrease (Table 12.4).

Because the effect of magnesium ions on chemical properties of soils is similar to that of calcium ions, in the past it was customary to use pH-0.5p(Ca+Mg) but not pH-0.5pCa to represent lime potential (Schofield and Taylor, 1955a; Turner and Nichol, 1962a; Webster and Harward, 1959). However, according to the definition of lime potential mentioned in Section 12.1.1, it is apparent that it would be more reasonable to use pH-0.5pCa to represent lime potential. In view of the advancement

Table 12.3 Effect of Degree of Mg-Saturation on Lime Potential of Soils

		pH		pH-0.5pCa	
Soil	Mg-Saturation (%)	$10^{-4} M^a$	$10^{-3} M$	$10^{-4} M$	$10^{-3} M$
Ali-Haplic	20	4.39	3.96	2.28	2.00
Acrisol	40	4.50	4.12	2.43	2.12
	60	4.57	4.32	2.59	2.10
	80	4.68	4.39	2.78	2.31
Rhodic	20	4.57	4.13	2.29	2.27
Ferralsol	40	4.57	4.23	2.37	2.32
	60	4.63	4.32	2.56	2.38
	80	4.66	4.46	2.68	2.44

[a]Concentration of $CaCl_2$ solution.

Table 12.4 Lime Potential of Soils with Different Adsorbed Ca:Mg
Ratios

Soil	Ca:Mg	pH−0.5pCa			
		25[a]	50	75	100
Ali-Haplic	2:8	2.38	2.88	3.25	4.24
Acrisol	4:6	2.61	2.90	3.31	4.36
	6:4	2.64	2.90	3.89	4.63
	8:2	2.77	2.98	3.99	4.70
Ferrali-Haplic	2:8	2.50	2.90	3.30	4.04
Acrisol	4:6	2.58	3.03	3.65	4.36
	6:4	2.66	3.05	4.01	4.64
	8:2	2.85	3.11	4.08	4.89
Rhodic Ferralsol	2:8	2.88	3.49	4.12	4.31
	4:6	2.93	3.51	4.25	4.53
	6:4	3.01	3.62	4.31	4.69
	8:2	3.15	3.85	4.41	4.86

[a]Apparent degree of base saturation (%).

in analytical techniques, this is not only feasible, but is even more convenient.

Besides magnesium ions, potassium, ammonium and sodium are also
commonly found cations in variable charge soils. These adsorbed ions,
although not large in quantity, are easy to hydrolyze and may have a
remarkable effect on the soil pH. This would affect the lime potential of the
soil. As can be seen from Table 12.5, when sodium ions account for a few
percent of the total base cations, the lime potential rises markedly.
Potassium ions have a similar effect, although not as pronounced as that of
sodium ions.

When variable charge soils are submerged, both the pH (cf. Chapter 11)
and the quantity of base ions, including calcium ions, would increase (Yu,
1985). Therefore, the lime potential increases constantly with the prolongation of submergence, until a nearly steady value is reached. Figure 12.16
shows such an example. For the paddy soil derived from Ferrali-Haplic
Acrisol and containing a considerable amount of easily-decomposable
organic matter, both pH and pH−0.5pCa changed quickly after submergence. If the original content of organic matter was low, the addition of

Table 12.5   Effect of Sodium Ions and Potassium Ions on Lime Potential
of Soils

Ions	S value[a]	pH-0.5pCa		
		Ali-Haplic Acrisol	Ferrali-Haplic Acrisol	Rhodic Ferralsol
Na	0	4.70	4.81	4.38
	0.04	4.98	5.20	5.35
	0.08	5.06	5.23	5.37
	0.12	5.11	5.24	5.39
K	0	4.70	4.81	4.38
	0.08	4.75	4.78	4.62
	0.12	5.02	4.89	4.65

[a]Symmetry value. The S value of Mg is 0.2; remaining value is Ca.
The S value of Ca is 0.8 when the S value of Na or K is 0.

easily-decomposable organic matter such as green manure could also induce
a similar change. However, if the amount of green manure added to the soil
was very high, the large amount of organic acids produced during the

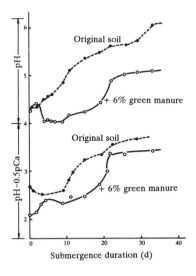

Fig. 12.16. Change in pH and pH-0.5pCa after submergence of a paddy
soil derived from Ferrali-Haplic Acrisol

vigorous decomposition of organic matter might become a factor in inducing the decrease in the pH. Thus, the magnitude of increase in lime potential became smaller.

From the above discussions it may be concluded that, in addition to the ratio of hydrogen (aluminum) ions to calcium ions that affects lime potential directly, other cations can also have effects on lime potential of the soil. These factors differ from the solid phase of the soil in behavior in that their amounts are easily changeable with the change in environmental conditions. This is one of the important reasons why lime potential of soils is apt to change. As shall be seen in the next section, because of such a reason the lime potential of different fields of the same soil type or even different depths of the same profile may differ markedly.

12.3 LIME POTENTIAL OF DIFFERENT SOILS UNDER FIELD CONDITIONS

12.3.1 Eluviation and Accumulation of Calcium

Under conditions of high temperature and high rainfall, variable charge soils have been intensively leached. Therefore, the content of calcium is very low, and the pCa value in soil solution is frequently larger than 5. Even though some calcium ions may be incorporated into the soil by some means, they are leached out rapidly. For example, it can frequently be observed that calcium ions applied to the cultivated layer of paddy soils with fertilizers and irrigation water are leached to the plowpan layer within a short time, with the result that the pH and pH-0.5pCa of this layer are higher than those of the cultivated layer. Some comparisons in this respect are shown in Table 12.6. If the duration of eluviation is long, calcium ions may be leached to deeper layers. The distribution pattern of pCa and pH within the paddy soil profile shown in Fig. 12.17 is quite representative in large areas of variable charge soils.

On the other hand, calcium can be enriched into the upper layer through absorption from the subsoil by plant roots and then accumulation as fallen leaves. If the vegetation is dense and the duration of enrichment is long, the content of calcium in the upper layers, particularly the litter layer, may be much higher than that of the subsoil. For example, Fig. 12.18 shows that for Ferrali-Haplic Acrisols under natural forest the pCa of the subsoil is as high as 5-5.5, whereas it is 3.8-4.5 in the surface layer. This means that the content of exchangeable calcium of the surface layer may be higher than that of the subsoil by one order of magnitude. This difference in calcium content is especially remarkable for Xanthic Ferrali-Haplic Acrisol with a dense vegetation. For soils under economic forest, the biological accumulation of calcium in the surface layer is also apparent (Fig. 12.19). Of course,

Table 12.6 Comparison of pCa and pH–0.5pCa Between Cultivated Layer and Plowpan of Paddy Soils Derived from Acrisols (Xishuangbanna) (Wang and Yu, 1983)

Field	Utilization	Horizon	pH	pCa	pH–0.5pCa
1	Fallow	Cultivated	4.89	3.08	3.35
		Plowpan	6.96	3.01	5.45
2	Fallow	Cultivated	4.85	3.50	3.10
		Plowpan	5.32	3.27	3.68
3	Fallow	Cultivated	5.23	3.62	3.42
		Plowpan	5.82	3.38	4.13
4	Rice	Cultivated	6.43	3.34	4.76
		Plowpan	6.53	2.95	5.05

for soils in tea and rubber plantations the effect of fertilization is also important.

When a forest soil is changed to an upland soil through slash-and-burn cultivation, because of the release of calcium from organic matter, the content of exchangeable and water-soluble calcium in the upper layer may be much higher than that of low-lying layers, as shown in Fig. 12.20.

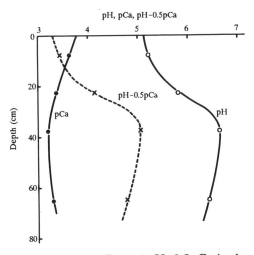

Fig. 12.17. Distribution of pH, pCa and pH–0.5pCa in the profile of paddy soil (Xishuangbanna) (Wang and Yu, 1983).

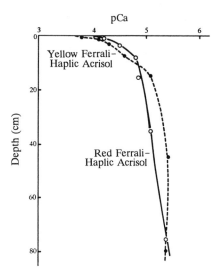

Fig. 12.18. pCa of soils under natural forest (Xishuangbanna) (Wang and Yu, 1983).

It can be concluded from the above examples that in variable charge soils under field conditions there are both biological accumulation and chemical eluviation of calcium ions. The intensity of both processes is high. The calcium content of a given soil is determined by the relative intensity

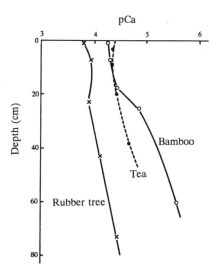

Fig. 12.19. pCa of soils under economic forest (Xishuangbanna) (Wang and Yu, 1983).

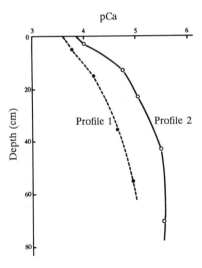

Fig. 12.20. pCa of upland soils (Xishuangbanna) (Wang and Yu, 1983).

between these two opposite processes. If the influence of agricultural measures is involved, the situation would be more complicated, as shall be shown in the next section.

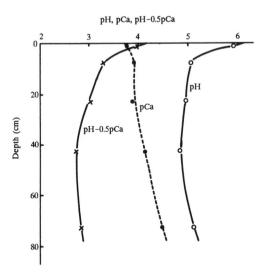

Fig. 12.21. Distribution of pH, pCa and pH–0.5pCa in the profile of Ferrali-Haplic Acrisol under rubber trees (Wang and Yu, 1983).

12.3.2 Effect of Agricultural Measures on Lime Potential

The effect of agricultural measures on lime potential of soils is caused chiefly by the application of fertilizers. This kind of effect may be reflected in soils under rubber trees, upland soils, and paddy soils.

The distribution patterns of pCa, pH, and pH–0.5pCa within the profile of a Ferrali-Haplic Acrisol under rubber trees are shown in Fig. 12.21. The high content of calcium ions in the upper layers is caused mainly by fertilization. Correspondingly, the pH and pH–0.5pCa of the upper layers are relatively high. This pattern of distribution of lime potential within the soil profile under rubber trees is representative, irrespective of the parent material from which the soil is derived (Fig. 12.22).

When the soil is cultivated for upland crops, the effect of fertilization can also be reflected by lime potential. As seen in Fig. 12.23, the lime potential of the cultivated layer of two upland soils is higher than that of low-lying layers.

If the soil is cultivated for rice, the effect of agricultural measures on lime potential is a little complicated. On one hand, there are large amounts of calcium ions and other base materials entering the soil with fertilizers and irrigation water. On the other hand, strong eluviation under submerged conditions makes calcium ions leached from the cultivated layer to lower layers. In Fig. 12.23, the lime potential of two paddy soil profiles is higher than that of the upland soil of the same region, whereas it is much lower in the cultivated layer than in subsoils within the profile, reflecting the

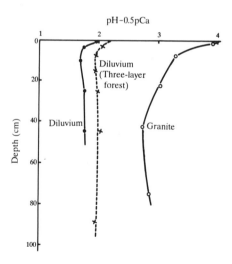

Fig. 12.22. Lime potential of Ferrali-Haplic Acrisols under rubber trees derived from different parent materials (Wang and Yu, 1983).

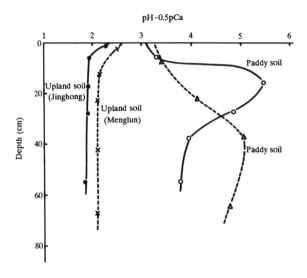

Fig. 12.23. Lime potential of agricultural soils (Xishuangbanna) (Wang and Yu, 1983).

combined effects of the two factors.

Compared to natural biological processes, agricultural measures are generally stronger in intensity and more frequent in operation. Therefore, the variation range of lime potential among different agricultural soils is generally wider than that of soils under natural vegetation.

12.3.3 Lime Potential of Soils under Natural Vegetation

It has been mentioned in the previous section that the lime potential of a soil under natural vegetation is determined by relative intensit of the chemical eluviation of base cations, including calcium ions, and the biological accumulation of these elements. The intensities of these two factors are different under different environmental conditions. Hence, the lime potentials will also differ considerably. Nevertheless, it can be found that in these soils there are certain regularities in the distribution pattern of lime potential.

For variable charge soils that are already strongly acid in reaction, owing to biological accumulation under natural vegetation, the lime potential within the upper depth is usually higher than that in lower horizons, caused by the high amount of calcium ions. The distribution patterns of lime potential in three Ferrali-Haplic Acrisol profiles under natural forest shown in Fig. 12.24 all reflect such a tendency.

On the other hand, if the upper layers that have been influenced by

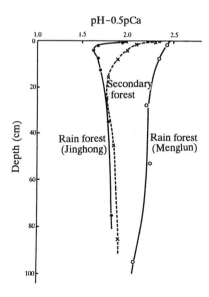

Fig. 12.24. Lime potential of Ferrali-Haplic Acrisols under natural forest (Xishuangbanna) (Wang and Yu, 1983).

biological accumulation are examined in detail, a complex situation can be found. In Fig. 12.24, for two Ferrali-Haplic Acrisols, lime potential is the lowest at a certain depth in the middle of the profile. This pattern of distribution is quite often in variable charge soils, as can be seen for other

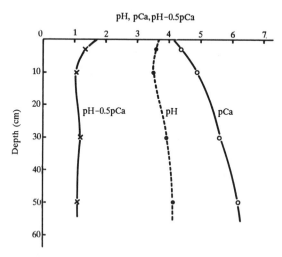

Fig. 12.25. Lime potential of hydrated Ferrali-Haplic Acrisols under natural forest (Dinghu Mountain).

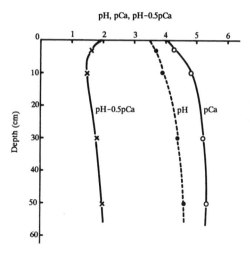

Fig. 12.26. Lime potential of Shallow Ferrali-Haplic Acrisols under natural forest (Dinghu Mountain).

two Ferrali-Haplic Acrisol profiles in Guangdong Province shown in Figs. 12.25 and 12.26, respectively.

The phenomenon that the lime potential is the lowest at a certain middle depth of the profile is especially remarkable in Xanthic-Haplic Acrisols. Figure 12.27 shows that the pH–0.5pCa value at a middle depth is lower

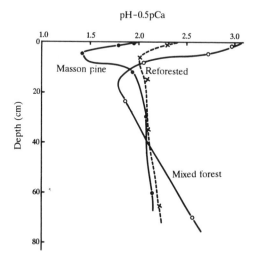

Fig. 12.27. Lime potential of Xanthic-Haplic Acrisols under natural forest (Guizhou) (Wang and Yu, 1983).

than that of the upper or lower depths by 0.3–1.1 units.

This phenomenon is caused by the important role of organic matter in affecting the pH. According to the principle mentioned in Section 12.2, for a given soil there should exist a certain relationship between pH and pCa, that is, the smaller the pCa value of a soil, the higher the pH. However, the establishment of such a principle requires a basic condition; that is, the soil system must be in a condition of dynamic chemical equilibrium. Under conditions of high temperature and high moisture, particularly in the litter layer, the large amount of organic acids produced during the vigorous decomposition of organic matter may make the limited amount of mineral part of the soil insufficient to establish a chemical equilibrium with these acids. Or, the speed of the establishment of chemical equilibrium may be too low. This would result in the presence of free organic acids in the solution. Therefore, the pH may be rather low. In the expression of lime potential, hydrogen ions are related to calcium ions in a manner of pH vs. $0.5pCa$, but not pH vs. pCa. Thus, in this middle layer, lime potential may be rather low, although the amount of calcium ions is quite high. The relationships among pH, pCa and pH–$0.5pCa$ in the second layer and the third layer (1–5 cm) of the profile of a Xanthic-Haplic Acrisol shown in Fig. 12.28 are a reflection of this complex situation. Apparently, the higher the amount of organic matter and the more moist the soil, the greater the possibility of the existence of free organic acids. This is the reason why the presence of a horizon with the lowest pH at a certain middle depth is more frequent and more remarkable in Xanthic-Haplic soils than in Rhodic

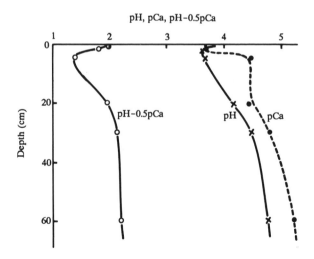

Fig. 12.28. pH, pCa, and pH–$0.5pCa$ in a Xanthic-Haplic Acrisol profile (Guizhou) (Wang and Yu, 1983)

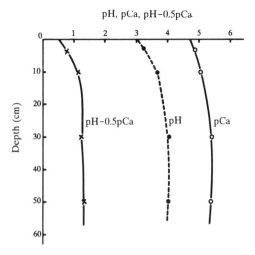

Fig. 12.29. Lime potential of Xanthic-Haplic Acrisols under natural forest (Dinghu Mountain).

Ferralsols and Ferrali-Haplic Acrisols. Under certain conditions, lime potential may be the lowest at the uppermost layer of the profile. As seen in Fig. 12.29, the pH and pCa at a depth of 3 cm in the profile of a Xanthic-Haplic Acrisol may be as low as 3.25 and 4.9, respectively. Therefore, the pH–0.5pCa value is only 0.8. This is the lowest numerical value of lime potential that has been found in variable charge soils.

From the data presented in this section is can be concluded that under the combined influences of chemical eluviation, biological accumulation and artificial measures, lime potential of variable charge soils may vary in a range from as low as 1 under strongly acid conditions to as high as 5.5 under neutral conditions. This variation is larger than that of pH (3.3–7.0) by about one unit. Thus, it would be possible to characterize the acidity intensity of different variable charge soils according to the difference in lime potential.

12.4 GRADATION OF LIME POTENTIAL

Based on its relationship with plant growth and required agricultural measures, the acidity intensity of variable charge soils may be characterized as strongly acid, acid, or neutral. The pH–0.5pCa ranges of these three categories are about <3, 3–4.5 and 4.5–6, respectively. Because the relationship between hydrogen ions and calcium ions in various types of soils is not in a straightforward manner, the corresponding pH and pCa of the three threshold values of lime potential–3, 4.5, and 6–may vary within a certain range. For pH the ranges are about 5-5.5, 6.0-6.5, and 7.0-7.5, respectively, and for pCa, they are 4-5, 3-4, and 2-3, respectively. When lime potential is lower than 3, the toxicity of aluminum ions and the insufficiency of calcium ions become two important factors for the poor growth of the majority of crops on strongly acid soils.

BIBLIOGRAPHY

Aitken, R. L. and Moody, P. W. (1991) *Aust. J. Soil Res.*, 29:483-491.

Aslyng, H. C. (1954) Royal Veterinary Agricultural College of Copenhagen Yearbook. pp. 1-50.

Beckett, P. H. T. (1964) *J. Soil Sci.*, 15:1-8.

Coleman, N. T. and Thomas, C. W. (1967) in *Soil Acidity and Liming* (R. W. Pearson and F. Adams, eds.). American Society of Agronomy, Madison, WI, pp.1-34.

Conyers, M. K., Munns, D. N., Helyar, K. R., and Poile, G. J. (1991) *J. Soil Sci.*, 42:599-606.

Gillman, G. P. and Bell, L. C. (1976) *Aust. J. Soil Res.*, 14:351-360.

Larsen, S. (1965) *J. Soil Sci.*, 16:275-278.

Krupsky, N. K. et al. (1975) *Agrochem.*, no. 3, 133-138.

Marshall, C. E. (1964) *The Physical Chemistry and Mineralogy of Soils.* vol. 1. John Wiley & Sons, New York, pp. 211-259.

Mattson, S. and Wiklander, L. (1940) *Soil Sci.*, 49:109-154.

Nichol, W. E. and Turner, R. C. (1957) *Can. J. Soil Sci.*, 37:96-101.

Odahara, K. and Wada, S. I. (1992) *Japanese J. Soil Sci. Plant Nutr.*, 63:64-71.

Schofield, R. K. (1947) *Proc. 11th Congr. Pure Appl. Chem.*, 3:257-261.

Schofield, R. K. (1955) *Soils and Fert.*, 18:373-375.

Schofield, R. K. and Taylor, A. W. (1955a) *Soil Sci. Soc. Am. Proc.*, 19:164-167.

Schofield, R. K. and Taylor, A. W. (1955b) *J. Soil Sci.*, 6:136-146.

Sumner, M. E. and Marques, J. M. (1968) *Agrochimica*, 12:191-195.

Turner, R. C. (1965a) *Soil Sci.*, 99:88-92.

Turner, R. C. (1965b) *Soil Sci.*, 100:14-19.

Turner, R. C. and Clark, J. S. (1965) *Soil Sci.*, 99:194-199.

Turner, R. C. and Nichol, W. E. (1958) *Can. J. Soil Sci.*, 38:63-68.

Turner, R. C. and Nichol, W. E. (1962a) *Soil Sci.*, 93:374-382.

Turner, R. C. and Nichol, W. E. (1962b) *Soil Sci.*, 94:58-63.

Turner, R. C., Nichol, W. E., and Brydon, J. E. (1963) *Soil Sci.*, 95:186-191.

Ulrich, B. (1961a) *Die Wechselbeziehung von Boden und Pflanze in Physikalisch-chemischer Betrachtung.* Ferdinard Enke, Stuttgart.

Ulrich, B. (1961b) *Lands. Forsch.*, 14:225-228.

Wang, J. H. (1986) in *Current Progress in Soil Research in People's*

Republic of China (Soil Science Society of China). Jiangsu Science and Technology Publishers, Nanjing, pp. 78-84.

Wang, J. H. and Yu, T. R. (1981) *Z. Pflanzenernähr. Bodenk.*, 144:514-523.

Wang, J. H. and Yu, T. R. (1983) *Acta Pedol. Sinica*, 20:286-294.

Wang, J. H. and Yu, T. R. (1986) *Z. Pflanzenernähr. Bodenk.*, 149:598-607.

Webster, G. R. and Harward, M. E. (1959) *Soil Sci. Soc. Am. Proc.*, 23:446-451.

Yu, T. R. (1976) *Electrochemical Properties of Soils and Their Research Methods*. Science Press, Beijing.

Yu, T. R. (1985) *Physical Chemistry of Paddy Soils*. Science Press/Springer Verlag, Beijing/Berlin.

Yu, T. R. (ed.) (1987) *Principles of Soil Chemistry*. Science Press, Beijing.

Yu, T. R. and Zhang, Z. N. (1983) in *Red Soils of China* (C. K. Li, ed.). Science Press, Beijing, pp. 74-90.

Yu, T. R. and Wang, Z. Q. (eds.) (1988) *Soil Analytical Chemistry*. Science Press, Beijing.

Yu, T. R. and Chen, Z. C. (eds.) (1990) *Chemical Processes in Soil Genesis*. Science Press, Beijing.

13

OXIDATION–REDUCTION REACTIONS

Z. G. Liu, C. P. Ding, Y. X. Wu, S. Z. Pan and R. K. Xu

Oxidation–reduction reactions are chemical reactions caused by the transfer of electrons between two substances. These reactions occur actively in variable charge soils. This is because that under conditions of high temperature and high precipitation both the accumulation and the decomposition of organic matter proceed rapidly. The decomposition products of organic matter may release electrons, providing the necessary condition for the occurrence of reduction reactions. In particular, because the soil may have a high content of water during seasonal rainy periods, the presence of a strongly reducing condition is possible. Furthermore, large areas of variable charge soils have been cultivated for rice production. For these paddy soils there are always intensive oxidation–reduction reactions proceeding alternately. Variable charge soils have a high content of iron oxides. The content of manganese is also higher than that of constant charge soils. Thus, the soil itself possesses plenty of electron–acceptors. Besides, the high concentration of hydrogen ions in variable charge soils is favorable for the occurrence of reduction reactions. Therefore, as shall be seen in this chapter, contrary to the belief that the significance of oxidation–reduction reactions is confined chiefly to submerged soils, these reactions may play an important role in soil genesis and soil fertility for variable charge soils even under well-aerated conditions.

In this chapter, after discussions on factors affecting the intensity of oxidation–reduction and interactions among various oxidation–reduction substances, the oxidation–reduction regimes of variable charge soils under different utilization conditions will be presented. Ferrous and manganous ions, two important inorganic reducing substances in soils, shall be dealt with in the next chapter.

13.1 INTENSITY OF OXIDATION–REDUCTION

13.1.1 Electron Activity and Oxidation–Reduction Potential

The oxidation–reduction intensity of a substance is determined by its ability to liberate or accept electrons. Therefore, electron activity in an equilibrium system may be used as an index for expressing its reduction strength. An electron has a radius of only approximately 1/20,000 of that of a hydrogen atom. Its large charge–to–size ratio prevents it from persisting in free form in aqueous systems. The ephemeral "hydrated electron" has a half–life of less than 1 msec (Bartlett and James, 1993). As a species with a potential of -2.7 V vs. the standard potential of H^+/H_2, it is a powerful reducing agent. Electron activity represents the contribution of electrons to the partial molar free energy of that system. Just as the pH value represents the negative logarithm of the activity of protons, the pe value may be used to represent the negative logarithm of the activity of electrons:

$$pe = -\log a_e \tag{13-1}$$

For a system in which the oxidized form (oxidant) and the reduced form (reductant) coexist at equilibrium,

$$(\text{Oxidant}) + ne \rightleftharpoons (\text{Reductant}) \tag{13-2}$$

the equilibrium constant K would be

$$K = \frac{(\text{Reductant})}{(\text{Oxidant})(e)^n} \tag{13-3}$$

Under standard conditions where the ratio of (oxidant) to (reductant) is equal to 1, when $n = 1$ we obtain

$$\log K = pe^0 \tag{13-4}$$

According to the definition in physical chemistry, the equilibrium constant of a chemical reaction is related to the change in standard free energy (ΔG^0) as

$$\Delta G^0 = -RT \ln K = -nFE^0 \tag{13-5}$$

Therefore,

$$E^0 = \frac{RT}{nF} \ln K \tag{13-6}$$

where E^0 is the standard potential of the system. If $n = 1$, at 25°C we have

$$E^0 = 0.059 \log K = 0.059 pe^0 \tag{13-7}$$

If the ratio of (oxidant) to (reductant) is not equal to 1, it will be related to the potential as

$$Eh = E^0 + \frac{RT}{nF} \ln \frac{(\text{Oxidant})}{(\text{Reductant})} \tag{13-8}$$

In the equation, Eh is called the *oxidation–reduction potential* of the system, taking the potential of the standard hydrogen electrode as zero. This potential represents the change in free energy when one mole of electrons is transferred from H_2 molecules to the oxidant. Apparently, when $n = 1$, this potential is related to the activity of electrons:

$$pe = \frac{F}{2.303RT} Eh \tag{13-9}$$

and at 25°C we have

$$pe = \frac{Eh}{0.059} \tag{13-10}$$

Oxidation–reduction potential (Eh) has long been used widely as an index for expressing the intensity of the oxidation–reduction of a system because it can be measured directly. It can been seen from equation (13–8) that this potential is determined by both the nature of the system and the ratio of the oxidized form to the reduced form. The concept pe was introduced much later in chemistry. However, because pe is directly related to the activity of electrons, it is more meaningful in physics.

13.1.2 Effect of Proton Activity

It has been shown in equations (11–9) to (11–13) of Chapter 11 that during the oxidation–reduction reactions there is invariably the participation of protons. Therefore, the activity of protons has a direct effect on the reaction equilibrium. For the following oxidation–reduction reaction:

$$(\text{Oxidant}) + ne + mH^+ \rightleftharpoons (\text{Reductant}) + xH_2O \tag{13-11}$$

the relationship at 25°C should be

$$Eh = E^0 + \frac{0.059}{n} \log \frac{(\text{Oxidant})}{(\text{Reductant})} - 0.059\frac{m}{n}\text{pH} \quad (13\text{-}12)$$

This means that the extent of the effect is determined by the ratio m/n. When m/n is equal to 1, we obtain

$$Eh = E^0 + \frac{0.059}{n} \log \frac{(\text{Oxidant})}{(\text{Reductant})} - 0.059\,\text{pH} \quad (13\text{-}13)$$

In this case a change in pH by one unit would induce a change in Eh ($\Delta Eh/\Delta \text{pH}$) by -59 mV at 25°C. This is what happens for the oxygen system and many organic systems. For the oxygen system we have

$$O_2 + 4H^+ + 4e \rightarrow 2H_2O \quad (13\text{-}14)$$

$$Eh = 1.23 + 0.015 \log P_{O_2} - 0.059\,\text{pH} \quad (13\text{-}15)$$

For some systems, the ratio m/n in the oxidation–reduction reaction may be larger than 1. For example, for the reduction of Fe_2O_3 we have

$$Fe_2O_3 + 6H^+ + 2e \rightarrow 2Fe^{2+} + 3H_2O \quad (13\text{-}16)$$

$$Eh = 0.728 - 0.059 \log(Fe^{2+}) - 0.177\text{pH} \quad (13\text{-}17)$$

Thus, a change in pH by one unit can induce a change in oxidation–reduction potential by -177 mV at 25°C.

There are a variety of oxidation–reduction reactions with different m/n ratios in soils. This makes the value $\Delta Eh/\Delta \text{pH}$ vary within a wide range in different soils. It has been found that, for well–aerated variable charge soils, because oxygen plays the dominant role in determining the oxidation–reduction potential, the $\Delta Eh/\Delta \text{pH}$ value was generally close to the theoretical value of oxygen (60 mV) (Yu and Li, 1957). On the other hand, for reduced soils, because both ferrous iron and organic reducing substances may be the principal reducing substances, the $\Delta Eh/\Delta \text{pH}$ value is frequently related to the ratio between these two substances, being larger for systems with a higher ratio. Table 13.1 shows such an example. When the soil was submerged after the addition of a large amount of green manure so that the concentration of ferrous iron attained a range of 10^{-3}–10^{-2} cmol L^{-1}, the $\Delta Eh/\Delta \text{pH}$ value might be as large as the theoretical value of the iron system.

The practical significance of the effect of pH on Eh lies in the fact that the Eh value of a soil with the same oxidation–reduction intensity may be

Table 13.1 Relationship Between $\Delta Eh/\Delta pH$ and the Ratio Ferrous Iron/Organic Reducing Substances in Paddy Soil Derived from Acrisol (Yu, 1985)

Treatment	Ferrous (A) ($\frac{1}{2}$cmol kg^{-1})	Organic (B) ($1/z$ cmol kg^{-1})	A/B	$\Delta Eh/\Delta pH$
+5% Rapeseed	3.35	0.81	4.13	104
+5% Vetch	1.31	0.93	1.41	104
+5% Milk vetch	1.76	2.32	0.76	102
+3% Milk vetch	0.77	1.75	0.44	93
+3% Rapeseed	0.19	1.22	0.16	83

different at different pH. This phenomenon can also be observed when comparing the oxidation–reduction status of different types of soils under field conditions. It has been found that for the substratum of soils with little organic matter the pH and Eh for Ferrali–Haplic Acrisols of the Guangdong Province were 4.0–4.5 and 650–720 mV, respectively, while the corresponding values for Xanthic–Haplic Acrisols of the Sichuan Province, Cambsols of the Jiangsu Province, and alluvial soils of the Beijing region were 4.5–5.0 and 630–690 mV, 5.5–6.0 and 550–600 mV, and 7.0–7.5 and 460–510 mV, respectively (Yu and Ding, 1990). Since the oxidation–reduction potential of the substratum of these soils is determined chiefly by oxygen, this difference in Eh among different types of soils should be closely related to the change in Eh for the oxygen system with the change in pH. Therefore, generally speaking, the Eh of a variable charge soil would be higher than that of a constant charge soil, except podzolized soil, with an equal oxidation–reduction intensity.

In order to compare the oxidation–reduction intensity among various soils, it is a frequent practice to convert the measured Eh value to a common pH basis, such as Eh_7 or Eh_5, assuming that the $\Delta Eh/\Delta pH$ value equals 59 mV in calculations, although it is apparent that this assumption is not strictly valid, as has been discussed in the previous sections.

13.1.3 Poising in Oxidation–Reduction

The variation of the activity of electrons in soils may range from pe −7 to 12, corresponding to an Eh range of −420 to 720 mV. Because the production and loss of electrons are very rapid, the oxidation–reduction status would be in a condition of constant and drastic change. Thus, it would be unfavorable to organisms if the soil cannot resist such drastic changes. Fortunately, like the buffering action against acidification–alkalinization, a soil also has the ability to buffer against oxida-tion–reduction. This is called *poising*.

The poising of a substance is the ability of that substance to retard the change in *Eh* when small amounts of oxidants or reductants are added to the system. This can be explained through the following derivations.

For equation (13–8), if it is assumed that the activity of the oxidant is x and the total activity of the oxidant plus the reductant is A, the activity of the reductant would be $A - x$. When the activity of the oxidant is increased to a small extent, the increase in oxidation–reduction potential should be

$$\frac{dEh}{dx} = \frac{RT}{nF} \cdot \frac{A}{x(A - x)} \tag{13-18}$$

If the reciprocal of dEh/dx is used as an index for expressing the poising of a substance, called *poising index*, it can be derived that

$$\frac{dx}{dEh} = \frac{nF}{RT} \cdot \frac{x(A - x)}{A} = \frac{nF}{RT} \cdot x\left(1 - \frac{x}{A}\right) \tag{13-19}$$

It can be seen from the equation that the higher the total concentration A of an oxidation–reduction substance, the stronger the poising, and that at a constant A the poising is strongest when $A = 2x$, that is, when the ratio

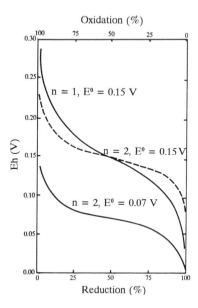

Fig. 13.1. Relationship between *Eh* and degree of oxidation or reduction (%) for different oxidation–reduction systems.

the oxidant to reductant equals 1. For different oxidation–reduction substances, the poising would be stronger when the numerical value of n is larger. These relationships are illustrated in Fig. 13.1. At the two extremes of the curve the addition of a small amount of oxidant or reductant can induce a remarkable change in Eh, while the change in Eh approaches zero when the proportions of the oxidant and the reductant all equal 50%. This situation is similar to the buffering action discussed in Chapter 11.

There are two complicating factors in actual soil conditions. Generally a soil does not contain merely one kind of oxidation–reduction substance, but is composed of a variety of such substances. In such a mixed system the Eh value is related not only to the relative proportions of these substances, but also to the reaction rates. In particular, the reaction rate of some organic substances participating in the oxidation–reduction reactions may be very low while their quantity may be quite high, thus inducing a hysteresis effect in poising. Another complicating factor is that the oxidation–reduction reactions in soils may proceed not only in solution but also at the interface between the solid phase and the solution. Thus, the overall reaction rate would be further lowered. These two complicating factors make it difficult to apply equation (13-19) to soils in a simple way. Nevertheless, if experimental conditions are kept identical, it would be possible to make comparisons among soils with different oxidation–reduction status.

In Fig. 13.2, the change in Eh during the titration of extracts from paddy soils with an oxidizing agent is shown. At the initial stage, the addition of a small amount of oxidizing agent can induce a marked increase in Eh.

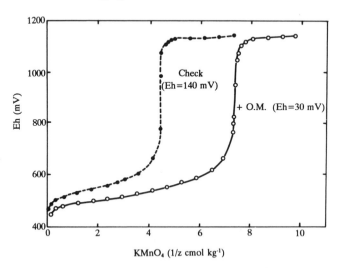

Fig. 13.2. Change in Eh during titration of $Al_2(SO_4)_3$ solution extract of paddy soils derived from Acrisols with $KMnO_4$ (Yu, 1985).

Then, there appears a distinct poising range where the change in *Eh* is quite small. Finally, near the titration end-point, because the potential becomes to be determined by the oxidizing agent (KMnO₄) gradually, the *Eh* increases sharply to a nearly constant value. The pattern of the whole titration curve is similar to that of the theoretical curve shown in Fig. 13.1. If the quantity of the consumed oxidizing agent at the break-point of *Eh* change is taken as the poising capacity against oxidation, it can be found that the capacity corresponds to 4.6 and 7.4 cmol kg⁻¹ of soil for the two soils with different oxidation-reduction regimes. It is noticeable that on the curves there is only one poising range. This is because the reducing substances are dominated by ferrous iron, and therefore the behavior of other substances cannot be clearly characterized.

If the titration is carried out directly with soil suspension, the pattern of *Eh* change is similar to that of the soil extract (Fig. 13.3). However, because in this case some substances of the solid phase (organic matter) can also participate in the oxidation-reduction reaction, the plateau region in the titration curve is not as distinct as in the case of solution.

There are reducible substances such as iron and manganese oxides present in the soil. Therefore, a soil also possesses a poising action against reducing agents. It can be seen from Fig. 13.4 that the poising capacity increases greatly after the addition of MnO₂ to a Ferrali-Haplic Acrisol, whereas the poising vanishes nearly completely for a Rhodic Ferralsol after the removal of the oxides of iron and manganese. When different soils are compared it can be seen from Fig. 13.5 that the poising capacity against reduction is of the order Rhodic Ferralsol > Ali-Haplic Acrisol >

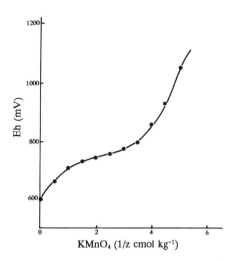

Fig. 13.3. Change in *Eh* during titration of suspension of paddy soil derived from Acrisol with KMnO₄.

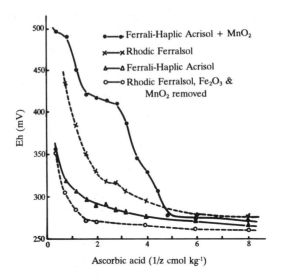

Fig. 13.4. Change in *Eh* during titration of soils of different iron-manganese status with ascorbic acid (sodium acetate solution, pH 4.5).

Xanthic-Haplic Acrisol, in keeping with the order of the contents of ironand manganese oxides for the three soils. Thus, it may be inferred that generally the poising capacity of variable charge soils against reduction is larger than that of constant charge soils, although the poising capacity

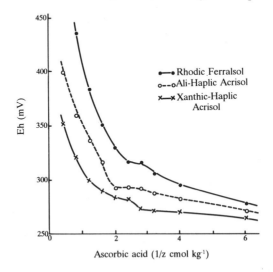

Fig. 13.5. Change in *Eh* during titration of three variable charge soils with ascorbic acid (in sodium acetate solution).

against oxidation may not necessarily be of such an order due to the involvement of organic matter.

13.1.4 Depolarization Reaction at an Electrode

In the oxidation–reduction reactions discussed above, only soil solution or the interface between the solution and the solid phase of the soil is involved. Although an electrode is used in the measurement of oxidation–reduction potential of soils, the electrode (usually a platinum electrode) is used only to indicate the activity of electrons of the system. Actually, it is also possible to induce an oxidation or reduction reaction at the interface between the electrode and the solution by removing electrons from or supplying electrons to the solution through the electrode. During this kind of reaction the concentration of oxidation–reduction substances in the interface layer would change, thus inducing a change in the electrode potential. This process is called *electrode polarization* (Yu and Ji, 1993). However, after the interruption of the polarization reaction, oxidation–reduction substances adjacent to the interface layer would diffuse to this layer due to the presence of a concentration gradient, gradually restoring the electrode potential to its original value. This phenomenon is called *depolarization*. Apparently, the higher the concentration of relevant substances in the bulk solution, the shorter the time required for the complete depolarization of the electrode. Thus, the depolarization reaction of the electrode by soil solution is in reality a reflection of a poising action of the solution when it loses electrons to or gains electrons from an external source.

In the experiment one can apply a potential higher than the equilibrium potential of the electrode so that electrons can be removed through the electrode. After the interruption of polarization, the change in electrode potential caused by depolarization with time is recorded. Thus, an anodic depolarization curve can be obtained. Similarly, a cathodic depolarization curve can be obtained after the interruption of the application of a potential lower than the equilibrium electrode potential.

Figure 13.6 shows the depolarization patterns of two soils with different reduction intensities at a platinum electrode. If the electrode potential is plotted against the logarithm of time, three portions of the curve can be distinguished. Within the initial 20 seconds of depolarization (1), the electrode potential changed rapidly. Within the time interval of 20 seconds to about 3 minutes (2), the depolarization curve gradually became nearly linear. Afterward (3), the slope of the linear portion of the curve became smaller. Because the changes of the anodic and cathodic depolarization curves are of opposite directions, two straight lines with an intersection at a certain point can be obtained if the electrode potential is plotted against

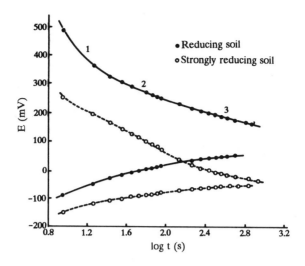

Fig. 13.6. Change in electrode potential with time for polarized platinum electrode during depolarization by two soils (Liu and Yu, 1984).

the logarithm of time within the interval of 3 minutes to a certain period. The electrode potential corresponding to the intersection point should be the potential of the depolarized electrode.

The depolarization reaction of a soil at the electrode is related to both the surface properties of the polarized electrode and the kind and concentration of depolarizers (oxidation–reduction substances) of the soil. Because the concentration of various oxidation–reduction substances invariably increases with the decrease in pH, irrespective of the oxidation–reduction status of the soil, it may be expected that for the same soil the required time for depolarization would be short if the pH is low. Figure 13.7 reflects this situation for a Rhodic Ferralsol and a Xanthic–Haplic Acrisol.

During depolarization reaction of the electrode by oxidation–reduction substances a diffusion process of these substances is involved. Therefore, water content, size of aggregates, and looseness in structure of the soil may all affect the reaction rate (Liu and Yu, 1984).

Thus, it is not only possible to judge the relative concentration of oxidation–reduction substances in the soil from the pattern of depolarization of polarized platinum electrode, because, as shown in Fig. 13.6, the concentration in the strongly reducing soil is apparently higher than that in the moderately reducing soil, but in practice it is also possible to estimate the oxidation–reduction potential of the soil from two depolarization curves. The reasoning for this is that the intersection point between the extrapolated lines of the two depolarization curves should be the potential of the

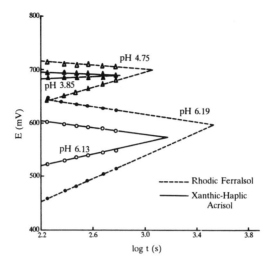

Fig. 13.7. Effect of soil pH on depolarization reaction (Liu and Yu, 1984).

depolarized electrode, and this potential in principle should also correspond to the equilibrium electrode potential in the soil-that is, the *Eh*. This point is of practical significance. It is now known that contrary to the previous belief that platinum electrode is an inert electrode that merely functions to transfer electrons, actually it can form platinum oxide films at its surface and can also adsorb other substances from the medium. The consequence

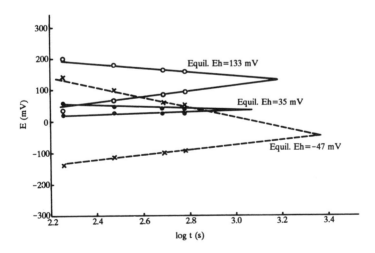

Fig. 13.8. Determination of oxidation–reduction potential of soils by extrapolated depolarization curve method (Yu, 1985).

of these reactions is that it requires a long time for the electrode to establish an equilibrium potential that truly reflects the oxidation–reduction status of the soil after the contact of the electrode with a soil, particularly if the poising of the soil is very weak (Bartlett and James, 1993; Bohn, 1971; Bottcher and Strebel, 1988; Cogger et al., 1992; Matia et al., 1991; Peiffer et al., 1992; Yu, 1976, 1985). In contrast, it is possible to determine the Eh of a soil within a half hour by determining the depolarization curves, as shown in Fig. 13.7 for example. Experimental data show that the Eh value determined in this way generally deviates from the equilibrium potential of the platinum electrode by less than 10 mV (Fig. 13.8). Many Eh data cited in this chapter were determined by this method. If the measuring apparatus is used in conjunction with a computer and the result is printed out automatically (Fang and Liu, 1987), the method would be more convenient to use.

13.2 REACTIONS AMONG OXIDATION–REDUCTION SUBSTANCES

13.2.1 Sequential Reduction

In Section 13.1, the relationship between the oxidized form and the reduced form of one kind of oxidation–reduction substance was discussed. In soils the general case is that there are a variety of oxidation–reduction substances existing together. When the oxidation–reduction status of one substance is changed, this substance may react with another oxidation–reduction substance, with the direction and extent of the reaction depending on the nature of the two species involved.

For example, when an equilibrium is established between two reacting oxidation–reduction substances we obtain

$$a(\text{Oxidant})_1 + b(\text{Reductant})_2 \rightleftharpoons c(\text{Reductant})_1 + d(\text{Oxidant})_2 \quad (13\text{-}20)$$

Because at equilibrium the Eh of the two substances are equal to each other, assuming that the values of n of the two substance are the same, then,

$$
\begin{aligned}
Eh &= E_1^0 + \frac{RT}{nF} \ln \frac{(\text{Oxidant})_1^a}{(\text{Reductant})_1^c} \\
&= E_2^0 + \frac{RT}{nF} \ln \frac{(\text{Oxidant})_2^d}{(\text{Reductant})_2^b}
\end{aligned}
\quad (13\text{-}21)
$$

$$E_1^0 - E_2^0 = \frac{RT}{nF} \ln \frac{(\text{Oxidant})_2^d (\text{Reductant})_1^c}{(\text{Reductant})_2^b (\text{Oxidant})_1^a}$$

$$= \frac{RT}{nF} \ln K \qquad\qquad (13-22)$$

It can be seen that the larger the difference in E^0 between the two substances, the larger the K value at equilibrium. If the sign of the difference is positive, the larger the difference, the more complete the reaction in equation (13-20) would proceed to the right. The same is true if the sign of the difference is negative. Thus, whether an oxidation-reduction reaction between two substances can proceed and to what extent the reaction would go is dependent on their E^0.

The E^0 and pe^0 of some oxidation-reduction substances commonly found in soils are given in Table 13.2.

In soils, electrons come chiefly from organic reducing substances produced during the decomposition of organic matter. Under aerated conditions, oxygen is the principal acceptor of electrons. Once the supply of oxygen is cut off due to submergence, manganese oxides or nitrate are reduced first, with the sequence of reduction between the two oxides depending on the pH, and then iron oxides and sulfate are reduced. This kind of serial reduction in a certain order is called *sequential reduction* (Patrick and Jugsujinda, 1992; Yu, 1976, 1985). However, if the difference in E^0 between two substances is not sufficiently large, the two reduction processes may overlap to some extent, caused chiefly by the difference in the oxidation-reduction rates.

For most variable charge soils, the most important oxidation-reduction

Table 13.2 E^0 and pe^0 of Some Oxidation-Reduction Systems Commonly Found in Soils (25°C)

System	E^0 (V) Standard	E^0 (V) at pH 7	pe^0 ($=\log K$)
$\frac{1}{4}O_2 + H^+ + e \rightleftharpoons \frac{1}{2}H_2O$	1.23	0.814	20.8
$\frac{1}{2}MnO_2 + 2H^+ + e \rightleftharpoons \frac{1}{2}Mn^{2+} + H_2O$	1.23	0.401	20.8
$Fe(OH)_3 + 3H^+ + e \rightleftharpoons Fe^{2+} + 3H_2O$	1.06	-0.185	17.9
$\frac{1}{2}NO_3^- + H^+ + e \rightleftharpoons \frac{1}{2}NO_2^- + \frac{1}{2}H_2O$	0.85	0.54	14.1
$\frac{1}{8}SO_4^{2-} + 5/4H^+ + e \rightleftharpoons \frac{1}{8}H_2S + \frac{1}{2}H_2O$	0.30	-0.214	5.1
$\frac{1}{8}CO_2 + H^+ + e \rightleftharpoons \frac{1}{8}CH_4 + \frac{1}{4}H_2O$	0.17	-0.244	2.9
$H^+ + e \rightleftharpoons \frac{1}{2}H_2$	0	-0.413	0

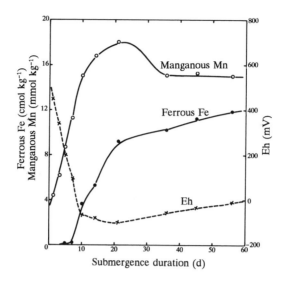

Fig. 13.9. Changes in *Eh*, ferrous ions and manganous ions with time for Rhodic Ferralsol during submergence.

reactions are those between iron or manganese compounds and organic reducing substances. In the following sections, the properties of these two kinds of substances and their interactions will be discussed.

13.2.2 Properties of Iron Ions and Manganese Ions

It can be seen from Table 13.2 that when the oxidized forms of iron and manganese are present together it is manganese that is reduced first at a given pH. This is the actual case in soils. Figure 13.9 shows the different trends in the reduction of manganese and the reduction of iron when a Rhodic Ferralsol was submerged. In the figure, the reduction of manganese and the change of *Eh* proceeded nearly synchronously. Following the decrease in *Eh*, namely following the intensification of reducing condition, the amount of reduced manganese increased rapidly. At the tenth day of submergence both *Eh* and the amount of manganous Mn attained a nearly steady value. On the other hand, the reduction of iron compounds showed a marked hysteresis phenomenon. The amount of ferrous Fe only accounted for one-third of the maximum value at the tenth day of submergence and only attained a nearly steady value on the twenty-first day. These data qualitatively illustrate the difference in behavior between iron and manganese with respect to the oxidation–reduction. As shall be seen in Chapter 14, this difference between the two elements also exists in field soils.

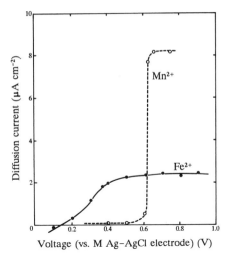

Fig. 13.10. *I-V* curve of ferrous ions and manganous ions at pH 6 (Yu, 1985).

Conversely, when reduced iron and manganese coexist, it is the ferrous iron that will be oxidized first. The *I-V* curves of the two ion species are shown in Fig. 13.10. At pH 6, ferrous iron begins to undergo oxidation at the electrode at about 0.2 V vs. *M* Ag/AgCl electrode, and the diffusion current attains a maximum value at about 0.4 V, whereas manganous ions can be oxidized only at the electrode at about 0.6 V. Therefore, if the electrode current at 0.35 V is taken as an index for expressing the quantity of *strongly reducing substances*, ferrous ions should belong to this kind of substance, whereas manganous ions do not. However, manganous ions can be oxidized if the voltage applied to the electrode is increased to 0.7 V. Substances that can be oxidized only at a comparatively higher electrode potential, such as manganous ions, may be called *weakly reducing substances* (Ding et al., 1982).

13.2.3 Properties of Organic Reducing Substances

Organic reducing substances are of tremendous importance in soils, although at present we do not know much about their nature. These substances are the principal electron donors in soils and are the main cause for inducing the reduction of other substances such as iron, manganese and sulfur. It can be seen from Fig. 13.11 that, because of the lack of easily decomposable organic matter, there was practically no reduction reaction within 2–3 days of submergence for the subsoil of an Ali–Haplic Acrisol. On the other hand, the *Eh* dropped rapidly on submergence when

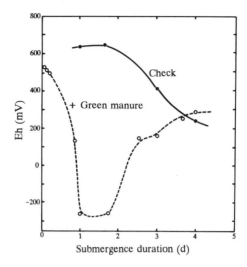

Fig. 13.11. Change in *Eh* for Ali–Haplic Acrisol during submergence (Yu, 1985).

some green manure was added. Therefore, it can be concluded that the production of organic reducing substances is the first step in the reduction process in soils. This is the case both under submerged conditions and when the soil is in contact with atmospheric air.

The composition of organic reducing substances in soils should be very complex because they are the decomposition products of various organic materials. From the viewpoint of oxidation–reduction reaction, what we are most interested in is the reduction intensity.

The *I–V* curves of the anaerobic decomposition products of two organic materials are shown in Fig. 13.12. For the decomposition products of vetch the concentration of the reducing substances is relatively high, in which four groups with half–wave potentials of 0.22, 0.33, 0.47, and 0.65 V vs. S.C.E. can be distinguished. The concentration of the decomposition products of rice straw is relatively low. It is also difficult to ascertain the half–wave potentials of various groups. Nevertheless, it can be seen from the figure that these latter products also include a variety of substances with properties varying from weakly reducing to strongly reducing. Generally, the quantity of reducing substances is the highest at the stage of vigorous decomposition of organic matter.

Some of these organic reducing substances can form complexes with ferrous and manganous ions. According to analytical data for 17 soil solutions, the complexing capacity (C_M) of these substances is related to their concentration (C_R) as follows (Yu and Ding, 1990):

$$C_R = (-0.123 \times 10^{-3}) + 7.544 C_M$$

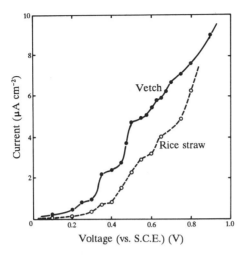

Fig. 13.12. *I-V* curve of anaerobic decomposition products of two organic materials (Yu, 1985).

Most of the organic reducing substances carry negative charge. As can be seen in Fig. 13.13, for the anaerobic decomposition products of tree leaves there appeared an oxidation peak at about 0.5 V. This peak remained unchanged after passing the solution through a column of cation–exchange resin. It disappeared after passing through anion–exchange resin. Because of the presence of negative charge, these reducing substances can be adsorbed by soils and iron oxides. According to an experiment, under comparable conditions the amounts of organic reducing substances adsorbed by goethite, Rhodic Ferralsol, Ali–Haplic Acrisol and Cambisol were 8.15, 4.08, 2.26, and 1.86 cmol kg^{-1}, respectively (Ding et al., 1990), showing a close relationship between the amount of adsorbed organic reducing substances and that of iron oxide. This point is of significance, because the first step of interactions between these substances and the oxides of iron or manganese is probably adsorption process, and it is only after this adsorption that it becomes possible for these substances to transfer their electrons to iron and manganese to reduce them. This will be further discussed in the following section.

According to equation (13–21), the lower the E^0 of the organic reducing substance, the faster the reduction of the oxides of iron and manganese. It has been found in a simulating experiment that, for the following oxidation–reduction substances

$$C_6H_4O_2 \text{ (Quinone)} + 2H^+ + 2e \rightleftharpoons C_6H_6O_2(\text{Hydroquinone}) \quad (13-23)$$
$$\text{Dehydroascorbic acid} + 2H^+ + 2e \rightleftharpoons \text{Ascorbic acid} \quad (13-24)$$

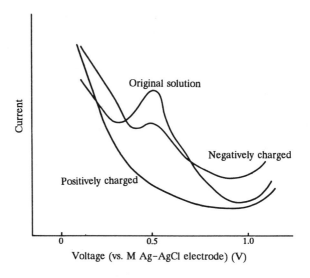

Fig. 13.13. *I–V* curve of organic reducing substances carrying different electric charges (Ding et al., 1990).

because the E^0 values are 0.699 and 0.40 V, respectively, the reaction rate of ascorbic acid with iron oxides was much higher than that of hydroquinone (Fig. 13.14). These data can in a qualitative way illustrate the difference in chemical reactivity among organic reducing substances with different reduction intensities in soils. As shall be seen in the following section, when

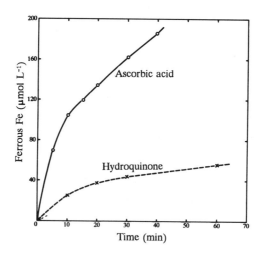

Fig. 13.14. Comparison of reduction rate of iron oxides in paddy soils derived from Acrisols by two organic reducing substances (pH 4.0).

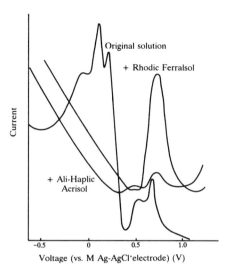

Fig. 13.15. Change in I-V pattern of organic reducing substances after reaction with variable charge soils (Ding et al., 1990).

these organic substances coexist, it is the one with a stronger reduction intensity that reacts with the oxides of iron and manganese first.

13.2.4 Oxidation–Reduction Reactions in Soils

When organic reducing substances are in contact with variable charge soils, two reactions may occur. First, those substances that carry negative and positive charges can be adsorbed by positive and negative surface charge sites of the soil, respectively. Second, if the reduction intensity of these substances is sufficiently high, these substances can transfer their electrons to the oxides of iron and manganese, inducing an oxidation–reduction reaction (McBride, 1987; Shindo and Huang, 1984; Stone and Morgan, 1984; Ukrainczyk and McBride, 1992). As has been discussed in the previous section, when the oxides of the two elements are present together, it is the oxide of manganese that should be reduced first.

The change in I-V pattern for the anaerobic decomposition products of pine leaves after reaction with two variable charge soils is shown in Fig. 13.15. In the original decomposition products there were several groups of reducing substances with peak potentials of −0.05, +0.13, +0.23, +0.53, and +0.69 V, respectively. After reaction with the soil, the former three groups with comparatively stronger reduction intensities disappeared. The quantity of those substances with a peak potential of 0.53 V also decreased markedly. Correspondingly, a large amount of manganous ions with a peak

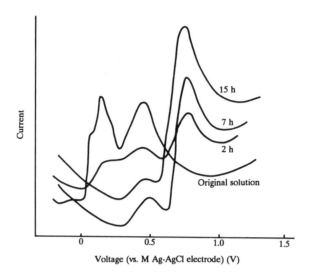

Fig. 13.16. Change in I-V pattern of organic reducing substances after reaction with Ali–Haplic Acrisol for different durations (Ding et al., 1990).

potential of 0.75 V appeared, especially for the Rhodic Ferralsol.

The reaction rate of different organic reducing substances with soils varies remarkably. It can be seen from Fig. 13.16 that those substances with peak potentials of 0.1 V and 0.2 V disappeared after 2 and 7 hr, respectively, whereas those with a peak potential of 0.5 V only decreased in amount. As expected, the amount of reduced manganese increased with time. The difference in reaction activity among organic reducing substances with different reduction intensities shown above is in conformity with the principle of sequential oxidation.

These data show that there are both adsorption and oxidation–reduction processes during the reaction of organic reducing substances with soils.

For variable charge soils containing both iron oxides and manganese oxides, as has been seen in Fig. 13.15 and Fig. 13.16, only the reduction of manganese oxides was observed. In order to compare the behavior of the two kinds of oxides, one can observe their reaction with organic reducing substances when they are present separately. In the case of manganese oxides, organic substances with various reduction intensities disappeared rapidly, and correspondingly large amounts of manganese chelates with a peak potential of 1.2 V were produced (Fig. 13.17). In contrast, when organic reducing substances reacted with iron oxides, it was only after 24 hr that a small amount of ferrous iron with a peak potential of 0.0 V appeared, although the amount of these organic substances decreased constantly with time (Fig. 13.18). Thus, it should be reasoned that the

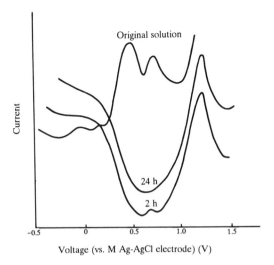

Fig. 13.17. Change in I-V pattern of organic reducing substances after reaction with manganese oxide (Ding et al., 1990).

decrease in amount of organic reducing substances in solution was caused chiefly by adsorption on goethite. The difference in activity between the two oxides when reacting with organic reducing substances is also in agreement with the principle of sequential reduction.

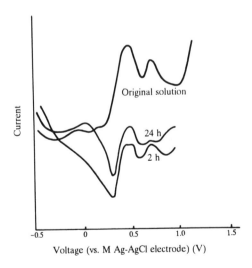

Fig. 13.18. Change in I-V pattern of organic reducing substances after reaction with goethite (Ding et al., 1990).

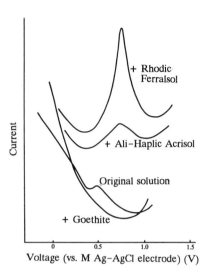

Fig. 13.19. Change in I-V pattern of organic reducing substances carrying negative charges after reactions with variable charge soils and goethite (Ding et al., 1990).

It has been mentioned that most of the organic reducing substances carry negative charge. The reaction patterns between negatively charged organic reducing substances separated from the anaerobic decomposition products of tree leaves and variable charge soils and goethite are shown in Fig. 13.19. It is understandable for the phenomenon that for the Rhodic Ferralsol system the amount of manganous ions with a peak potential of 0.75 V was larger than that for the Ali–Haplic Acrisol system, because the manganese content of the former soil is much higher. It can also be seen that organic reducing substances with a peak potential of 0.5 V disappeared after reaction with goethite, although there was no ferrous iron formed during the reaction. Apparently, the disappearance of these negatively charged organic substances was a result of the adsorption of them by goethite carrying positive charges.

From the above discussions it can be concluded that organic reducing substances can both be adsorbed by variable charge soils and be oxidized by the oxides of iron and manganese of the soil. These two processes are interrelated, but may not be necessarily synchronous. The adsorption of these substances should be helpful in increasing their stability against air–oxidation. On the other hand, the reduction of iron oxides and manganese oxides would result in the production of ferrous and manganous ions. Therefore, under natural conditions variable charge soils invariably contain both organic and inorganic reducing substances, mainly ferrous and manganous ions.

13.3 OXIDATION-REDUCTION REGIMES OF SOILS UNDER FIELD CONDITIONS

13.3.1 General Description

Under field conditions the oxidation-reduction regime of a soil is determined chiefly by the contents of organic matter and water. Organic matter is the principal source of electrons. Water content affects aeration, whereas atmospheric oxygen is an inexhaustible supply of oxygen. Once aeration is hindered, because the soil is in a closed or semiclosed state, electrons can only be transferred among various oxidation-reduction substances contained by the soil themselves.

For variable charge soils, because the temperature is high and the rainfall is abundant, biological activities and hence oxidation-reduction reactions are very active. In the following, according to the pattern of utilization, the oxidation-reduction regimes of these soils are described.

13.3.2 Soils Under Natural Forest

The distribution of oxidation-reduction potential in the profile of four Ferrali-Haplic Acrisols is shown in Fig. 13.20. The Eh_7 of the surface soil

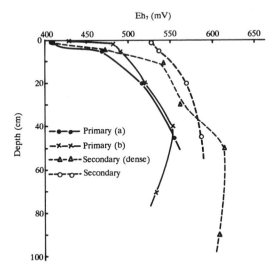

Fig. 13.20. Oxidation-reduction potentials in profile of Ferrali-haplic Acrisols under natural forest (Xishuangbanna) (Ding et al., 1984).

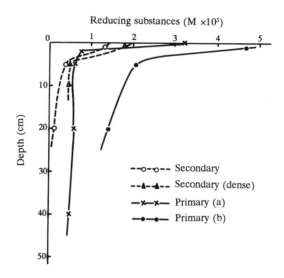

Fig. 13.21. Amounts of reducing substances in profile of Ferrali-Haplic Acrisols under natural forest (Xishuangbanna) (Ding et al., 1984).

may be lower by 100-200 mV than that of low-lying horizons even during the dry season (April) of the year. The numerical value of Eh_7 is as low as 400-450 mV. The Eh is lower in soils under primary forest than in those under reforested forest. The difference in Eh between soil a (annual rainfall 1400 mm) and soil b (annual rainfall 1000 mm), both under primary forest, is caused chiefly by the difference in vegetation density. For soils under reforested forest the Eh is also comparatively lower when the vegetation is denser.

The lower Eh in the surface layer of the four soils is caused by the presence of large amounts of reducing substances. A comparison between Fig. 13.21 and Fig. 13.22 can show that the larger the amount of reducing substances in a soil, the lower the Eh. Within the uppermost several millimeters where organic matter is abundant, the concentration of reducing substances can be as high as corresponding to $1-5 \times 10^{-5}$ mol L^{-1} of Mn^{2+}. In contrast, at a depth below 50 cm where the Eh_7 is higher than 550 mV, reducing substances generally cannot be detected.

Within the same climatic region, the moisture in Xanthic-Haplic Acrisols is generally higher than that in Ferrali-Haplic Acrisols and Ali-Haplic Acrisols. Hence, the oxidation-reduction regime is also different. Table 13.3 shows this difference. On the other hand, for Xanthic-Haplic Acrisols developed due to local poor drainage, the oxidation-reduction regime during dry season may not be remarkably different from other types of Acrisols of the same region (Yu et al., 1957).

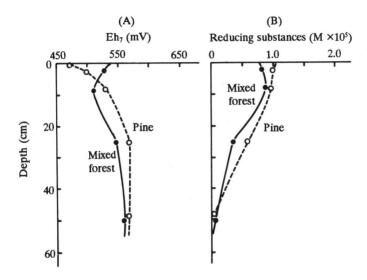

Fig. 13.22.   Oxidation–reduction potentials and amounts of reducing substances in Xanthic–Haplic Acrisol profiles (Guizhou) (Ding et al., 1984).

In subtropical regions where typical zonal Xanthic–Haplic Acrisols distribute, such as in the Guizhou Province of China, because the temperature is lower and the humidity is higher than those in the Guangdong Province, reducing substances can accumulate to a deeper depth in the soil profile, as shown in Fig. 13.22. The Eh_7 range in low-lying horizons is 550-570 mV, narrower than that in Xanthic–Haplic Acrisols of tropical regions (540-620 mV). The difference in Eh_7 between upper horizons and lower horizons, about 100 mV, is also comparatively small.

Table 13.3   Oxidation–Reduction Potentials in the Profile of a Ferra-li-Haplic Acrisol and a Xanthic–Haplic Acrisol (Guangdong) (Yu et al., 1957)

	Eh (mV)	
Depth (cm)	Ferrali–Haplic Acrisol	Xanthic–Haplic Acrisol
0–3	500	–
3–6	620	490
6–25	650	550
25–	670	560

13.3.3 Soils Under Economic Forest

In the surface layer of variable charge soils under economic forest there is also the accumulation of organic matter, although the amount is smaller than that for soils under natural forest. Correspondingly, the reduction condition is generally not as pronounced as for the latter type of soil. As can be seen in Fig. 13.23, for a Ferrali–Haplic Acrisol under bamboo the oxidation–reduction potential is high and the content of reducing substances is low, due to dry conditions of the soil and a low content of organic matter. In soils under rubber trees, more organic matter can accumulate. If some medicinal plants are grown in the rubber plantation so that a litter layer with a depth of about 2 cm can accumulate, a moderately reducing condition can develop in the soil. However, generally speaking, the difference in oxidation–reduction condition between the surface layer and the lower horizons is smaller than that for soils under natural forest.

13.3.4 Agricultural Soils

When variable charge soils are utilized for upland crops, except for the cultivated layer where a small amount of reducing substances can accumulate, the whole profile is in an oxidizing state (Fig. 13.24). When utilized for rice, the oxidation–reduction status is determined by the water regime. In cases where there is a glei horizon (G) in the profile, the amount of

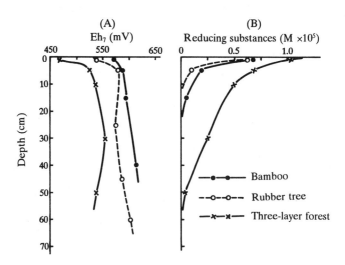

Fig. 13.23. Oxidation-reduction potentials and amounts of reducir substances in profile of Ferrali-Haplic Acrisols under economic fore (Xishuangbanna) (Ding et al., 1984).

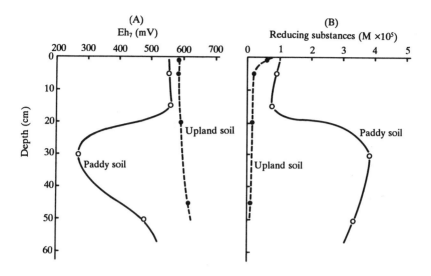

Fig. 13.24. Oxidation-reduction regimes of Ferrali-Haplic Acrisol under upland crop and rice-growing conditions (Xishuangbanna) (Ding et al., 1984).

reducing substances in this horizon may be very high even in seasons when no rice crop is planted. For an example shown in Fig. 13.24, the amount corresponds to 4×10^{-5} M Mn^{2+} in this horizon. Correspondingly, the Eh in this horizon is very low. When the soil is submerged for the cultivation of rice, the amount of reducing substances in the cultivated layer may be as high as corresponding to 1.3×10^{-4} M Mn^{2+}. The Eh_7 in this layer is generally lower than 100 mV.

It has been mentioned in Section 13.2 that those substances that can be oxidized at the electrode at an applied voltage of 0.35 V vs. M Ag/AgCl electrode may be regarded as strongly reducing substances, and that those oxidized only at 0.70 V may be regarded as weakly reducing substances. Thus, the current density at the relevant voltages may be used as a relative index to express the concentration of the two groups of reducing substances. The oxidation-reduction regime of the cultivated layer of paddy soils under different oxidation-reduction conditions is shown in Table 13.4. It can be seen that for the same type of paddy soil the stronger the reducing intensity, the larger the amount of reducing substances. The differences between the oxidizing soil and the strongly reducing soil may be as large as two orders of magnitude. Besides, the stronger the reduction intensity of the soil, the larger the proportion of strongly reducing substances in the total reducing substances.

Table 13.4 Oxidation-Reduction Potentials and Amounts of Reducing Substances in the Cultivated Layer of Paddy Soils Derived from Acrisols Under Different Oxidation-Reduction Conditions (Ding et al., 1982)

Soil Type	Redox Status	Eh (mV)	Reducing Substances[a](A)		Percentage of A	
			0.35V	0.7V	0.35V	0.7V
Young	Oxidizing	450	0.08	0.77	10.4	89.6
	Moderate reducing	110	1.83	12.5	14.6	85.4
	Strongly reducing	20	7.03	20.2	34.8	65.2
Developed	Weakly reducing	400	0.39	3.11	12.5	87.5
	Moderate reducing	150	1.41	5.27	26.8	73.2
	Strongly reducing	-10	11.7	16.3	71.8	28.2

[a]In unit of (μA cm^{-2}).

13.3.5 Relationship Between Intensity Factor and Capacity Factor of Oxidation-Reduction

In soils under field conditions there are various oxidation-reduction substances existing together in different proportions and in different oxidized or reduced forms. Therefore, one cannot expect to directly relate Eh to the quantity of any oxidation-reduction substance by applying the simple equation (13-8). On the other hand, it has been found, based on a large amount of analytic data, that for aerated soils there exists a noticeable relationship between the Eh_7 and the total concentration (C) of the reducing substances (Ding et al., 1984) (Fig. 13.25), following the equation

$$Eh_7 = 653 - 159 \log C$$

For paddy soils, within the wide Eh range of -200 to 600 mV, this kind of relationship is also remarkable (Yu, 1985). It is thus seen that even for soils with a complex composition the intensity factor and the quantity factor of oxidation-reduction are closely interrelated.

From the above discussions it can be concluded that for variable charge soils under field conditions, reduction processes can develop to a considerable extent even when the soil is in contact with atmospheric air. In the past, it was generally supposed that the significance of reduction reaction is limited mainly to submerged soils where the soil has been isolated from

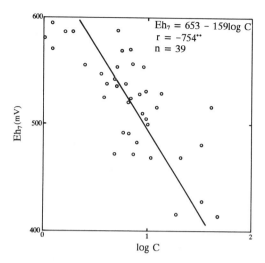

Fig. 13.25. Relationship between Eh_7 and concentration (C) of reducing substances in soils (C in $M \times 10^6$) (Ding et al., 1984).

the atmosphere. Actually, this is not the case. It can be said that wherever there is the presence of organic matter in the soil there is the condition for the production of electrons. Besides, within a soil there are microregional sites where the oxidation–reduction potential may be low due to the presence of organic substances in the rhizosphere secreted by plant roots or to poor aeration in the interior of soil aggregates (Yu, 1985). For variable charge soils in particular, oxidation–reduction reactions are very active.

The occurrence of active oxidation–reduction reactions in variable charge soils is of practical significance. The reduction and thus the mobilization of some elements, such as iron and manganese, make them available to plants. Because the contents of iron and manganese oxides are high, in some cases the concentration of these reduced elements is so high that they may be toxic to plants. In soil genesis, the migration and segregation of iron and manganese within the profile is an important criterion of soil development.

BIBBLIOGRAPHY

Ahmad, A. R. and Nye, P. H. (1990) *J. Soil Sci.*, 41:395–409.
Bartlett, R. J. and James, B. R. (1993) *Adv. Agron.*, 50:152–208.
Bohn, H. L. (1971) *Soil Sci.*, 112:39–45.
Bottcher, J. and Strebel, O. (1988) *Z. Pflanzenernaehr. Bodenk.*, 151: 363–368.
Cogger, C. G., Kennedy, P. E. and Carlson, D. (1992) *Soil Sci.*, 154:50–58.

Ding, C. P., Liu, Z. G., and Yu, T. R. (1982) *Soil Sci.*, 134:252-257.

Ding, C. P., Liu, Z. G., and Yu, T. R. (1984) *Geoderma*, 32:287-295.

Ding, C. P., Wu, Y. X., and Yu, T. R. (1990) *Trans. 14th Intern. Congr. Soil Sci.*, II:62-67.

Eary, L. E. and Rai, D. (1991) *Soil Sci. Soc. Am. J.*, 55:676-683.

Fang, J. A. and Liu, Z. G. (1987) *Anal. Instr.*, 3:16-19.

Faulkner, S. P., Patrick, W. H., and Cambrell, R. P. (1989) *Soil Sci. Soc. Am. J.*, 53:883-890.

Komada, M. (1990) *Trans. 14th Intern. Congr. Soil Sci.*, II:44-49.

Liu, Z. G. (1990) *Soils*, 22:237-240.

Liu, Z. G. (1986) in *Current Progress in Soil Research in People's Republic of China* (Soil Science Society of China). Jiangsu Science and Technology Publishers, Nanjing, pp. 92-99.

Liu, Z. G. and Yu, T. R. (1984) *J. Soil Sci.*, 35:469-479.

Mansfeldt, T. (1993) *Z. Pflanzenernahr. Bodenk.*, 156:287-292.

Matia, L., Rauret, G., and Rubio, R. (1991) *Z. Anal. Chem.*, 339:455-462.

McBride, M. B. (1987) *Soil Sci. Soc. Am. J.*, 51:1466-1472.

Miller, D. M., Tang, T., and Paul, D. W. (1993) *Soil Sci. Soc. Am. J.*, 57:356-360.

Patrick, W. H. and Jugsujinda, A. (1992) *Soil Sci. Soc. Am. J.*, 56:1071-1073.

Peiffer, S., Klemm, O., Pecher, K., and Hollerung, R. (1992) *J. Contam. Hydrol.*, 10:1-18.

Sadana, U. S., Naygar, V. K., and Takkar, P. N. (1990) *Trans. 14th Intern. Congr. Soil Sci.*, II:56-61.

Shindo, H. and Huang, P. M. (1984) *Soil Sci. Soc. Am. J.*, 48:927-934.

Stone, A. T. and Morgan, J. J. (1984) *Environ. Sci. Technol.*, 18:450-456.

Ukrainczyk, L. and McBride, M. B. (1992) *Clays Clay Miner.*, 40:157-166.

Willett, I. R. (1990) *Trans. 14th Intern. Congr. Soil Sci.*, II:38-43.

Yu, T. R. (1976) *Electrochemical Properties of Soils and Their Research Methods.* Science Press, Beijing.

Yu, T. R. (1985) *Physical Chemistry of Paddy Soils.* Science Press/Springer Verlag, Beijing/Berlin.

Yu, T. R. (ed.) (1987) *Principles of Soil Chemistry.* Science Press, Beijing. pp.365-406.

Yu, T. R. and Li, S. H. (1957) *Acta Pedol. Sinica*, 5:97-110.

Yu, T. R., Ling, Y. X., Ding, C. P., Mu, Y. S., and Liu, Z. G. (1957) *Science Bull.*, 11:338-339.

Yu, T. R. and Ding, C. P. (1990) in *Soils of China* (Y. Hseng and C. K. Li, ed.). Science Press, Beijing, pp. 533-553.

Yu, T. R. (1992) *Adv. Agron.*, 48:205-250.

Yu, T. R. and Ji, G. L. (1993) *Electrochemical Methods in Soil and Water Research.* Pergamon Press, Oxford.

14

FERROUS AND MANGANOUS IONS

X. M. Bao

It has been seen in the previous chapter that, when organic reducing substances are in contact with soils, iron and manganese oxides of the soil can be reduced, producing ferrous and manganous ions. This reduction process is of particular significance for variable charge soils. This is because the contents of iron and manganese oxides in these soils are higher than those in constant charge soils, and therefore there is the basis for the production of large amounts of ferrous and manganous ions. On the other hand, the low pH of variable charge soils is one of the favorable factors in the reduction of iron and manganese. Hence, the possibility exists for the production of the reduced form of these two elements exists even under well–aerated conditions, provided that the soil contains organic matter. It is for these reasons that iron and manganese can actively participate in chemical reactions in variable charge soils in the form of ions, and they are of great significance in both soil genesis and plant nutrition. As a matter of fact, many reports have shown that the amount of manganous ions in strongly acid soils is so high that the toxicity of manganese may be one of the causes for the poor growth of plants.

In this chapter, after the elucidation of the reduction and dissolution of iron and manganese, emphasis of discussions will be placed on the chemical equilibria of the two kinds of ions. Finally, the regimes of the two ions in variable charge soils under field conditions will be described.

14.1 REDUCTION AND DISSOLUTION OF IRON AND MANGANESE

14.1 Chemical Properties of Iron and Manganese

The solubility of iron oxides in soils is extremely low. The solubility product of $Fe(OH)_3$ is about 10^{-38}. Therefore, the concentration of Fe^{3+} ions in solution is only 10^{-8} M even at a pH of 4, if the concentration is determined solely by the solubility of the oxides. Similarly, the quantity of Mn^{4+} ions due to the dissolution of manganese oxides is also very small. On the other hand, because the solubility products of $Fe(OH)_2$ and $Mn(OH)_2$ are about 10^{-16} and 10^{-13}, respectively, after reduction, ferrous and manganous (in this chapter the latter is simply referred to as manganese) ions can exist stably in soils.

Let us examine the reduction of iron oxide:

$$Fe_2O_3 + 6H^+ + 2e \rightleftharpoons 2Fe^{2+} + 3H_2O \tag{14-1}$$

$$K = \frac{[Fe^{2+}]^2}{[H^+]^6[e]^2} \tag{14-2}$$

$$pFe^{2+} = pe + 3pH + 0.5pK \tag{14-3}$$

where K is the equilibrium constant in equation (14-1). The value of K is about 10^{25}.

Therefore, we can get

$$pFe^{2+} = pe + 3pH - 12.5 \tag{14-4}$$

For the reduction of manganese oxide, the following equations can be derived:

$$MnO_2 + 4H^+ + 2e \rightleftharpoons Mn^{2+} + 2H_2O \tag{14-5}$$

$$pMn^{2+} = 2pe + 4pH - 41.7 \tag{14-6}$$

Thus, at pH 7 we have

$$pFe^{2+} = pe + 8.5 \tag{14-7}$$

$$pMn^{2+} = 2pe - 13.7 \tag{14-8}$$

and at pH 5 we obtain

$$pFe^{2+} = pe + 2.5 \tag{14-9}$$

$$pMn^{2+} = 2pe - 21.7 \tag{14-10}$$

It is thus seen that under the same oxidation–reduction condition the lower the pH, the more favorable the chances for the reduction of iron and manganese; and at the same pH the stronger the reducing condition (the smaller the pe value), the higher the concentration of the two ion species in solution. These are the principal factors in controlling the quantity of the two ion species in soils. Because the composition and chemical reactions in soils are very complex, it is not possible to directly calculate the quantities of ferrous and manganese ions in soils with the above equations. Nevertheless, the above discussions indicate that the quantities of the two ion species are affected by the pH and oxidation–reduction conditions and that manganese is easier to reduce and dissolve than iron. This is the actual case in soils. The quantities of iron and manganese that were reduced by hydroquinone for different horizons of a Haplic Acrisol and a paddy soil derived from it are shown in Table 14.1. With the same concentration of hydroquinone, the quantity of reduced manganese was always larger than that of iron. For manganese oxides, as high as 70% can be reduced, whereas for iron oxides it was less than 1^. This phenomenon can be used to explain why the segregation of manganese is much more pronounced than that of iron in variable charge soils and paddy soils derived from them.

Table 14.1 Contents of Easily–Reducible Iron and Manganese in Different Horizons of a Paddy Soil Derived from Acrisol (Yu, 1985)

Horizon	Concentration of hydroquinone (%)	Reducible (mg kg^{-1})		% of Total	
		Fe	Mn	Fe	Mn
Cultivated	0.02	8.0	10.6	0.27	23.7
	0.05	9.3	18.0	0.32	40.1
	0.1	11.0	19.0	0.37	42.4
	0.2	13.5	18.3	0.46	40.9
Illuvial	0.02	1.3	60.4	0.04	68.0
	0.05	2.0	62.8	0.07	70.8
	0.1	1.9	62.8	0.07	70.8
	0.2	2.9	–	0.10	–
Parent Acrisol	0.02	2.0	16.4	0.06	42.3
	0.05	2.1	16.5	0.06	42.4
	0.1	2.8	–	0.07	–
	0.2	3.5	18.0	0.09	46.4

14.1.2 Reduction of Iron

During the reaction of organic reducing substances with iron oxides of the solid phase of the soil, the electrons liberated by organic substances migrate to the surface of the iron oxides, rendering Fe^{3+} ions changed to Fe^{2+} ions through accepting electrons. Then, Fe^{2+} ions diffuse into the solution. In the process several steps are involved. Among these steps, the release of Fe^{2+} ions from the surface of solid iron oxides may be the rate-determining one. Therefore, the reduction of iron oxides involving two phases is a rather slow process. Apparently, the stronger the reduction intensity of organic substances, that is, the larger the amount of electrons, the more rapid the reaction. The change in proportions of organic reducing substances and ferrous iron with time during submergence of a paddy soil derived from Xanthic-Haplic Acrisol is shown in Fig. 14.1. It can be seen that with the prolongation of time, organic reducing substances were consumed gradually, and correspondingly ferrous ions were produced. At the end of 3 weeks, more than 80% of the organic form of reducing substances was consumed for the transformation of iron. It should be noted that for organic reducing substances coming from the decomposition products of rice straw, it required 10 days to transfer half of their electrons to iron oxides, whereas for those coming from vetch it required only 5 days. It has been shown that in the latter case the reduction intensity of organic substances is stronger.

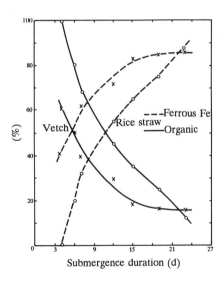

Fig. 14.1. Change in proportions of ferrous iron and organic reducing substances with time for paddy soil derived from Xanthic-Haplic Acrisol after addition of different organic materials and submergence (Yu, 1985).

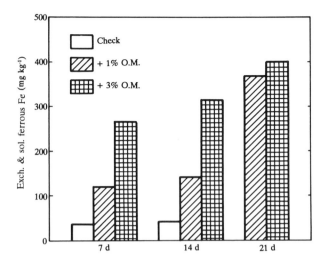

Fig. 14.2. Change in amounts of water–soluble and exchangeable ferrous iron with time for Ferrali–haplic Acrisol (Dinghu Mountain) after addition of different amounts of organic matter (water content 40%).

It can be expected that for the same soil the higher the content of organic matter, the more the amount of ferrous iron released from the solid phase, due to the higher concentration of organic reducing substances produced. Figure 14.2 shows such an example. The quantity of water–soluble and exchangeable ferrous iron in the surface layer of a Ferrali–Haplic Acrisol without the addition of organic matter was low. With the increase in the amount of added organic matter, the quantity of ferrous iron increased.

According to equation (14–4) describing the relationship between the quantity of ferrous iron and the activity of electrons, there should be a linear correlation between pFe^{2+} and pe with a slope of 1. Figure 14.3 shows that the quantities of water–soluble ferrous iron and water–soluble+exchangeable ferrous iron are closely related to the activity of electrons, although the correlation slope between $pFe_{(water-soluble)}$ or $pFe_{(water-soluble + exchangeable)}$ and pe is much smaller than 1. The small slope is assumed to be caused by other factors that control the distribution of different forms of ferrous iron, as shall be discussed in a later section.

14.1.3 Reduction of Manganese

Similar to the case for iron oxides, in the reduction of manganese oxides it is also the slow release of manganese ions from the solid phase that is the rate–determining step of the whole process. Various mechanisms, including

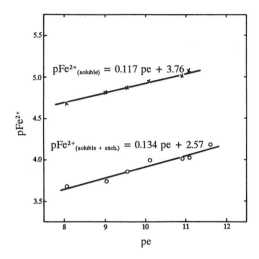

Fig. 14.3. Relationship between amount (mol kg⁻¹) of ferrous iron and pe for the surface soil of Ferrali–haplic Acrisols (Dinghu Mountain).

chemical and microbial (Bartlett and James, 1993; Ritchie, 1989; Willett, 1990), have been suggested to account for the reduction.

In variable charge soils, iron oxides and manganese oxides generally exist together. According to the principle of sequential reduction, when reduction

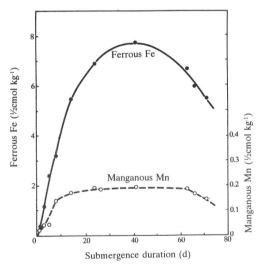

Fig. 14.4. Change in amounts of ferrous and manganous ions in cultivated layer of a paddy soil derived from Acrisol after submergence (Jiangxi) (Yu, 1985).

occurs, manganese should be reduced first. However, because the rate of reduction of manganese may not be sufficiently high, the reduction of manganese may be accompanied by a simultaneous reduction of iron during a certain period of reaction under field conditions. The production of manganese ions and ferrous ions for a paddy soil derived from Acrisol after submergence is shown in Fig. 14.4. Within the first several days both of the two ion species increased. After 10 days, manganese ions did not increase further, because at this time all the reducible manganese had been reduced, whereas the newly produced organic reducing substances can react continuously with the large amount of iron oxides. These data illustrate qualitatively the difference in oxidation–reduction behavior between manganese and iron oxides in soils. Therefore, within a certain period of submergence the amount of Mn^{2+} may exceed that of Fe^{2+}, although eventually the reverse is true due to the presence of a much higher content of iron oxides than that of manganese oxides in most of the variable charge soils (Pan et al., 1994).

Like the case for iron, the reduction of manganese would be closely related to the reduction strength of the soil. Water content should be an important factor in this respect. The amounts of manganese ions for two soils under different water regimes are shown in Table 14.2. The amounts of both water–soluble and exchangeable manganese ions increased with the increase in water content and the corresponding decrease in Eh. The increase in water–soluble manganese was more remarkable. Under field

Table 14.2 Effect of Water Content on Amount of Manganese Ions in Soils

Soil	Water (%)	pH	Eh (mV)	Mn (mg kg^{-1}) Sol.[a]	Exch.[b]	% of Total Sol.	Exch.
Ali–Haplic	30	4.32	475	6.4	15.0	29.9	70.1
Acrisol	40	4.66	340	12.7	28.7	30.7	69.3
	50	4.76	287	20.0	37.0	35.1	64.9
	60	5.08	126	25.0	44.1	36.2	63.8
Ferrali–Haplic	30	5.11	544	9.3	177	5.0	95.0
Acrisol	40	5.21	470	13.3	219	5.7	94.3
	50	5.21	418	15.6	254	5.8	94.2
	60	5.27	276	19.2	301	6.0	94.0

[a]Soluble. [b]Exchangeable.

Table 14.3 Effect of Organic Matter on Amount of Manganese Ions in
Soils

Soil	Depth (cm)	O. M. (%)	Mn (mg kg⁻¹)			
			Sol.[a]	Exch.[b]	Compl.[c]	Sum
Rhodic Ferralsol	0–15	5.3	1.09	140	80.8	221.9
(Xuwen)	100+	0.6	0.13	43.8	69.8	113.7
Fer.-Haplic Acrisol	0–20	6.58	0.33	3.71	38.9	42.9
(Liuzhou)	60+	2.50	0.80	2.01	11.5	14.3

[a]Soluble; [b]Exchangeable; [c]Complexed.

conditions the water regime of the soil may change constantly, especially in areas where there is a distinct dry-wet cycle; hence, the amount of manganese in a soil may change constantly during the year. For different soils, the higher the content of organic matter, the larger the amount of organic reducing substances and hence the amount of reduced manganese. It can be seen from Table 14.3 that the amounts of manganese ions in the surface layer of two soils are much larger than those of the subsoil.

14.2 CHEMICAL EQUILIBRIA OF FERROUS IONS

Ferrous iron exists in soils in forms of water-soluble, exchangeable, complexed, and precipitated. A part of water-soluble ferrous iron can be chelated by organic ligands (Yu, 1985). There are a series of dynamic equilibria among these forms, with the equilibrium status determined by physicochemical reactions, including dissolution, ion exchange, chelation, and complexation. These shall be discussed in the following sections.

14.2.1 Dissolution and Chelation

The possibility for the formation of precipitates of ferrous compounds in well-drained variable charge soils is very small. For the chemical equilibrium

$$Fe^{2+} + 2OH^- \rightleftharpoons Fe(OH)_2 \qquad (14\text{-}11)$$

the solubility product of $Fe(OH)_2$ cited in the literature differs greatly. This

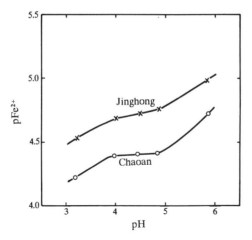

Fig. 14.5. Relationship between amount of water–soluble ferrous iron and pH for two Ferrali–Haplic Acrisols.

is probably caused by differences in precipitation conditions and in degree of ageing. If it is assumed that the value is 10^{-16}, it can be calculated that at pH 6 the concentration of Fe^{2+} ions in solution would be 1 mol L^{-1}, a high value that actually cannot be achieved in soils. Therefore, the quantity of water–soluble ferrous iron should be determined chiefly by other factors.

The relationship between the quantity of water–soluble ferrous iron and the pH for two Ferrali–Haplic Acrisols is shown in Fig. 14.5. Within the pH range of 4 to 5 the quantity is not affected by the pH. The increase in quantity when the pH is lower than 4 may reflect the enhancing effect of hydrogen ions on reduction and dissolution of iron oxides, whereas the decrease in quantity when the pH is higher than 5 may be related to the intensification of complexation and adsorption.

When a soil is submerged, because of the presence of a large amount of sulfide ions, ferrous ions can combine with these ions to form ferrous sulfide. The increase in the pH under submerged conditions is favorable to this reaction. Figure 14.6 shows the results of a simulating experiment in which different amounts of ferrous iron and sulfide ions were added to a paddy soil derived from Ali–Haplic Acrisol, taking the solubility product of FeS as 3.7×10^{-19} in calculations. In the figure, the relationship between pFe^{2+} and the pH for two submerged paddy soils is also shown. It can be seen that two pH ranges can be distinguished. When the pH is lower than about 5.3, the effect of pH on the quantity of ferrous iron is rather small. At a higher pH, the results of the simulating experiment can be expressed by the following equation:

$$pFe^{2+} = 1.47pH - 4.88 \qquad (14\text{-}12)$$

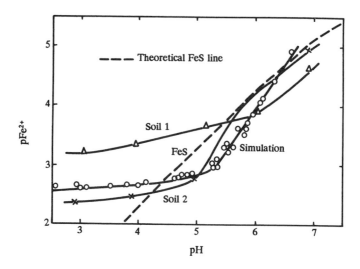

Fig. 14.6. Relationship between amount of water–soluble ferrous iron and pH for two variable charge paddy soils during submergence (Yu, 1985).

The curves for the two soils lie close to that of the simulating experiment.

The above data indicate that, for submerged soils, the dissolution of ferrous sulfide may be an important factor in controlling the quantity of water–soluble ferrous iron.

Organic ligands present in soil solution can form chelates with ferrous ions, with the relative proportions of ionic form and chelated form

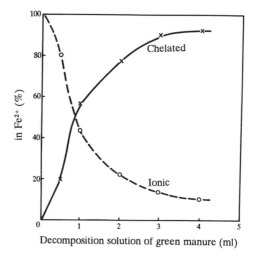

Fig. 14.7. Relative proportions of chelated and ionic ferrous iron in relation to amount of chelating agent (Yu, 1985).

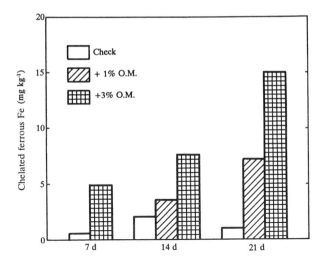

Fig. 14.8. Effect of organic matter on amount of chelated ferrous iron in soil (Ferrali-Haplic Acrisol).

determined by the relative quantities of the two reactants (Fig. 14.7). Therefore, generally speaking, the higher the content of organic matter and hence the larger the amount of ligands, the larger the amount of chelated ferrous iron (Fig. 14.8). It has been found that in variable charge soils, chelated ferrous iron can amount to tens percents of the total water–soluble ferrous iron. This point is of significance because the mobility will be increased and the stability against oxidation strengthened when ferrous ions are chelated. This is the reason why ferrous ions can be present in the surface layer of variable charge soils even under aerated conditions when the content of organic matter is high. It is also for this reason that the color of the surface layer of well-developed variable charge soils is not reddish brown, but gray, caused by the leaching loss of a large part of iron in water–soluble form during the long–term genesis process.

The stability constant of chelated ferrous iron is related to the nature of the ligands. In a complex soil system, because there are a variety of ligands present together, the stability constant as determined by ordinary methods is an apparent mean value. It was observed that the stability constant increased considerably after the aeration of the soil solution (Table 14.4), caused probably by the oxidation of the components with a low chelating stability, particularly those with a small molecular weight, because it has been known that the stability constant for the ligands with a molecular weight smaller than 500 in the decomposition products of plants is lower than those with a molecular weight larger than 1000 by about one order of magnitude when forming chelates with ferrous iron (Bao, 1987).

Table 14.4 Effect of Oxidation of Chelating Agents on Stability Constant of Ferrous Chelates (Bao and Yu, 1987)

Soil[a]	log K	
	Original Solution	Aerated
Xanthic–Haplic Acrisol	4.2	4.7
Xanthic–Haplic Acrisol	3.1	5.6
Litter layer of Xanthic Acrisol	4.0	4.1
Ali–Haplic Acrisol	3.6	4.3
Ferrali–Haplic Acrisol	3.5	4.2
Ferrali–Haplic Acrisol	4.4	4.6
Mean	3.8	4.6

[a]Incubated for 1 week with 20% water.

14.2.2 Ion Exchange

In Chapter 3 we have seen the effect of the adsorption of ferrous ions on the adsorption of potassium and calcium ions in variable charge soil. It has been observed that the larger the quantity of negative surface charge carried by the soil, the larger the amount of ferrous ions adsorbed (Yu, 1985). The competitive adsorption of ferrous ions with other cations

Table 14.5 Exchange of Calcium Ions by Ferrous Ions in Paddy Soils (Yu, 1985)

Parent Soil	Apparent CEC (cmol kg^{-1})	Fe^{2+} (S value)	Adsorbed (A) ($\frac{1}{2}$cmol kg^{-1})	Replaced (B) ($\frac{1}{2}$cmol kg^{-1})	A/B
Xanthic–Haplic Acrisol	8.82	1.0	3.10	3.52	0.88
		2.0	3.44	3.67	0.94
Ali–Haplic Acrisol	9.32	1.0	3.10	2.85	1.09
		2.0	3.47	3.23	1.07
Haplic Acrisol	7.36	2.5	4.14	4.39	0.94
		5.0	5.17	5.17	1.00

proceeds in a stoichiometric manner. As can be seen in Table 14.5, when ferrous ions were used to replace adsorbed calcium ions, the ratio of adsorbed ferrous iron to replaced calcium was close to 1.

The relationship between the amount of exchangeable ferrous iron and the pH for three soils is shown in Fig. 14.9. Within the pH range of 4.5 to 6.0 the effect of the pH is not great. The increase in exchangeable ferrous iron at pH lower than 4.5 probably reflects the increase in the amount of total ferrous iron, whereas the decrease in exchangeable ferrous iron at higher pH will be caused by the intensification of complexation of ferrous ions with organic matter. This is because according to the principle of complexation, the higher the pH, the larger the stability constant of the complexes of ferrous iron (Bao and Yu, 1987).

14.2.3 Complexation

The term *complexation* referred to here is used to denote the reaction between ferrous ions and ligands carried by organic matter of the solid part of a soil, so that it can be distinguished from the reaction between ferrous ions and organic ligands in soil solution, although these two reactions are both mainly chelation in nature. Apparently, for different soils the relative importance of this kind of complexation would be dependent on the content of organic matter. Indeed, experiments have shown that complexed ferrous iron did not exist after the removal of organic matter from the soil, whereas the amount of complexed ferrous iron was linearly related to the content

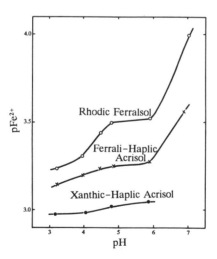

Fig. 14.9. Relationship between amount of exchangeable ferrous iron and pH in variable charge soils.

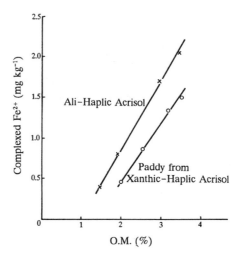

Fig. 14.10. Relationship between amounts of complexed ferrous iron and organic matter in a paddy soil derived from Xanthic–Haplic Acrisol and an Ali–Haplic Acrisol (Yu, 1985).

of organic matter when the soil was incubated after the addition of different amounts of organic materials (Fig. 14.10). Statistical data also showed that

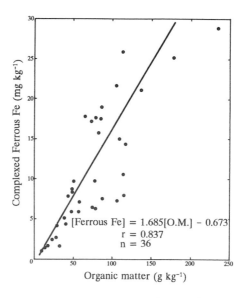

Fig. 14.11. Correlation between amounts of complexed ferrous iron and organic matter in variable charge soils.

there is a close relationship between the amount of complexed ferrous iron and the content of organic matter for various variable charge soils (Fig. 14.11). It can be calculated from the slope in the figure that one gram of soil organic matter can complex about 0.16 mg of ferrous iron.

14.2.4 Chemical Equilibria Among Different Forms of Ferrous Iron

In soils, there is a dynamic equilibrium among different forms of ferrous iron. All factors capable of affecting dissolution, chelation, adsorption, and complexation of ferrous iron, including the amount ferrous iron, the surface charge of soil colloids, and the status of organic matter, will exert influences on this kind of equilibrium. Among environmental factors the pH plays a special role, because, as has been discussed in the previous sections, pH can affect each of the above-mentioned chemical equilibria.

The distributions of different forms of ferrous iron in two Ferrali-Haplic Acrisols at different pH are shown in Figs. 14.12 and 14.13, respectively. According to the above discussions, within the pH range of 3.2 to 7.0 the existence of precipitated ferrous hydroxide would be unlikely. It can be seen from the figures that for the same soil the percentages of water-soluble and exchangeable forms decreased while that of complexed form increased with the increase in the pH. This is in conformity with the principle of the effect of the pH on the chemical equilibria of ferrous iron. It is noticeable that the distribution pattern of various forms of ferrous iron differs markedly with the kind of the soil. For the Ferrali-Haplic Acrisol of the Xishuangbanna region with an organic matter content of as high as 7.84% due to dense forest vegetation, complexed iron may account for more than 70% of the total ferrous iron. For the Ferrali-Haplic Acrisol of the Chaoan region the organic matter content is 4.48%. Correspondingly, the percentage of complexed form in total ferrous iron ranges from 10% to 40%. These data mean that organic matter plays an important role in determining the forms of ferrous iron in variable charge soils.

Water-soluble ferrous ions can account for several percent of the total ferrous iron at pH 5. Following the decrease in the amount of ferrous ions in solution, these ions will be liberated from the exchangeable form and the complexed form. Therefore, under field conditions, iron would be constantly leached out from the surface layer of variable charge soils when the rainfall is high.

14.3 CHEMICAL EQUILIBRIA OF MANGANOUS IONS

14.3.1 General Discussion

It has been stated in Section 14.1 that in soils where iron and manganese

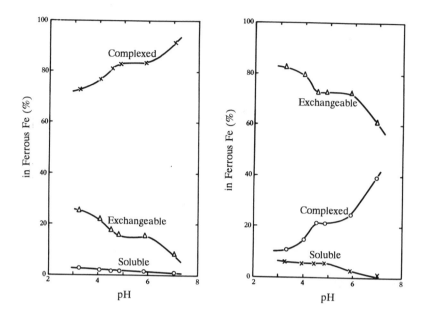

Fig. 14.12. (Left) Distribution of various forms of ferrous iron in relation to pH in the surface layer of a Ferrali–Haplic Acrisol (Xishuangbanna).

Fig. 14.13. (Right) Distribution of various forms of ferrous iron in relation to pH in the surface layer of a Ferrali–Haplic Acrisol (Chaoan).

oxides exist together it is the latter that is easier to be reduced. However, except for horizons where iron and manganese have been accumulated, the content of manganese oxides of variable charge soils is generally only 0.1–0.3%, although higher than that of constant charge soils. Therefore, in soils where there is a reduction condition it is mainly the insufficient quantity of manganous ions that limits the full proceeding of some chemical reactions. The solubility product of $Mn(OH)_2$ is about 10^{-13}, larger than that of $Fe(OH)_2$ by several orders of magnitude. Hence, the possibility of the formation of the precipitate of manganese hydroxide in soils under ordinary pH conditions is very small. Besides, the stability constants of complexes or chelates of manganese ions with organic ligands in soils are smaller than those of ferrous ions by about one order of magnitude (Bao and Yu, 1987). This makes the binding stability between manganous ions and organic matter of the solid phase of soils comparatively weak. These factors are the reasons why manganous Mn exists chiefly in the form of free ions in soils (Ritchie, 1989).

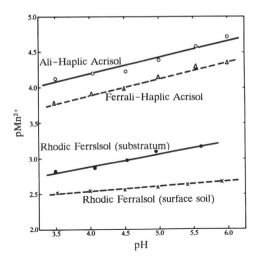

Fig. 14.14. Relationship between amount of exchangeable (including water-soluble) manganese and pH in variable charge soils (Mn in unit mol kg^{-1}).

14.3.2 Dissolution

According to equation (14-6), at the same reduction intensity the quantity of reduced manganese is determined mainly by pH. Therefore, there are extensive data in the literature showing that the amount of manganese ions in a soil is generally proportional to that of hydrogen ions, although the proportional constant varies with some other factors. The relationship between the amount of manganous ions, including the water-soluble form and the exchangeable form, and pH for four variable charge soils is shown in Fig. 14.14. The amount of manganese ions in different soils at the same pH varies markedly. This is caused by differences in the amounts of reducible manganese and organic matter. The amount of reducible manganese in the Rhodic Ferralsol is the largest. In particular, because the organic matter content of the surface layer of this soil is as high as 4.0%, the amount of manganese ions is high.

There is a dynamic equilibrium between adsorbed manganese ions and Mn ions in solution. It can be observed that the amount of water-soluble manganese is also linearly related to the pH (Fig. 14.15).

The amount of manganese ions in a soil determines the availability of that element to plant. Figure 14.16 shows that the manganese content of a pea plant is linearly related to the amount of manganese ions in soil within the pH range of 4.5-6. Thus, when the pH of a soil is sufficiently low, the toxicity of manganese to plants may become a realistic problem (Foy, 1984).

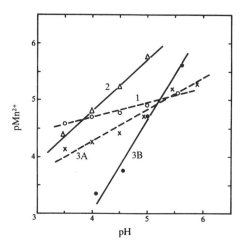

Fig. 14.15. Relationship between amount of water–soluble manganese and pH in variable charge soils. 1, Ali–Haplic Acrisol; 2, Ferrali–Haplic Acrisol; 3A, surface layer of Rhodic Ferralsol; 3B, substratum of same soil.

14.3.3 Chelation

As in the case for ferrous ions, manganese ions can form chelates with

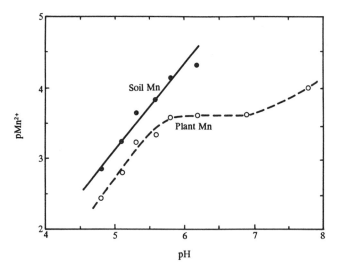

Fig. 14.16. Relationship between content of manganese in a pea plant and amount of manganese ions in soil (Manganese is expressed in moles per kilogram of soil or dry plant) (Yu et al., 1958).

ligands in the decomposition products of plant materials, although the stability constants of the chelates are smaller than those of the ferrous ions. Ligands with a larger molecular weight can form more stable chelates with manganese ions (Table 14.6). Some of the ligands in soil solution possess a reducing property, and the concentration of these ligands may be closely related to that of reducing substances (Bao et al., 1983). The stability constants of Mn-chelates increased after aeration of the solution (Table 14.7). Most of the ligands carry negative charges and can be adsorbed by iron oxides of the soil (Yu, 1985).

14.3.4 Ion Exchange

In the adsorption of manganese ions by variable charge soils, both electrostatic attraction and specific reaction are involved. Therefore, this adsorption is closely related to the surface properties of the soil. The amount of adsorbed manganese ions in two variable charge soils is shown in Table 14.8. When the amount of manganese ions added to the soil was the same, the amount of adsorbed manganese for the Hyper-Rhodic Ferralsol was two times that of the Ferrali-Haplic Acrisol in both the potassium-saturated and calcium-saturated soils, although the quantity of negative surface charge carried by the latter soil is larger than that of the former soil. The amount of adsorbed manganese was much larger for the potassium-saturated soil than that for the calcium-saturated soil, in conformity with the principle of ion exchange. These data indicate that at a pH of about 7, both ion exchange and specific adsorption of manganese

Table 14.6 Relationship Between Stability Constant of Mn-Chelates and Molecular Weight of Ligands of Decomposition Products of Plant Materials (Bao, 1987)

	log K		
Source of Ligands	<500[a]	500–1000	>1000
Rice straw	3.0	3.1	3.6
Vetch	2.8	3.3	4.6
Milk vetch	2.8	3.0	3.8
Grass	2.8	3.1	4.1
Mean	2.9	3.1	3.9

[a]Molecular weight.

Table 14.7 Effect of Oxidation of Chelating Agents of Soil Solution on Stability Constant of Mn-Chelates (Bao and Yu, 1987)

Soil[a]	log K	
	Original Solution	Aerated
Xanthic Acrisol (Guizhou)	3.2	3.8
Xanthic Acrisol (Jiangxi)	4.1	4.5
Litter layer	2.0	3.7
Ali-Haplic Acrisol (Jiangxi)	3.3	3.7
Ferrali-Haplic Acrisol (Yunnan)	3.0	3.3
Ferrali-Haplic Acrisol (Yunnan)	3.1	3.8
Mean	3.1	3.8

[a]Incubated for 1 week with 20% water.

Table 14.8 Adsorption of Manganese Ions by Variable Charge Soils

Soil	Mn Added (cmol kg^{-1})	Mn Adsorbed (cmol kg^{-1})	
		K-Saturated	Ca-Saturated
Ferrali-Haplic	0.45	0.23	0.05
Acrisol[a]	0.89	0.45	0.10
	1.79	0.58	0.19
	2.67	0.74	0.54
	3.57	1.04	0.92
	4.46	1.45	1.34
Hyper-Rhodic	0.45	–	0.38
Ferralsol[b]	0.89	0.64	0.63
	1.79	–	1.04
	2.69	2.05	1.66
	3.57	2.50	2.10

[a]pH of K-saturated and Ca-saturated soils, 6.93 and 6.59, respectively.
[b]pH of K-saturated and Ca-saturated soils, 6.89 and 6.60, respectively.

ions can occur in variable charge soils. It has been mentioned in Chapter 4 that iron oxide is the principal adsorbent for hydrolytic cations in variable charge soils (Barah et al., 1992; McKenzie, 1980; Paterson et al., 1990). The Hyper-Rhodic Ferralsol, with an iron oxide content of as high as 25%, should adsorb more manganese ions than do other types of soils.

The content of manganese oxides in soils is of the order of 0.1-0.3% in most cases. In soil chemistry, the contribution of manganese oxides to ion adsorption is generally not considered. Actually, these compounds with a very low point of zero charge carry a large amount of negative surface charge under ordinary soil pH conditions. Therefore, their capacity for adsorbing cations is much larger than that of the oxides of iron and aluminum, if compared on a weight unit basis. In particular, manganese oxides possess a very strong affinity for manganese ions (McKenzie, 1980). This can be seen from a comparison of the replacing power between the manganese ions and the sodium or aluminum ions for replacing potassium or calcium ions from artificially prepared manganese oxides shown in Fig. 14.17 and Fig. 14.18. The replacing power of manganese ions is even much stronger than that of aluminum ions. This indicates that in the interaction between manganese oxides and manganese ions a specific force plays a very important role.

Adsorbed manganese ions can be replaced by other cations. However, for iron- and manganese-rich soils, because the adsorption affinity for manganese ions is strong, the replacement is difficult. It can be seen from Table 14.9 that only a very small amount of manganese ions was replaced

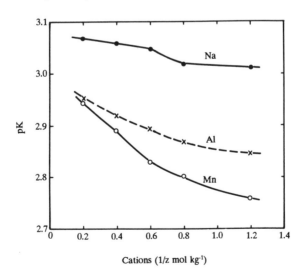

Fig. 14. 17. pK value in suspension of K-saturated manganese oxide added with different cations (data of T. R. Yu, F. L. Wang, and P. M. Huang)

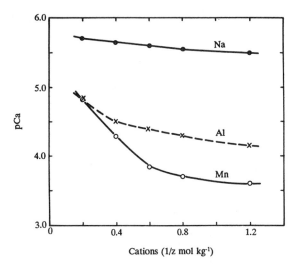

Fig. 14.18. pCa value in suspension of Ca-saturated manganese oxide added with different cations (data of T. R. Yu, F. L. Wang, and P. M. Huang)

by potassium or calcium ions, even when 20 $1/z$ cmol kg^{-1} of these replacing ions were added to a Mn-saturated Rhodic Ferralsol, although the replaced amount by calcium ions was a little larger than that by potassium ions. This phenomenon means that Rhodic Ferralsols possess a very strong adsorbing strength for manganese ions. This strong adsorbing strength for manganese

Table 14.9 Exchange of Potassium Ions and Calcium Ions with Mmanganese Ions in a Rhodic Ferralsol

Soil	K or Ca Added ($1/z$ cmol kg^{-1})	Mn Replaced (½cmol kg^{-1})	
		K	Ca
Surface soil	0.02	–	0.01
	0.2	0.09	0.19
	2	0.14	0.43
	20	0.21	0.48
Substratum	0.02	0.01	0.01
	0.2	0.04	0.12
	2	0.17	0.59
	20	0.31	0.68

ions is one of the characteristics of variable charge soils.

14.4 FERROUS AND MANGANOUS IONS IN SOILS UNDER FIELD CONDITIONS

The pH of variable charge soils under natural vegetation is generally lower than 6. In the surface layer of the profile, owing to the presence of a large amount of organic matter, there is a condition for the reduction of iron and manganese oxides even during dry seasons. Therefore, in this layer there is invariably the presence of ferrous and manganous ions. Actually, a large part of reducing substances in variable charge soils, such as those in Ferrali–Haplic Acrisols of the Xishuangbanna region cited in the previous chapter, consists of ferrous and manganese ions.

Table 14.10 Contents of Ferrous Iron in Surface Layer of Soils of Dinghu Mountain Region and Hainan Island Under Field Conditions

Soil	Vegetation	pH	Eh_7	Fe^{2+} (mg kg^{-1}) Soluble	Exch.
Hydrated Ferrali–Acrisol	Broadleaf	3.57	400	1.12	6.32
Ferrali–Haplic Acrisol	Broadleaf	3.59	500	0.93	6.32
Stony Ferrali–Acrisol	Mixed forest	3.89	420	0.49	4.07
Ferrali–Haplic Acrisol	Masson pine	3.76	400	0.21	2.20
Ferrali–Haplic Acrisol	Mixed forest	3.79	378	0.34	17.3
Ferrali–Haplic Acrisol	Tea	4.58	500	0.30	1.85
Ferrali–Haplic Acrisol	Sparse needle	4.55	520	0.33	3.07
Xanthic–Haplic Acrisol	Rain forest	4.50	440	0.83	21.2
Ferrali–Haplic Acrisol	Olive	5.30	541	tr.	2.39
Ferrali–Haplic Acrisol	Coffee	4.99	520	0.79	1.13
Ferrali–Haplic Acrisol	Oil palm	5.37	570	0.34	1.75
Ferrali–Haplic Acrisol	Rubber tree	5.12	550	0.30	3.03
Ferralsol	Pepper	5.19	570	1.12	0.31
Xanthic–Haplic Acrisol	Reforested	4.78	540	0.79	6.35
Xanthic–Haplic Acrisol	China fir	4.77	560	0.14	4.05
Xanthic Ferralsol	Rain forest	5.29	590	0.43	2.03
Xanthic–Haplic Acrisol	Primeval forest	4.67	543	2.46	2.32
Xanthic–Haplic Acrisol	Rain forest	4.45	480	0.46	10.17
Xanthic–Haplic Acrisol	Mixed forest	4.50	490	0.31	3.01

The contents of ferrous iron in the surface layer of some variable charge soils under natural vegetation are given in Table 14.10. In this layer the amount of water–soluble and exchangeable ferrous iron is generally several milligrams to 20 milligrams per kilogram of soil. The amount is closely related to the oxidation–reduction status of the soil. This kind of relationship for the surface layer of soils with a dense vegetation is shown in Fig. 14.19.

Ferrous iron may also be present in low–lying horizons of the soil, provided that the pH is sufficiently low (below 4.5), although the amount is smaller than that of the surface layer (Yu et al., 1957).

Ferric ions can also exist in variable charge soils. The amount may attain 5 mg kg^{-1} (Yu et al., 1957). However, this existence is only confined to the surface layer with a high content of organic matter. These ferric ions are probably in forms of complexes or chelates (James and Bouldin, 1986; McBride et al., 1983; Paterson et al., 1990).

When a soil is cultivated for rice, the amount of ferrous iron would be much larger than that in soils under natural vegetation. In this case, the amount is chiefly determined by the water regime of the soil as shown in Table 14.11 for three paddy soils derived from Acrisols.

At present no material is available concerning the amount of manganese ions in variable charge soils under field conditions. However, because manganese ions can be detected even after air–drying of these soils, it can be expected that there would be the presence of these ions under field conditions when the soil contains a certain amount of water. If submerged,

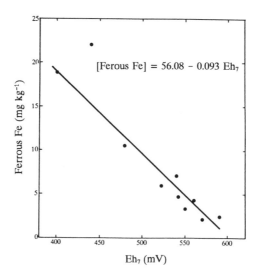

Fig. 14.19. Correlation between amount of ferrous iron and Eh_7 in the surface layer of variable charge soils (dense vegetation).

Table 14.11 Contents of Ferrous Iron in Different Types of Paddy Soils
Derived from Acrisols (Hunan)

| Soil | Depth (cm) | Ferrous Iron (mg kg^{-1}) | |
		Water–Soluble	Exchangeable
APC type	0–16	3.0	348
	16–31	0.2	4.8
	31+	0	0
G type	0–25	8.0	242
	25–36	25.0	225
	36+	0.5	125
AG type	0–3	0	0.5
	3–25	2.3	135
	25+	0.5	250

as has been shown in Fig. 14.4, all the reducible manganese can be changed
into ionic form.

From the discussions made in this chapter it can be concluded that iron
and manganese are not only important components of the solid part of
variable charge soils, but their ionic forms also play an important role in
reactions in soil solution and at the interface between solution and the solid
part. These elements are of significance in both soil genesis and plant
nutrition. They may even be toxic to plants under certain circumstances.
Therefore, they do not belong to inert materials as was previously thought.
On the contrary, they are important soil components. This is especially true
for variable charge soils where the contents of these elements are comparatively high and there are favorable conditions for their mobilization due to
the low pH and active oxidation–reduction processes.

BIBLIOGRAPHY

Bao, X. M. (1987) *Acta Pedol. Sinica*, 24:313–317.

Bao, X. M. and Yu, T. R. (1986) *Acta Pedol. Sinica*, 23:44–47.

Bao, X. M. and Yu, T. R. (1987) *Biol. Fert. Soils*, 5:88–92.

Bao, X. M., Ding, C. P., and Yu, T. R. (1983) *Z. Pflanzenernähr. Bodenk.*,
146:285–294.

Barah, D. K., Rattan, R. K., and Barerjee, N. K. (1992) *J. Indian Soc. Soil
Sci.*, 40:283–288.

Bartlett, R. J. and James, B. R. (1993) *Adv. Agron.*, 50:152–208.

Bryant, R. B. and Macedo, J. (1990) *Soil Sci. Soc. Am. J.*, 54:819–821.

Collins, J. F. and Buol, S. W. (1970) *Soil Sci.*, 110:111–117.

Foy, C. D. (1984) in *Soil Acidity and Liming* (F. Adams, ed.). American Society of Agronomy, Madison, WI, pp. 57–97.

Geering, M. R., Hodson, J. R., and Sdano, C. (1969) *Soil Sci. Soc. Am. Proc.*, 33:81–85.

James, B. R. and Bouldin, D. R. (1986) *Commun. Soil Sci. Plant Anal.*, 17:1185–1201.

McBride, M. B. (1987) *Soil Sci. Soc. Am. J.*, 51:1466–1472.

McBride, M. B. (1989) *Adv. Soil Sci.*, 10:1–56.

McBride, M. B., Goodman, B. A., Russell, J. D., Fraser, A. R., Farmer, V. C., and Dickson, D. P. E. (1983) *J. Soil Sci.*, 34:825–840.

McKenzie, R. M. (1980) *Aust. J. Soil Res.*, 18:61–73.

Miller, D. M., Tang, T., and Paul, D. W. (1993) *Soil Sci. Soc. Am. J.*, 57:356–360.

Norvell, W. A. and Lindsay, W. L. (1982) *Soil Sci. Soc. Am. J.*, 46:710–715.

Pan, S. Z., Sun, H. Y., Xu, R. K., and Wu, Y. X. (1994) *Pedosphere*, 4: 217–224.

Paterson, E., Goodman, B. A., and Farmer, V. C. (1990) in *Soil Acidity* (B. Ulrich and M. E. Sumner, eds.). Springer–Verlag, Berlin, pp. 97–124.

Ponnamperuma, F. N. (1972) *Adv. Agron.*, 24:29–96.

Ponnamperuma, F. N. (1981) in *Proceedings of the International Symposium on Paddy Soils*. Science Press/Springer Verlag, Beijing/Berlin, pp. 59–94.

Patrick, W. H. and Jugsujinda, A. (1992) *Soil Sci. Soc. Am. J.*, 56:1071–1073.

Patrick, W. H. and Reddy, C. N. (1978) in *Soil and Rice*. International Rice Research Institute, Los Benos, pp. 361–380.

Ritchie, G. S. P. (1989) in *Soil Acidity and Plant Growth* (A. D. Robson, ed.). Academic Press, Sydney, pp. 1–60.

Sanders, J. R. (1983) *J. Soil Sci.*, 34:315–323.

Scheffer, F. and Schachtschabel, P. (1992) *Lehrbuch der Bodenkunde.* Ferdinand Enke, Stuttgart.

Schwab, A. P. and Lindsay, W. L. (1983) *Soil Sci. Soc. Am. J.*, 47:201–205.

Ukrainczyk, L. and McBride, M. B. (1992) *Clays Clay Miner.*, 40:157–166.

Warden, B. T. and Reisenauer, H. M. (1991) *Soil Sci. Soc. Am. J.*, 55:345–349.

Willett. I. R. (1990) *Trans. 14th Intern. Congr. Soil Sci.*, II:38–43.

Yu, T. R. (1985) *Physical Chemistry of Paddy Soils*. Science Press/Springer Verlag, Beijing/Berlin, pp. 69–91.

Yu, T. R., Ling, Y. X., Ding, C. P., Mu, R. S., and Liu, Z. G. (1957) *Science Bull.*, 11:338–339.

Yu, T. R., Ling, Y. X., Mu, R. S., and Liu, W. L. (1958) *Soils Bull.*,

33:16–30.
Zinder, B., Furrer, G., and Stum, W. (1986) *Geochim. Cosmochim. acta*, 50:1861–1869.

INDEX

502